IMAGE REGISTRATION FOR REMOTE SENSING

Image registration employs digital image processing in order to bring two or more digital images into precise alignment for analysis and comparison. Accurate registration algorithms are essential in supporting Earth and planetary scientists as they mosaic remote sensing satellite images and track changes of the planet's surface over time for environmental, political and basic science studies. The book brings together invited contributions by 36 distinguished researchers in the field to present a coherent and detailed overview of current research and practice in the application of image registration to satellite imagery. The chapters cover the problem definition, theoretical issues in accuracy and efficiency, fundamental algorithms used in its solution, and real-world case studies of image registration software applied to imagery from operational satellite systems.

This book is an essential reference for Earth and space scientists who need a comprehensive and practical overview on how to obtain optimal georegistration of their data, an indispensable source for image processing researchers interested in current research, and the ideal text for teaching a special topic university graduate course.

JACQUELINE LE MOIGNE is the Assistant Chief for Technology in the Software Engineering Division at NASA Goddard Space Flight Center where she leads the strategic vision and the development of goals and objectives for advanced software and information system technologies. During her 20 years experience at NASA, Dr. Le Moigne has performed significant work in the processing and the analysis of remote sensing data. She has become an international expert in image registration, especially as it relates to the use of wavelet analysis, high-performance and onboard processing. She has published over 120 refereed papers and has been an Associate Editor for the *IEEE Transactions on Geoscience and Remote Sensing* and for the journal *Pattern Recognition*.

NATHAN S. NETANYAHU is an Associate Professor in the Department of Computer Science at Bar-Ilan University, Israel, and is also affiliated with the Brain Research Center at Bar-Ilan University and the Center for Automation Research at the University of Maryland, College Park. He has previously worked for the Israeli Ministry of Defense, the Space Data and Computing Division, NASA Goddard Space Flight Center, and for the Center for Excellence in Space Data and Information Sciences (CESDIS) at NASA Goddard. Professor Netanyahu's main research interests are in the areas of algorithm design and analysis, computational geometry, image processing, pattern recognition, remote sensing, and robust statistical estimation. He has coauthored nearly 70 refereed papers that appeared in journals, international conference proceedings, and book chapters, and has served as Associate Editor for *Pattern Recognition*.

ROGER D. EASTMAN is an Associate Professor of Computer Science at Loyola University Maryland, with over 25 years of experience in image matching and registration for medical, robotic and Earth science applications. Professor Eastman has collaborated with NASA-Goddard researchers in Earth science registration on techniques for generalizing and evaluating algorithms, and for robust subpixel registration, and with NIST-Gaithersburg researchers on advanced sensors for manufacturing robotics for general assembly. He regularly reviews articles on image registration for the *IEEE Transactions on Geoscience and Remote Sensing* and other remote sensing venues.

IMAGE REGISTRATION FOR REMOTE SENSING

Edited by

JACQUELINE LE MOIGNE
NASA Goddard Space Flight Center, USA

NATHAN S. NETANYAHU
Bar-Ilan University, Israel, and University of Maryland, USA

ROGER D. EASTMAN
Loyola University Maryland, USA

CAMBRIDGE
UNIVERSITY PRESS

CAMBRIDGE
UNIVERSITY PRESS

University Printing House, Cambridge CB2 8BS, United Kingdom

One Liberty Plaza, 20th Floor, New York, NY 10006, USA

477 Williamstown Road, Port Melbourne, VIC 3207, Australia

4843/24, 2nd Floor, Ansari Road, Daryaganj, Delhi - 110002, India

79 Anson Road, #06-04/06, Singapore 079906

Cambridge University Press is part of the University of Cambridge.

It furthers the University's mission by disseminating knowledge in the pursuit of education, learning and research at the highest international levels of excellence.

www.cambridge.org
Information on this title: www.cambridge.org/9781108445757

© Cambridge University Press 2011

First published 2011
4th printing 2015
First paperback edition 2017

A catalogue record for this publication is available from the British Library

Library of Congress Cataloging in Publication data
Image registration for remote sensing / edited by Jacqueline Le Moigne, Nathan S. Netanyahu, Roger D. Eastman.
p.   cm.
Includes bibliographical references and index.
ISBN 978-0-521-51611-2
1. Image registration.   2. Image analysis.   3. Remote sensing.
4. Image processing – Digital techniques.
I. Le Moigne, Jacqueline.   II. Netanyahu, Nathan S.
III. Eastman, Roger D.   IV. Title.
TA1632.I4826   2010
621.36 78 – dc22   2010041514

ISBN 978-0-521-51611-2 Hardback
ISBN 978-1-108-44575-7 Paperback

To my mother, Noëlie Le Moigne, for teaching me
that all dreams can be achieved through determination.
To Gavin, Lauriane and Gordon,
for all the hours I spent on this endeavor.

Jacqueline Le Moigne

To my precious children, Aviv and Yovel, and to
my beloved partner and closest friend, Ella Aviram,
for their love, support, and endless patience!

Nathan S. Netanyahu

To my wife Michele and son Daniel
for their support and understanding.

Roger D. Eastman

We dedicate this book to the memory of Professor Azriel Rosenfeld, who inspired us with a love of image processing

Contents

Contributors

SIMON BAILLARIN, *CNES (Centre National d'Etudes Spatiales), Toulouse, France*

DANIEL G. BALDWIN, *Colorado Center for Astrodynamics Research, Aerospace Engineering Science Department, University of Colorado at Boulder, Colorado*

JÓN A. BENEDIKTSSON, *IEEE Geoscience and Remote Sensing Society and University of Iceland, Reykjavik, Iceland*

MARC BERNARD, *Spot Image, Toulouse, France*

AURÉLIE BOUILLON, *IGN (Institut Géographique National), Saint-Mandé, France*

JAMES L. CARR, *Carr Astronautics, Washington, DC*

RAMA CHELLAPPA, *Center for Automation Research, University of Maryland at College Park, Maryland*

QIN-SHENG CHEN, *Hickman Cancer Center, Flower Hospital, ProMedica Health System, Sylvania, Ohio*

ARLENE COLE-RHODES, *Electrical and Computer Engineering Department, Morgan State University, Baltimore, Maryland*

R. IAN CROCKER, *Colorado Center for Astrodynamics Research, Aerospace Engineering Science Department, University of Colorado at Boulder, Colorado*

ROGER DAVIES, *Department of Physics, The University of Auckland, New Zealand*

DAVID J. DINER, *Jet Propulsion Laboratory, California Institute of Technology, Pasadena, California*

ROGER D. EASTMAN, *Loyola University, Baltimore, Maryland*

WILLIAM J. EMERY, *Colorado Center for Astrodynamics Research, Aerospace Engineering Science Department, University of Colorado at Boulder, Colorado*

A. ARDESHIR GOSHTASBY, *Department of Computer Science and Engineering, Wright State University, Dayton, Ohio*

VENU M. GOVINDU, *Department of Electrical Engineering, Indian Institute of Science, Bangalore, India*

VELJKO M. JOVANOVIC, *Jet Propulsion Laboratory, California Institute of Technology, Pasadena, California*

CHARLES S. KENNEY, *Department of Electrical and Computer Engineering, University of California at Santa Barbara, California*

JACQUELINE LE MOIGNE, *NASA Goddard Space Flight Center, Greenbelt, Maryland*

B. S. MANJUNATH, *Department of Electrical and Computer Engineering, University of California at Santa Barbara, California*

JEFFREY MORISETTE, *U.S. Geological Survey (USGS) Fort Collins Center, Colorado; formerly Hydrospheric and Biospheric Sciences Laboratory, NASA Goddard Space Flight Center, Greenbelt, Maryland*

DAVID M. MOUNT, *Department of Computer Science, University of Maryland at College Park, Maryland*

NATHAN S. NETANYAHU, *Bar-Ilan University, Israel, and University of Maryland at College Park, Maryland*

MASAHIRO NISHIHAMA, *Raytheon at Terrestrial Information Systems Branch, NASA Goddard Space Flight Center, Greenbelt, Maryland*

FREDERICK S. PATT, *Science Applications International Corporation (SAIC), NASA Goddard Space Flight Center, Greenbelt, Maryland*

SAN RATANASANYA, *formerly Department of Computer Science, University of Maryland at College Park, Maryland*

KAUSHAL SOLANKI, *Department of Electrical and Computer Engineering, University of California at Santa Barbara, California*

HAROLD S. STONE, *NEC Research Laboratory Retiree, New Jersey*

JAMES STOREY, *SGT (Stinger Ghaffarian Technologies) at USGS Center for Earth Resources Observation and Science (EROS), Sioux Falls, South Dakota*

SYLVIA SYLVANDER, *CNES (Centre National d' Etudes Spatiales), Toulouse, France*

BIN TAN, *Earth Resources Technology Inc., Annapolis Junction, Maryland*

PRAMOD K. VARSHNEY, *Electrical Engineering and Computer Science Department, Syracuse University, New York*

ROBERT E. WOLFE, *Terrestrial Information Systems Branch, NASA Goddard Space Flight Center, Greenbelt, Maryland*

CURTIS WOODCOCK, *Department of Geography and Environment, Boston University, Massachussets*

MIN XU, *Department of Electrical Engineering and Computer Science, Syracuse University, New York*

ILYA ZAVORIN, *formerly Goddard Earth Science Technology (GEST) Center at NASA Goddard, University of Maryland Baltimore County, Maryland*

MARCO ZULIANI, *Department of Electrical and Computer Engineering, University of California at Santa Barbara, California*

Foreword

In recent years, image registration has become extremely important in remote sensing applications. Image registration refers to the fundamental task in image processing to match two or more pictures which have been taken of the same object or scene, for example, at different times, from different sensors, or from different viewpoints.

The main reason for the increased significance of image registration in remote sensing is that remote sensing is currently moving towards operational use in many important applications, both at social and scientific levels. These applications include, for example, the management of natural disasters, assessment of climate changes, management of natural resources, and the preservation of the environment; all of which involve the monitoring of the Earth's surface over time. Furthermore, there is an increasing availability of images with different characteristics, thanks to shorter revisiting times of satellites, increased flexibility of use (different acquisition modalities) and the evolution of sensor technologies. Therefore, a growing need emerges to simultaneously process different data, that is, remote sensing images, for information extraction and data fusion. This includes the comparison (integration or fusion) of newly acquired images with previous images taken with different sensors or with different acquisition modalities or geometric configurations – or with cartographic data. The remote images can, therefore, be multitemporal (taken at different dates), multisource (derived from multiple sensors), multimode (obtained with different acquisition modalities), or stereo-images (taken from different viewpoints).

The different images are initially in different coordinate systems. The registration process spatially aligns them by considering one of the images as a reference and transforming the remaining images one at a time. Therefore, a selection of corresponding structures/elements (e.g., pairs of good control points, linear features, etc.) in the reference and in each of the other images is necessary to determine an appropriate transformation. After the completion of the registration process, the images can be processed for information extraction. The registration procedure can both be manual and automatic. A wide variety of situations requires diverse registration techniques, spanning from quite simple to very complex and flexible ones, depending also on the degree of heterogeneity of the images and on the level of accuracy needed by the user or by the next computerized analysis stages to which the registration results are addressed. A number of approaches can be put under the umbrella of image registration. Geolocation and geometric correction are examples of such techniques.

Although a few books have been written on image registration in general and several for specific application fields, like medical imaging, in particular, no book has until now been available on image registration research in remote sensing. Therefore, this book edited by Dr. Jacqueline Le Moigne, Professor Nathan S. Netanyahu, and Professor Roger D. Eastman, is very welcome and is of great importance to researchers in remote sensing. The editors are renowned experts in the field of image registration of remote sensing data, and they have selected a group of outstanding authors to cover the most important topics in image registration for remote sensing.

The book is very well organized and split into four main parts. The first part gives an overview of image registration in remote sensing and discusses its importance. The next two parts discuss specific topics in the image registration chain, i.e., similarity metrics and feature matching. Finally, examples on several important applications and systems are given in part four. The book has the significant advantage that it is written in such a way that it is suitable not only for those who are advanced in processing of remote sensing data but also for those who are new to the field, including students. Newcomers to the field will get a clear understanding of what image registration for remote sensing is about after studying a few chapters in the book.

Professor Jón Atli Benediktsson
President
IEEE Geoscience and Remote Sensing Society
Pro-Rector of Academic Affairs
University of Iceland

Acknowledgements

The editors would like to acknowledge Dr. Harold Stone, Dr. Ilya Zavorin, and Professor Arlene Cole-Rhodes for their long-standing research collaboration and their help in preparing this book. They would also like to acknowledge Professor Arthur Goshtasby for his forthcoming advice on the notation used in the book and Dr. Jeffrey Morisette for his thoughtful insights on many Earth science issues. The editors are grateful to Professor Jón Benediktsson for writing the Foreword and Professor Sebastiano Serpico for his book endorsement. Finally, the editors are deeply indebted to all the contributors to this volume for their expertise, dedicated work, and patience.

Part I

The Importance of Image Registration
for Remote Sensing

1

Introduction

JACQUELINE LE MOIGNE, NATHAN S. NETANYAHU,
AND ROGER D. EASTMAN

Despite the importance of *image registration* to data integration and fusion in many fields, there are only a few books dedicated to the topic. None of the current, available books treats exclusively image registration of Earth (or space) satellite imagery. This is the first book dedicated fully to this discipline. The book surveys and presents various algorithmic approaches and applications of image registration in remote sensing. Although there are numerous approaches to the problem of registration, no single and clear solution stands out as a standard in the field of remote sensing, and the problem remains open for new, innovative approaches, as well as careful, systematic integration of existing methods. This book is intended to bring together a number of image registration approaches for study and comparison, so remote sensing scientists can review existing methods for application to their problems, and researchers in image registration can review remote sensing applications to understand how to improve their algorithms. The book contains invited contributions by many of the best researchers in the field, including contributions relating the experiences of several Earth science research teams working with operational software on imagery from major Earth satellite systems. Such systems include the Advanced Very High Resolution Radiometer (AVHRR), Landsat, MODerate resolution Imaging Spectrometer (MODIS), Satellite Pour l'Observation de la Terre (SPOT), VEGETATION, Multi-angle Imaging SpectroRadiometer (MISR), METEOSAT, and the Sea-viewing Wide Field-of-view Sensor (SeaWiFS).

We have aimed this collection of contributions at researchers and professionals in academics, government and industry whose work serves the remote sensing community. The material in this book is appropriate for a mixed audience of image processing researchers spanning the fields of computer vision, robotic vision, pattern recognition, and machine vision, as well as space-based scientists working in the fields of Earth remote sensing, planetary studies, and deep space research. This audience represents many active research projects for which the collaboration between image processing researchers and Earth scientists is essential, as the former try to solve the problems posed by the latter. A common language is not only appropriate but also needed. Our intent is to ensure that the material is accessible to both audiences. We have strived to provide a broad overview of the field, ranging from theoretical advanced algorithms to applications, while maintaining rigor by including basic definitions and equations.

In the Introduction we focus on the basic essence of image registration and the main rationale for its pursuit in the domain of remote sensing. The individual contributions in the rest of the book cover extensively various ways in which image registration is carried out. Specifically, we will describe applications for which accurate and reliable image registration is essential, and briefly review their corresponding challenges. We will then define remote sensing, describe how remote sensing data are acquired, and consider characteristics of these data and their sources. Finally, we will summarize the overall contents of the book, and provide definitions of selected general terms used throughout the chapters.

1.1 A need for accurate image registration

Earth science studies often deal with issues such as predicting crop yield, evaluating climate change over multiple timescales, locating arable land and water sources, monitoring pollution, and understanding the impact of human activity on major Earth ecosystems. To address such issues, Earth scientists use the global and repetitive measurements provided by a wide variety of satellite remote sensing systems. Many of these satellites have been launched (e.g., the Earth Observing System (EOS) AM and PM platforms), while the launch of others is being planned (e.g., the Landsat Data Continuity Mission (LDCM)). All these systems support multiple-time or simultaneous observations of the same Earth features by different sensors. Viewing large areas of the Earth at very high altitudes by spaceborne, remote sensing systems provide global measurements that would not be available using ground or even airborne sensors, although these global measurements often need to be complemented by local or regional measurements to complete a more thorough investigation of the phenomena being observed.

Image registration for the integration of digital data from such disparate satellite, airborne, and ground sources has become critical for several reasons. For example, image registration plays an essential role in spatial and radiometric calibration of multitemporal measurements for obtaining large, integrated datasets for the long-term tracking of various phenomena. Also, change detection over time or scale is only possible if multisensor and multitemporal data are accurately calibrated through registration. Previous studies by Townshend *et al.* (1992) and Dai and Khorram (1998) showed that even a small error in registration might have a large impact on the accuracy of global change measurements. For example, when looking at simulated data of MODIS at 250-m spatial resolution, a misregistration error of one pixel can produce a 50% error in the computation of the Normalized Difference Vegetation Index (NDVI). Another reason for integrating multiple observations is the resulting capability of extrapolating data throughout several scales, as researchers may be interested in phenomena that interact at multiple scales, whether spatial, spectral, or temporal. Generally, changes caused by human activity occur at a much faster rate and affect much larger areas. For all these applications, very accurate registration, that is, *exact pixel-to-pixel matching of two different images or matching of one image to a map*, is one of the first requirements for making such data integration possible.

More generally, image registration for remote sensing can be classified as follows:

(1) *Multimodal registration*, which enables the integration of complementary information from different sensors. This suits, for example, land cover applications, such as agriculture and crop forecasting, water urban planning, rangeland monitoring, mineral and oil exploration, cartography, flood monitoring, disease control, real-estate tax monitoring, and detection of illegal crops. In many of these applications, the combination of remote sensing data and Geographic Information Systems (GISs), see, for example, Cary (1994), and Ehlers (1995), shows great promise in helping the decision-making process.
(2) *Temporal registration*, which can be used for change detection and Earth resource surveying, including monitoring of agricultural, geological, and land cover features extracted from data obtained from one or several sensors over a period of time. Cloud removal is another application of temporal registration, when observations over several days can be fused to create cloud-free data.
(3) *Viewpoint registration*, which integrates information from one moving platform or multiple platforms navigating together into three-dimensional models. Landmark navigation, formation flying and planet exploration are examples of applications that benefit from such registration.
(4) *Template registration*, which looks for the correspondence between new sensed data and a previously developed model or dataset. This is useful for content-based or object searching and map updating.

Scientific visualization and virtual reality, which create seamless mosaics of multiple sensor data, are other examples of applications which are based on various types of registration, in particular, multimodal, temporal, and viewpoint registration.

1.2 What is image registration?

As a general definition, image registration is the process of aligning two or more images, or one or more images with another data source, for example, a map containing vector data. An image is an array of single measurements, and alignment is provided by a mathematical transformation between geometric locations in two image arrays. To be mutually registered, two images should contain overlapping views of the same ground features. In the basic case, one image may need to be translated, or translated and rotated, to align it with the other. The problem of image-to-image registration is illustrated in Fig. 1.1, which shows a reference image, extracted from an IKONOS scene over Washington, DC, with a corresponding translated and rotated image. In later chapters we will consider complex transformations, beyond translation and rotation, for alignment of the images.

Image registration involves locating and matching similar regions in the two images to be registered. In manual registration, a human carries out these tasks visually using interactive software. In automatic registration, on the other hand, autonomous algorithms perform these tasks. In remote sensing, automated procedures do not always offer the needed reliability and accuracy, so manual registration is frequently used. The user extracts from both images distinctive locations, which are typically called *control points* (CPs), tie-points, or reference points. First, the CPs in both images (or datasets) are interactively matched pairwise to achieve correspondence. Then, corresponding CPs are used to compute the parameters of a geometric transformation in question. Most available commercial systems follow this registration approach. Manual CP selection represents, however, a repetitive, laborious, and time-intensive task that becomes prohibitive for large amounts of data. Also, since the interactive choice of control points in satellite images is sometimes difficult to begin with, and since often too few points, inaccurate points, or ill-distributed points might be chosen, manual registration could lead to large registration errors. The main goal of image registration research, in general, is to improve the accuracy, robustness, and efficiency of fully automatic, algorithmic approaches to the problem. Specifically, the primary objective of this book is to review and describe the main research avenues and several important applications of automatic image registration in remote sensing.

Usually, automatic image registration algorithms include three main steps (Brown, 1992):

(1) Extraction of distinct regions, or *features*, to be matched.
(2) *Matching* of the features by searching for a transformation that best aligns them.
(3) *Resampling* of one image to construct a new image in the coordinate system of the other, based on the computed transformation.

Automatic approaches differ in the way they solve each step. One algorithm may extract simple features, but use a complex matching strategy, while another may use rather complex features, but then employ a relative simple matching strategy. Chapter 3 provides a survey of many current automatic image registration methods,

Figure 1.1. A reference image and its transformed image, extracted from an IKONOS scene acquired over Washington, DC. See color plates section. (IKONOS satellite imagery courtesy of GeoEye. Copyright 2009. All rights reserved.)

focusing mainly on their feature extraction and matching steps. Additional chapters discuss particular algorithmic approaches, and several other chapters describe ground control systems successfully implemented for satellite systems.

This book mainly deals with feature extraction and matching. While feature extraction and matching must be integrated, resampling is performed post-matching and can be handled relatively independently. For some applications, this step is replaced by an indexing of the incoming data into an absolute reference system, for example, a (latitude, longitude) reference system for Earth satellite data. Doing so preserves the original data values, which can be important for scientific applications. When several data sources are integrated, the resampling step can be replaced or supplemented by the fusion process. Finally, an automatic method may have two resampling stages. A temporary stage is used during matching to increase the similarity of the two images, but its results are discarded while a second, more accurate phase is used for the production of the final image product.

More generally, for all the applications described in Section 1.1, the main requirements from an image georegistration system are accuracy, consistency (i.e., robustness), speed, and a high level of autonomy that will facilitate the processing of large amounts of data in real time. With the goal of developing such a system, the purpose of this book is to examine the specific issues related to image registration in the particular domain of remote sensing, and to describe the methods that have been proposed to solve these issues. Before describing these methods, we first look at how remote sensing data are being acquired.

1.3 Remote sensing fundamentals

Remote sensing can be defined as the process by which information about an object or phenomenon is acquired from a remote location (e.g., an aircraft or a satellite). More specifically, satellite/sensing imaging refers to the use of sensors located on spaceborne platforms to capture electromagnetic energy that is reflected or emitted from a planetary surface such as the Earth. The Sun, like all terrestrial objects, is a source of energy. The sensors are either *passive* or *active*, that is, all energy which is observed by passive satellite sensors originates either from the Sun or from planetary surface features, while active sensors, such as radar systems, utilize their own source of energy to capture or image specific targets.

All objects give off radiation at all wavelengths, but the emitted energy varies with the wavelength and with the temperature of the object. A *blackbody* is an ideal object that absorbs and reemits all incident energy, without reflecting any. According to Stefan-Boltzman's and Wien's displacement laws (Lillesand and Kiefer, 1987; Campbell, 1996), a *dominant wavelength*, defined as the wavelength at which the total radiant exitance is maximum, can be computed for all blackbodies.

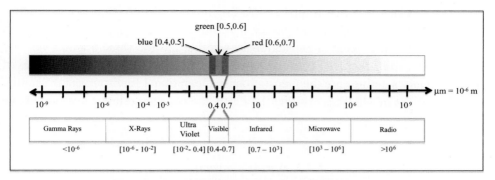

Figure 1.2. Electromagnetic spectrum. See color plates section.

Assuming that the Earth and the Sun behave like blackbodies, their respective dominant wavelengths are 9.7 μm (in the infrared (IR) portion of the spectrum) and 0.5 μm (in the green visible portion of the spectrum). This implies that the energy emitted by the Earth is best observed by sensors which operate in the thermal infrared and microwave portions of the electromagnetic spectrum, while Sun energy which has been reflected by the Earth predominates in the visible, near-infrared and mid-infrared portions of the spectrum. As a consequence, most passive satellite sensing systems operate in the visible, infrared, or microwave portions of the spectrum (Lillesand and Kiefer, 1987; Le Moigne and Cromp, 1999). See Fig. 1.2 for a summary of the above electromagnetic spectrum wavelengths definitions.

1.3.1 Characteristics of satellite orbits

Different orbiting trajectories may be chosen for a satellite depending on many requirements, including the characteristics of the sensors, the data acquisition frequency, the required spatial resolution, the necessary ground coverage, and the type of observed phenomenon. The two most common orbiting modes are usually referred to as *polar orbiting* and *geostationary* (or *geosynchronous*) satellites. A polar orbit passes near the Earth's North and South Poles. Some examples are the Landsat and SPOT satellites whose orbits are almost polar, passing above the two poles and crossing the Equator at a small angle from normal (e.g., 8.2° for the Landsat-4 and Landsat-5 spacecraft). If the orbital period of a polar orbiting satellite keeps pace with the Sun's westward progression compared to the Earth's rotation, these satellites are also called *Sun-synchronous*, that is, a Sun-synchronous satellite always crosses the Equator at the same local Sun time. This time is usually very carefully chosen, depending on the application of the sensing system and the type of features which will be observed with such a system. Atmospheric scientists prefer observations later in the morning to allow for cloud formation, whereas researchers performing land studies prefer earlier morning observations to minimize cloud cover. On the other hand, a geostationary satellite has the same angular velocity as the Earth, so its relative position is fixed with respect to the Earth. Examples of geostationary satellites are the Geostationary Operational Environmental Satellite (GOES) series of satellites orbiting the Earth at a constant relative position above the equator.

1.3.2 Sensor characteristics

Each new sensor is designed for specific types of features to be observed, with requirements that define its spatial, spectral, radiometric, and temporal resolutions. This term of *resolution* corresponds to the smallest unit of granularity that can be measured by the sensor. The spatial resolution corresponds to the area on the ground from which reflectance is integrated to compute the value assigned to each pixel. The spectral resolution relates to the bandwidths utilized in the electromagnetic spectrum, and the radiometric resolution defines the number of "bits" that are used to record a given energy corresponding to a given wavelength. Finally, the temporal resolution corresponds to the frequency of observation, defined by the orbit of the satellite and the scanning of the sensor.

One of the main characteristics of sensors is their signal-to-noise ratio (SNR), or the noise level relative to the strength of the signal. In this context, the term *noise* refers to variations of intensity which are detected by the sensor and that are not caused by actual variations in feature brightness. If the noise level is very high compared to the signal level, the data will not provide an optimal representation of the observed features. At a given wavelength λ, the SNR is a function of the detector quality, as well as the spatial resolution of the sensor and its spectral resolution. Specifically,

$$(S/N)_\lambda = D_\lambda \beta^2 (H/V)^{1/2} \Delta_\lambda L_\lambda, \qquad (1.1)$$

where

D_λ is the sensor detectivity (i.e., a measure of the detector's performance quality),
β is the instantaneous field of view,
H is the flying height of the spacecraft,
V is the velocity of the spacecraft,
Δ_λ is the spectral bandwidth of the channel (or spectral resolution), and
L_λ is the spectral radiance of the ground features.

Equation (1.1) demonstrates that maintaining the SNR of a sensor at an acceptable level often requires tradeoffs between the other characteristics of the sensor.

For example, to maintain the same SNR while improving the spatial resolution by a factor of four (i.e., decreasing β by a factor of two), we must degrade the spectral resolution by a factor of four (i.e., increase Δ_λ by a factor of four). Note that there are additional factors not accounted for in Eq. (1.1), such as atmospheric interactions, that also affect the signal-to-noise ratio.

Another way of characterizing Earth remote sensors is by the number of spectral bands of each sensor. In general, most Earth remote sensors are multispectral, that is, they utilize several bands to capture the energy emitted or reflected from Earth features. The addition of panchromatic imagery, which is usually of significantly higher spatial resolution than that of multispectral imagery in the visible part of the spectrum, provides higher quality detail information. Similarly, the number of bands in Landsat-4 and -5 was increased from four to seven (relative to Landsat-1 and -2) to include bands from the visible and thermal-IR range. The Landsat series was further extended with the introduction of Landsat-7, which contains an additional panchromatic band. Other sensors which provide coregistered, multispectral-panchromatic imagery are the Indian Remote Satellite-1 (IRS-1) sensor and SPOT.

Ideally, if a sensor had an infinite number of spectral channels, each observed area on the ground would be represented by a continuous spectrum and, therefore, could be identified from a database of known spectral response patterns. Practically, adding more bands and making each of them narrower is the first step towards realizing this ideal sensor. However, as previously explained by Eq. (1.1), it has been very difficult to increase the number of bands without decreasing the signal-to-noise ratio. But, due to recent advances in solid-state detector technology, it has become feasible to increase significantly the number of bands without decreasing the signal-to-noise ratio. This has led to the rise of new types of sensors, known as hyperspectral sensors. Usually, the criterion by which a sensor is regarded a multispectral or hyperspectral sensor is the number of bands (which can be as low as ten). Hyperspectral imaging refers typically to the simultaneous detection in hundreds to thousands of spectral channels, covering evenly a limited portion of the electromagnetic spectrum. The aim of hyperspectral sensors is to provide unique identification (or *spectral fingerprints*) capabilities for resolvable spectral objects. The NASA Earth Observing-1 (EOS-1) Hyperion sensor, launched in 2000 and still flying, is the first spaceborne civilian hyperspectral sensor. It spans 220 contiguous spectral bands from the visible to the infrared range (corresponding to a wavelength range of 0.4–2.5 μm). Hyperion data are currently used for scientific objectives related to land cover/land use activities, such as monitoring the global environment (e.g., deforestation) and climate change, disaster management, etc. Its targeting abilities complement those of other sensors, like MODIS, with a wider swath but lower spatial resolution.

Examples of different spectral and spatial resolutions are given in Table 1.1, with a focus on the operational Earth remote sensing systems described in Part IV of this book (Chapters 14–22). The table also provides information about the bandwidths and spatial resolutions of these sensors, whose spectral wavelength varies from the visible to the thermal-IR range.

Finally, another difference between sensors deals with their scanning mechanisms. Most Earth sensors utilize either across-track scanning or along-track scanning systems. Both types of scanning acquire data in scan lines that are perpendicular to the travel direction of the spacecraft. However, while cross-track scanners use a scanning mirror that rotates as it acquires the data, along-track scanners have a linear array of sensors that are "pushed along" the direction of travel (hence the term *pushbroom scanners*). Both types of scanning introduce errors that are corrected as part of the systematic correction step, although these corrections may include inconsistencies in the data radiometry, thereby creating errors in the registration process. Additional information about remote sensing can be found in several introductions to remote sensing, for example, Lillesand and Kiefer (1987), Campbell (1996), and Short (2009).

1.4 Issues involved with remote sensing image registration

Once the data are collected and packaged, they are transmitted to the ground where they are unpacked and processed in a ground processing station. Another scenario involves more processing on board the spacecraft but in any case, after transmission from the satellites, raw data are usually processed, calibrated, archived, and distributed with some level of processing. Most of NASA's satellite data products are classified according to the following data levels (Asrar and Dozier, 1994):

- *Level 0 data* are the reconstructed raw instrument data at full resolution.
- *Level 1A data* are reconstructed, time-reference raw data, with ancillary information including radiometric and geometric coefficients.
- *Level 1B data* are corrected Level 1A data (in sensor units).
- *Level 2 data* are derived geophysical products from Level 1 data, at the same resolution and location, e.g., atmospheric temperature profiles, gas concentrations, winds variables, etc.
- *Level 3 data* correspond to the same geophysical information as Level 2, but mapped onto a uniform space-time grid.
- *Level 4 data* are model output or results from prior analysis of lower-level data.

In Chapters 2–14 of this book, we will usually refer to image registration as performed on Level 1B data, which means that the spatial coordinates of image data have been computed according to a systematic correction using ancillary/ephemeris

Table 1.1 Spatial and spectral characteristics of all operational sensors described in Part IV (Chapters 14–22)

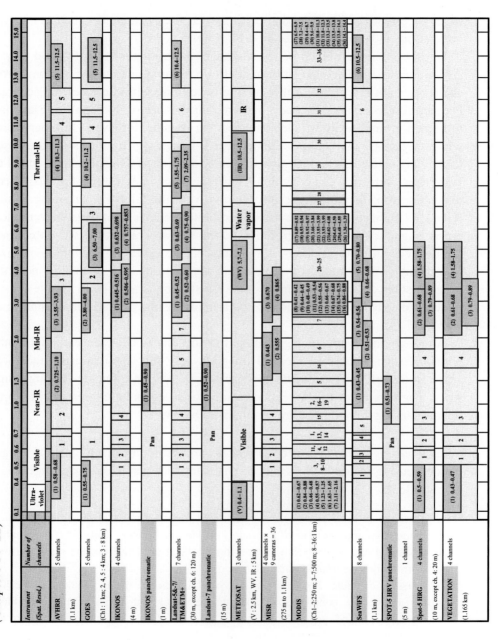

data from the spacecraft. In short, by determining where the satellite is pointing while acquiring an image, the image can be given approximate ground coordinates. This type of correction is also sometimes referred to as *navigation*, since it is based on a *navigation model* that takes into account parameters such as the type, orientation, and shape of a satellite orbit (Logsdon, 1997).

During satellite ground processing, image registration is typically used for *precision correction*. The navigation model may have systematic or random errors, and it does not report where the satellite is pointing within the desired accuracy. Precision correction is the process of correcting these errors by registering an image to known ground features (such as a specific coastline or river). In other words, while systematic correction is model-based, image registration is feature-based. Depending on the age and the type of remote sensing systems, the accuracy of the systematic correction can be within a few pixels up to a few tens of pixels. Recent navigation models that utilize information from the Global Positioning System (GPS) (El-Rabbany, 2002) are usually accurate within a few pixels. Nevertheless, errors might still occur, for example, during a spacecraft maneuver. In contrast, the desired, ultimate accuracy is typically on the order of fractions of a pixel. For example, it is crucial for registration applications such as change detection to reach subpixel accuracy. Thus, image registration is used in all remote sensing systems to refine the initial geolocation accuracy to the desired subpixel accuracy level.

Although many image registration methods have been developed in other domains, such as applied medical imaging, very few automatic methods – let alone an underlying, systematic approach – exist within a remote sensing framework. The reason for this is related mainly to issues that are very specific to the remote sensing domain, and are summarized below.

Remote sensing vs. medical or other type of imagery Compared to medical images, remote sensing imagery offers various characteristics that make image registration more difficult.

(1) *The variety in the types of sensor data and the conditions of data acquisition.* A technique that appears to work accurately on satellite imagery acquired at a given time over some given location may not perform as well on data from the same sensor at other times or over another location.

(2) *The size of the data.* For example, a typical Landsat scene is of size 7000 × 7000 pixels on average, containing 7 bands whose wavelength varies from the visible to the thermal infrared range. Handling such amounts of data in real-time must take into account computational requirements such as speed and memory. As a consequence, the implementation of such methods on parallel, distributed or even onboard computers must be considered.

(3) *The lack of a known image model.* Similarly to fiducial points, a very rough sketch of a city, containing a river or a network of roads, can be utilized as a global model to initiate the registration. However, this usually lacks the amount of detail and the degree of invariance to atmospheric conditions and seasonal variations that would be needed for subpixel registration accuracy.

(4) *The lack of well-distributed "fiducial points" resulting in the difficulty to validate image registration methods in the remote sensing domain.* Although, it is possible to use well-known landmarks such as the Washington Monument or the Tour Eiffel ("Eiffel Tower") as fiducial points, such landmarks are very rare, and are not evenly distributed around the globe. The key factor in any accuracy assessment of remote sensing data is linked to the ability to gather ground reference data independently of the remote sensing data themselves. The most reliable fashion would be to record actual GPS locations of various sites on the ground and link them to recorded image data. To be sufficiently accurate, though, millions of such locations should be recorded, and so this approach could become very tedious and prohibitively expensive. Additionally, depending on the time between the on-site ground reference gathering and the imaging of the area, the validity of the ground reference data may be lessened due to anthropogenic or natural influences. Another approach is to compare the digital image with other sources of ground reference data, such as air photos or appropriate reference maps, provided that the features of interest are detectable using these sources. The degree of correspondence between the ground reference data and the measurement derived from the sensor data can then be compared for accuracy assessment. Other assessment methods, including manual registration, the use of synthetic data, and round-robin measurements, are discussed in later chapters, in particular, Chapter 14.

Navigation error Several types of errors may occur in the navigation-based correction, thus resulting in registration errors of the Level 1B data. First, errors may be introduced in the input parameters during spacecraft maneuvers. Sometimes these errors are not detected and corrected until a few days or even a few weeks after the maneuver. Additional errors occur when the spacecraft and the sensor themselves age and perform differently from the way they were modeled in the navigation model. All these errors generally do not impact the data distributed at a later time, since regular checks are usually performed using *ground control points* (GCPs). These errors affect, however, data being used in real or near-real time, for example in efforts that support disaster relief. More generally, these errors affect all data being transmitted in a *direct readout* mode, that is, data transmitted almost immediately after acquisition to any receiving station within the satellite footprint.

Atmospheric and cloud interactions The atmospheric effects on data fidelity depend on the distance travelled by the radiation through the atmosphere and on the

Figure 1.3. Three Landsat images over Virginia acquired in August, October, and November 1999. (Courtesy: Jeffrey Masek, NASA Goddard Space Flight Center.)

magnitude of the energy signal. The two main atmospheric effects are *scattering* and *absorption*. Earth remote sensors usually concentrate their observations within some *atmospheric windows* that are defined outside of the wavelengths of maximum atmospheric absorption. For each sensor, the spectral bands or *channels* are defined within these atmospheric windows while focusing on the phenomena to be observed. Most of the atmospheric effects, including atmospheric humidity and the concentration of atmospheric particles, are corrected by physical models, although effects related to altitude or local and temporal weather during data acquisition are usually not included in these models.

Another issue related to registration of remote sensing images deals with cloud interactions. When performing image registration, recognizing and discarding cloud features is often considered an important preprocessing step.

Multitemporal effects We distinguish between natural effects and human-induced effects that occur over time. The former consist of, for example, different lighting conditions due to the change in the Sun angle during the year. Also, the viewing angle of the instrument can change from pass to pass. And with seasonal changes, the surface reflectance varies with weather conditions, and land cover is altered as crops appear in different stages and deciduous trees change. Figure 1.3 illustrates the above type of effects. It shows three Landsat images (with cloud cover) over the same area of Virginia taken at three different months in 1999.

To all these natural temporal effects must be added human-induced effects related to activities such as urban development, agricultural practices, and deforestation. Figure 1.4, which shows two Landsat images taken over Bolivia in 1984 and 1998, illustrates the human-induced changes that can be observed over time. In view of the above multitemporal effects, whether natural or human-induced, certain features may not be visible from one image to the next, and may thus induce registration errors.

Figure 1.4. Human-induced land cover changes observed by Landsat-5 in Bolivia in 1984 and 1998. See color plates section. (Courtesy: Compton J. Tucker and the Landsat Project, NASA Goddard Space Flight Center.)

Terrain/relief effect Another source affecting the registration of remote sensing imagery is the topography or the terrain. Various terrain features will be represented by variations in image brightness, in different ways, depending on the angle of illumination. This means that depending on the slope of the geographic relief, the characteristics of the sensor and the satellite orbit, and the time of the day, terrain relief effects might appear very differently in the images to be registered. Large topographic variations can be corrected using a terrain model but small local effects will still be present.

Multisensor (having different spatial and spectral resolutions) When dealing with multiple sensors, with different geometries and various spatial, spectral, radiometric and temporal resolutions (as described in Subsection 1.3.2), it is necessary to address the following image registration issues:

(1) Choice of geometric transformations that respond to various spatial resolutions and different scanning patterns.
(2) Extraction of image features that are invariant to radiometric differences due to multi-spectral and multitemporal resolutions.
(3) Choice of relevant channels when performing band-to-band registration (corresponding to approximate similar regions of the electromagnetic spectrum).

Figure 1.5 shows an example of terrain features observed at different times by the Landsat-7 Enhanced Thematic Mapper (ETM) and IKONOS over the Colorado Mountains. Note that some of the relief details are seen with IKONOS but not with ETM. This inconsistency is attributed to the different spatial resolution of the two sensors (i.e., 30 m for Landsat and 4 m for IKONOS).

Figure 1.5. Two near-infrared bands of Landsat-7 (left) and IKONOS (right) taken over the CASCADES area, Colorado. (Courtesy: Jeffrey Morisette, USGS and EOS Validation Core Sites.)

1.5 Book contents

Overall, the book consists of four main parts. In Part I (Chapters 1–3), image registration for remote sensing is defined, explained, and surveyed. Chapter 2 examines the effects of misregistration on validation efforts. Other important misregistration effects (not described in Chapter 2) are those affecting change detection (Townshend *et al.*, 1992; Dai and Khorram, 1998), and those that relate to weather forecasting, as well as political and legal issues, such as management of water resources or precise localization of property boundaries. Many of these topics are described in recent *IEEE Geoscience and Remote Sensing Symposium (IGARSS)* conferences. Chapter 3 provides a general overview of image registration, with special emphasis on the domain of remote sensing. In particular, it provides an extensive survey of current image registration methodologies, in the context of remote sensing.

Part II (Chapters 4–6) describes different possible choices of similarity metrics (Step (1) in Section 1.2), namely correlation, phase correlation, and mutual information. Methods based on these similarity measures are described from both a theoretical and practical standpoint, including their performance on synthetic and real data. Part III (Chapters 7–13) investigates Step (2) defined in Section 1.2. Specifically different choices of features (e.g., points, wavelets, contours, etc.) are discussed, and various feature-matching techniques and strategies, involving, for example, a hierarchical, multiresolution approach, robust feature matching, and different types of optimization. Finally, Part IV (Chapters 14–22) describes several remote sensing systems – most currently operational – with applications spanning many Earth remote sensing domains, such as land cover, meteorology, and geological and ocean studies, to name a few. The sensors that are studied in this part include IKONOS, Landsat, AVHRR, SPOT, and VEGETATION, which all focus mainly on land cover and urban studies; SeaWiFS, geared towards ocean color; GOES and METEOSAT, used for weather applications; and MISR and MODIS, which provide information about land surface and the atmosphere, at regional and global scales, for studies of the Earth climate.

1.6 Terminology

This section describes some of the terms used in the various chapters of the book, as the readers might find it useful to refer to this (alphabetically sorted) glossary when reading some of the subsequent chapters.

- *Along-track scanning*

   Uses a linear array of sensors that are "pushed along" the direction of travel (hence the term "pushbroom scanners").

- *Ancillary data*

   Refer to the data from sources other than remote sensing that are used to analyze remote sensing data.

- *Control point (CP) or ground control point (GCP)*

   A point on the Earth surface whose location is known very accurately and that is used to georeference image data.

- *Cross-track scanning*

   Uses a scanning mirror that rotates as it acquires the data (hence the name "whiskbroom scanners").

- *Distributed Active Archive Center (DAAC)*

   Centers that process, archive, and distribute data and products from NASA's past and current Earth-observing satellites and field measurement programs. There are currently 12 DAACs, each serving a specific Earth system science discipline.

- *Earth Observing System (EOS)*

   A coordinated series of polar-orbiting and low-inclination satellites for long-term global observations of the land surface, biosphere, solid Earth, atmosphere, and oceans.

- **EOS data levels**

  Level 0 data are the reconstructed raw instrument data at full resolution.

  Level 1A data are reconstructed, time-reference raw data, with ancillary information including radiometric and geometric coefficients.

  Level 1B data are corrected Level 1A data (in sensor units).

  Level 2 data are derived geophysical products from Level 1 data, at the same resolution and location, e.g., atmospheric temperature profiles, gas concentrations, or winds variables.

  Level 3 data correspond to the same geophysical information as Level 2, but mapped onto a uniform space-time grid.

  Level 4 data are model output or results from prior analysis of lower-level data.

- **Ephemeris data**

  A set of parameters that are acquired onboard the spacecraft and that can be used to calculate accurately the location of a satellite.

- **Field of view (FOV)**

  The angle over which the sensor observes and records data.

- **Geostationary (GEO) orbit**

  A constant and circular orbit above the Equator, in which the satellite travels, in the same direction and at the same speed as the Earth, thus appearing to be *stationary* with respect to a specific location on the Earth.

- **Geosynchronous orbit**

  An orbit around the Earth whose period matches the Earth's rotation period, i.e., each point on the Earth is observed at the same time every day. A geostationary orbit is a special case of a geosynchronous orbit.

- **Hyperspectral imaging**

  Simultaneous detection or sensing in hundreds to thousands of spectral channels, covering almost completely a limited portion of the electromagnetic spectrum.

- **Instantaneous field of view (IFOV)**

  The smallest solid angle through which a sensor is sensitive to radiation. It is also related to the spatial resolution of the sensor and to its altitude.

- **Line spread function (LSF)**

  A detector's response to light from an ideal light source; defines the apparent shape of an object as it appears in the output image.

- *Low-Earth orbit (LEO)*

An orbit of altitude lower than 2000 km.

- *Medium-Earth orbit (MEO)*

An orbit usually above a low-Earth orbit and below a geostationary orbit (i.e., above 2000 km and below 35 786 km).

- *Multispectral imaging*

Simultaneous detection in several bands (usually less than 10 or 20) covering separate portions of the electromagnetic spectrum.

- *Point spread function (PSF)*

A measure of the geometric performance of an optical system; defines the apparent shape of a point as it appears in the output image.

- *Polar orbit*

An orbit passing near the Earth's North and South Poles.

- *Pushbroom scanner*

See along-track scanning.

- *Radiometric resolution*

The number of "bits" used to record a given amount of energy corresponding to a given wavelength.

- *Spatial resolution*

The area on the ground utilized to compute the value assigned to each pixel.

- *Spectral resolution*

The bandwidths utilized in the electromagnetic spectrum.

- *Sun-synchronous satellite*

A polar orbiting satellite that keeps pace with the Sun's westward progression compared to the Earth rotation; a Sun-synchronous satellite always crosses the Equator at the same local Sun time.

- *Temporal resolution*

Corresponds to the observation frequency, defined by the orbit of the satellite and the scanning of the sensor.

- *Troposphere*

The lowest layer of the atmosphere; up to 10 km above the Earth and right below the stratosphere.

- *United States Geological Survey (USGS) Center for Earth Resources Observation and Science (EROS)*

Responsible for the processing systems that capture, correct, and distribute land remote sensing data, such as Landsat data products.

- *Whiskbroom scanner*

See cross-track scanning.

1.7 Conclusion

As stated at the beginning of the Introduction, this book is aimed at several types of readers, including Earth scientists, image processing researchers, engineers, instructors, and students. With contents evolving from general overview material to theoretical descriptions of the separate components of image registration to specific studies of operational systems, the book chapters can be read sequentially, similarly to a textbook, or any one chapter can be read at a time, out of order, depending on the needs of the readers. We hope that readers will enjoy the material presented, as much as we enjoyed putting it together.

References

Asrar, G. and Dozier, J. (1994). *EOS: Science Strategy for the Earth Observing System.* Woodbury, NY: AIP Press.

Brown, L. G. (1992). A survey of image registration techniques. *ACM Computing Surveys*, **24**(4), 325–376.

Campbell, J. B. (1996). *Introduction to Remote Sensing*, 2nd edn. New York: Guilford Press.

Cary, T. (1994). A world of possibilities: Remote sensing data for your GIS. *Geo Info Systems*, **4**(9), 38–42.

Dai, X. and Khorram, S. (1998). The effects of image misregistration on the accuracy of remotely sensed change detection. *IEEE Transactions on Geoscience and Remote Sensing*, **36**(5), 1566–1577.

Ehlers, M. (1995). Integrating remote sensing and GIS for environmental monitoring and modeling: Where are we? *Geo Info Systems*, 36–43.

El-Rabbany, A. (2002). *Introduction to GPS: The Global Positioning System.* Norwood, MA: Artech House, Inc.

Le Moigne, J. and Cromp, R. F. (1999). Satellite imaging and sensing. In J. G. Webster, ed., *The Measurement, Instrumentation and Sensors Handbook.* Boca Raton, FL: CRC Press, pp. 73.42–73.63.

Lillesand, T. M. and Kiefer, R. W. (1987). *Remote Sensing and Image Interpretation*, 2nd edn. New York: John Wiley & Sons.

Logsdon, T. (1997). *Orbital Mechanics*. New York: John Wiley & Sons.

Short, N. M. (2009). *The Remote Sensing Tutorial*. http://rst.gsfc.nasa.gov/; also NASA Reference Publication 1078 and Library of Congress Catalog Card No. 81–600117.

Townshend, J. R., Justice, C. O., Gurney, C., and McManus, J. (1992). The impact of misregistration on change detection. *IEEE Transactions on Geoscience and Remote Sensing*, **30**(5), 1054–1060.

2

Influence of image registration on validation efforts

BIN TAN AND CURTIS E. WOODCOCK

Abstract

Currently, several high level satellite-based data products relevant to Earth science system and global change research are derived from medium spatial resolution remote sensing observations. Because of their high temporal frequency and important spatial coverage, these products, also called biophysical variables, are very useful to describe the mass and energy fluxes between the Earth's surface and the atmosphere. In order to check the relevance of using such products as inputs into land surface models, it is of importance to assess their accuracy. Validation activity consists in evaluating by independent means the quality of these biophysical variables. Currently, these variables are derived from coarse-resolution remote sensing observations. Validation methods consist in generating a ground truth map of these products at high spatial resolution. It is produced by using ground measurements of the biophysical variable and radiometric data from a high spatial resolution sensor. The relationship between biophysical variable and radiometric data, called the transfer function, allows extending the local ground measurements to the whole high spatial resolution image. The resulting biophysical variable map is aggregated to be compared with the medium spatial resolution satellite biophysical products. Three important geometrical issues influence the validation results: (1) The localization error of the local ground measurements, (2) the localization error between high and medium resolution images, and (3) the *point spread function* (PSF) associated with the medium spatial resolution remote sensing observations. This work aims at analyzing the influence of these issues within the validation process. First, these problems are investigated at the field measurement scale. The localization error between ground measurement located by the Global Positioning System (GPS) and high spatial resolution pixel is modeled by a Gaussian random variable. For each possible relative position, the ground measurement is related to the surrounding area radiometric data weighted by the ground measurement spatial support area. A Monte Carlo simulation scheme accounting for the localization error provides the probability distribution function of the parameters of the transfer function. Second, registration error is investigated when an aggregated high spatial resolution image is compared with a coarse resolution image. The registration error is minimized by getting the best geometrical match between the two images using correlation techniques. Finally, the importance of the PSF associated with the medium spatial resolution biophysical products is evaluated using the Moderate Resolution Imaging Spectroradiometer (MODIS). This investigation is applied to several validation sites expressing a range of spatial heterogeneity. Results are discussed with emphasis on the possibility of achieving the validation at the original resolution of the medium resolution products.

2.1 Introduction

Satellite remote sensing is the most effective way for collecting global data continuously. Satellite data improve our understanding of global dynamics and processes occurring on the land, in the oceans, and in the lower atmosphere. With satellite data being widely used, assessing the quality of satellite data becomes a key issue. The accuracy of satellite data is determined primarily through validation efforts (Liang *et al.*, 2002; Morisette *et al.*, 2002; Privette *et al.*, 2002; Wan *et al.*, 2002; Wang *et al.*, 2004; Tan *et al.*, 2005). Validation is the process of assessing by independent means the accuracy of the data products derived from sensor system outputs. In general, reference data, which is assumed to represent the target value, is compared with the satellite sensor derived product; and in order to compare two datasets, they must be registered in a common coordinate system, normally represented with latitude/longitude. The geolocation of reference data is usually measured with a Global Positioning System (GPS), whose location accuracy without differential correction is about 10–15 m, since the deliberate selective availability error was removed from the U.S. GPS system after May 2, 2000. When the differential correction is applied, the GPS location accuracy is improved to within 2 m. This accuracy is sufficient to validate moderate or coarse resolution satellite data. However, the geolocation of satellite data is calculated from the satellite position by complex models. The accuracy of registering satellite data to a coordinate system is limited by both model accuracy and the nature of satellite sensors. In this chapter we focus on the effects of geolocation error on MODIS data and its implications for the validation of MODIS products.

2.2 Product and algorithm validation

There are (at least) two types of validation activities for satellite products. The first is to assess the satellite product accuracy independently of the observations from which it is derived. In this approach the products are viewed much like maps. They can have both attribute error and locational error and there is not necessarily any effort to determine the source of the errors. The errors could be due to a wide variety of sources, including imperfections in the observations, their registration, and imperfections in the algorithms used to generate the products. In this book, we refer to this approach as *product validation*. A second approach is to evaluate algorithm accuracy. This is an attempt to quantify the contribution to product error coming from imperfections in the algorithm. In this book, we refer to this approach as *algorithm validation.*

Product accuracy could be easily achieved through comparing the reference data and the satellite product directly. However, the estimate of the algorithm accuracy is not straightforward due to the entangling of imperfect algorithms, imperfect observations, and locational errors. Is there any analytical relationship between the results of product validation and algorithm validation? The answer is yes. Let us define the quantity R^2 as the square of the correlation coefficient for two variables, which is a common measure of the strength of the relationship between a dependent variable and one or more independent variables.

Let Y represent the reference data, X represent the ideal/perfect satellite data, and X' represent the actual/imperfect satellite data. Then, $R^2_{(Y,X)}$, $R^2_{(Y,X')}$, and $R^2_{(X,X')}$ represent the correlation coefficients between Y and X, Y and X', and X and X' respectively.

$R^2_{(Y,X)}$ could be considered as the result of algorithm validation, while $R^2_{(Y,X')}$ is the result of product validation. $R^2_{(X,X')}$ is equal to 1 if satellite observations are perfect, which means free of all kinds of errors and biases, such as up-stream algorithm errors, geolocation errors, etc. It is reasonable to assume the relationship between X and X' is a line relationship, i.e.,

$$X' = aX + b + \varepsilon,$$

where a and b are constant, and ε is a Gaussian distributed random error independent of both X and X'. In such a situation, the relationship among $R^2_{(Y,X)}$, $R^2_{(Y,X')}$, and $R^2_{(X,X')}$ could be derived as follows:

$$R^2_{(Y,X')} = \frac{\text{cov}^2(Y, X')}{\text{var}(Y) \times \text{var}(X')} = \frac{\text{cov}^2(Y, aX + b + \varepsilon)}{\text{var}(Y) \times \text{var}(aX + b + \varepsilon)}$$

$$= \frac{a^2\text{cov}^2(Y, X)}{a^2\text{var}(X) \times \text{var}(Y) + \text{var}(\varepsilon) \times \text{var}(Y)}.$$

Then,

$$\frac{1}{R^2_{(Y,X')}} = \frac{var(Y) \times var(X)}{cov^2(Y,X)} + \frac{var(Y) \times var(\varepsilon)}{cov^2(Y,X)} = \frac{1}{R^2_{(Y,X)}} + \frac{1}{R^2_{(Y,X)}} \left(\frac{1}{R^2_{(X',X)}} - 1 \right)$$

$$= \frac{1}{R^2_{(Y,X)} \times R^2_{(X',X)}}.$$

Therefore,

$$R^2_{(Y,X')} = R^2_{(Y,X)} \times R^2_{(X',X)}, \tag{2.1}$$

which means that the *product accuracy* could be expressed as a product of the algorithm accuracy and the accuracy of the remote sensing observations.

$R^2_{(Y,X')}$ can be estimated through product validation, that is, through comparison of reference data with map products. $R^2_{(Y,X)}$, the algorithm validation result, can be estimated if $R^2_{(X',X)}$, the relationship between perfect and imperfect satellite observation, is derivable according to

$$R^2_{(X',X)} = \frac{cov^2(X,X')}{var(X) \times var(X')}. \tag{2.2}$$

X' depends on the nature of the sensor, its calibration, atmospheric correction, the strategy for allocating satellite observations to a predefined grid system, and the heterogeneity magnitude of the land surface. Among all impact factors, the heterogeneity, which is usually measured with semivariance (Curran, 1988; Woodcock *et al.*, 1997; Tian *et al.*, 2002), is the most difficult one to remove/isolate because it is not a systematic bias and it varies from one pixel to another. As a result, estimation of $R^2_{(X',X)}$ can be more complicated than simply using the value of a variogram of the distance associated with the mean offset between imperfect and perfect observations. A better solution for this situation is needed and is a topic of current research.

2.3 Locational errors introduced in image registration

The locational error in gridded satellite products comes from two sources: (1) Assigning a coordinate (latitude/longitude) to an observation with a model, and (2) registering observations to a predefined grid system (MODIS data are registered into a sinusoidal grid). The accuracy of assigning a location to individual MODIS observations is known (Wolfe *et al.*, 2002). However, the mismatch between satellite observations and the predefined grid system could reduce the locational accuracy of satellite products.

The dimensions of a grid cell are normally the same as the dimensions of satellite observations (e.g., Landsat) or the same as the dimensions of the satellite

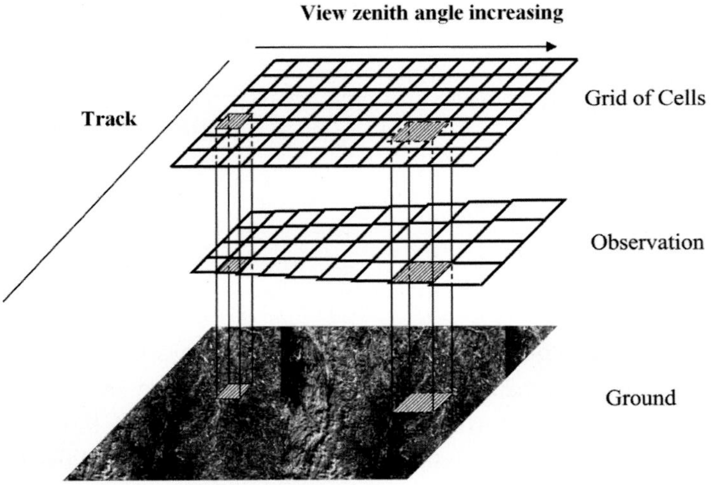

Figure 2.1. For wide-swath sensors, the observation dimension increases as the view zenith angle increases. The observation dimension is the same as the grid dimension at nadir, while the observation dimension could be eight times larger than the grid cell size at the end of the swath.

observation at nadir (e.g., AVHRR, MODIS, etc.) if the observation dimensions vary within a scene. Rarely do the boundaries of an observation coincide with the boundary of a predefined grid cell due to mismatches in size, location or alignment (Fig. 2.1). The nearest-neighbor method is widely used to allocate observations to predefined grid cells. The magnitude of the mismatch/misalignment following gridding is valuable to validation efforts as the reference data rarely correspond directly with the location from which the observations were derived. However, this mismatch between the grid cell location and the observations in the grid cell is normally ignored in validation efforts due to the difficulty or the impossibility of getting such information.

For wide-swath sensors (like AVHRR and MODIS) whose observation dimensions vary as a function of the view zenith angle (Fig. 2.1), the mismatch becomes worse towards the end of the swath. The observation dimensions are significantly greater than the grid cell dimension at the end of the swath, and one observation is allocated to as many as eight adjacent grid cells (Fig. 2.2). In addition, when the point spread function (PSF) of a sensor is a triangular function, as with the MODIS sensors in along-scan direction, the mismatch between observations and grid cells becomes even worse. Figure 2.3 presents the MODIS PSF in the along-scan direction. Only 75% of an observation comes from the nominal observation area. This means that, even when the grid cell and the nominal observation coincide perfectly in location (a very rare situation), the observation still cannot perfectly represent the grid area.

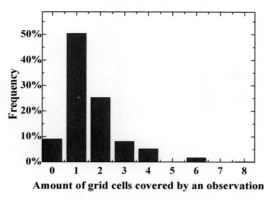

Figure 2.2. Frequency distribution of the number of grid cells to which single observations of MODIS are assigned. Approximately 41% of the observations are allocated to multiple adjacent cells. An observation can be allocated to as many as eight adjacent cells. Approximately 9% of the observations are not used in the output image because of the overlap between consecutive scan lines. A grid cell can be observed multiple times, but only one observation can be allocated to this cell. Therefore, some observations are not allocated to any cell.

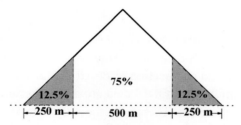

Figure 2.3. Triangular sensor point spread function (PSF) (examples at 500-m resolution). The nominal observation area contributes 75% of the actual observation. (Derived from Figure 2.6 in Nishihama *et al.*, 1997.)

2.4 Impact of locational error in validation efforts

The low correspondence between grid cells and the observations in them has particularly significant implications for validation of satellite data with reference data. Comparison of satellite data products with reference data for a cell can be misleading because they generally will not be from the same location. For product validation (defined in Section 2.1) this issue is not terribly relevant, as the mismatch between grid cells, reference data, and observations is one of many factors contributing to product inaccuracy. But for algorithm validation, it is important that reference data and observations do match. In order to evaluate quantitatively the impact of locational errors in validations, a set of simulated MODIS data is produced as an example. Both simulated MODIS data (500-m resolution) and reference data (500-m resolution) are generated from the same Landsat data (30-m resolution).

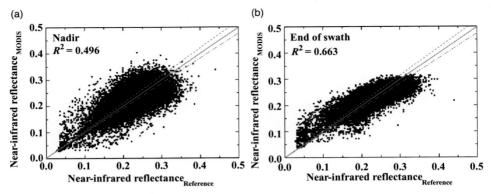

Figure 2.4. Near-infrared reflectance (NIR) of reference data versus NIR of simulated data: (a) Near nadir, and (b) at the end of the swath, where view zenith angle is high. The cell-by-cell correspondence between values can be expected to be poor due to the locational error in gridded data. The solid gray line is the 1:1 line. The dash gray line is the 95% confidence level. There are 53.8% and 47.9% points outside of the 95% confidence level at nadir and the end of swath, respectively. Reference data and simulated MODIS data were generated using the Landsat ETM+ data of the Konza Prairie LTER of April 4, 2000. (See Tan *et al.*, 2005.)

The reference data are simply averaged Landsat pixels which follow a 500-m grid, while the simulated MODIS data incorporate the effects of wide-swath, triangular PSF effect, pixel shift, and geolocation errors (for details, refer to Tan *et al.*, 2005). The reference and simulated MODIS data are atmospherically corrected with the same method and parameters. Therefore, the only source of difference between them is the locational error introduced in the image registration process, which is the process of transforming the reference and simulated MODIS data into one coordinate system. From the results shown in Fig. 2.4, it is clear that in general the reference data and MODIS data do not match. For example, for the near-infrared band about 50% of MODIS pixels are beyond 95% of the range with an error of $\pm 5\%$. (It is interesting to remark that the percentage within the 5% error threshold increases as the view angle increases.) A quantitative indicator of the locational error is *obscov* (Wolfe *et al.*, 2002), which is defined as the integrated response over the overlapped area of observation and grid cell divided by the integrated response over the area of the observation footprint (Yang and Wolfe, 2001; Tan *et al.*, 2005). Figure 2.5 shows the frequency distribution of *obscov* for 50 days of simulated MODIS data along with *obscov* for Collection 4 MODIS 500 m daily products for tile h18v04 and h12v04 from January 1 to February 19, 2004. The result shown in Fig. 2.5 confirms that the MODIS observations and predefined grid cells generally do not coincide. The average overlap between observations and their grid cells is less than 30%. The maximum possible overlap is only 75%

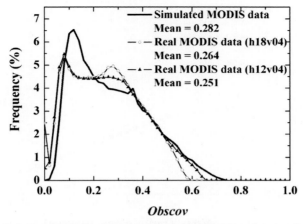

Figure 2.5. Frequency distribution of *obscov* for 50 days of simulated MODIS data and of *obscov* of the Collection 4 MODPTHKM product in tile h18v04 and h12v04 from January 1 to February 19, 2004. The similarity between the observed and simulated *obscov* distributions helps verify that the simulation is a good approximation of the real MODIS processing stream.

due to the triangular PSF (Fig. 2.3). Therefore, the direct comparison between reference data and MODIS data is not feasible. Other alternatives are required to perform cell-by-cell comparisons. One possibility is to use the geolocation data provided with MODIS products to take into account the effects of a pixel shift. This approach has the advantage of improving the locational accuracy of individual MODIS observations. But it is important to remember that this approach does not solve the problem entirely, as the effects of locational error will still exist (Wolfe *et al.*, 2002). Another option is to aggregate the data, which tends to improve *obscov* (see Fig. 2.6). It is worth noting that an average *obscov* of 80% can be achieved by aggregation to eight times the native resolution (or 2 km for the 250-m bands).

Another way to explore the effect of *obscov* on MODIS data quality is to use cell-by-cell comparison of Normalized Difference Vegetation Index (NDVI) values between the reference data and the simulated MODIS data (Fig. 2.7). The cells near nadir with *obscov* > 0.7 and *obscov* < 0.3 are compared with reference data separately. As expected, there is better correspondence between reference and simulated MODIS data ($R^2 = 0.83$) for large *obscov* values (Fig. 2.7(a)), and poor correlation ($R^2 = 0.43$) for small *obscov* values (Fig. 2.7(b)). In MODIS data, the *obscov* values of adjacent cells can be quite different. This spatial variability in *obscov* is due to misalignment and mismatch between observations and grid cells. The spatial distribution pattern of *obscov* differs from one place to another, and depends on the alignment between the orientation of observations and the grid cells. The comparison between reference data and simulated MODIS data at the

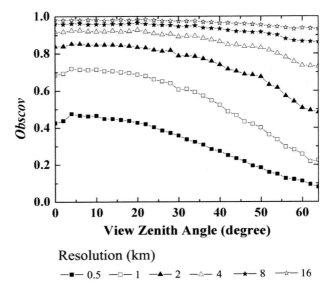

Figure 2.6. Impact of view zenith angle (VZA) and spatial resolution on *obscov*. This parameter increases as resolution decreases, and decreases with increasing VZA. The result shown here is the average over 50 days of simulations. Note that to have at least 80% overlap between the location of grid cells and their observations, MODIS data need to be aggregated by a factor of eight (from 500 m to 4 km in this example) and high VZAs avoided.

end of a swath is shown in Fig. 2.7(c). The correlation is poor ($R^2 = 0.54$). The NDVI dynamic range in the reference data is larger than in the simulated MODIS data. This "blurring" effect in MODIS is due to the growth of the observation dimensions near the end of the swath.

2.5 Summary

The image registration process unavoidably results in a pixel shift effect, which is moderate for nadir-view sensors (e.g., Landsat) and significant for wide-swath sensors (e.g., MODIS). The locational error introduced in the image registration process undermines the local spatial properties of satellite data. For algorithm validation, direct comparison of reference data on a pixel-by-pixel basis with satellite data products will introduce error unrelated to the algorithm (Figs. 2.4 and 2.7). Two alternatives are recommended to reduce this effect. The first is to utilize the detailed geolocation information about the satellite observations, including the *obscov* value, the dimensions of the observation, and the offset between the grid cell and the observation. The ground area covered by an observation can be more accurately located using this information than the location of the grid cell. In such a

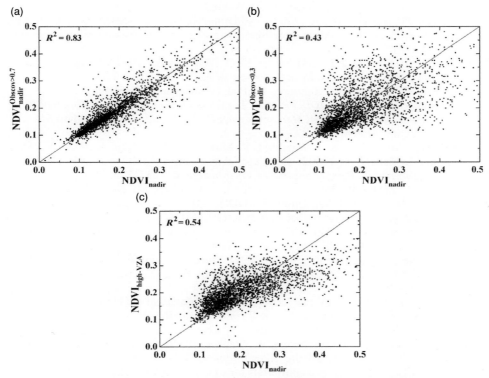

Figure 2.7. NDVI of reference data versus NDVI of simulated MODIS data: (a) Near nadir when *obscov* values are greater than 0.7, (b) near nadir when *obscov* values are less than 0.3, and (c) for high VZA regardless *obscov*. The cell-by-cell correspondence between values can be expected to be poor when *obscov* values are low. Reference data and simulated MODIS data were generated using the ETM+ data of the Konza Prairie LTER of April 4, 2000.

way, the comparison between field data and satellite data is more reliable. However, it is important to note that using this approach will not allow for perfect alignment of ground reference data and satellite observations, as geolocation error will still be present. The second approach, when detailed geolocation data are not available, is to compare them at a coarser resolution (e.g., patch level) rather than the native resolution (Tian *et al.*, 2002; Wang *et al.*, 2004; Tan *et al.*, 2005). For this approach, the validation site should be large to minimize the effect of the locational error and to minimize the noise associated with individual observations.

References

Curran, P. J. (1988). The semivariogram in remote sensing: An introduction. *Remote Sensing of Environment*, **24**, 493–507.

Liang, S., Fang, H., Chen, M., Shuey, C. J., Walthall, C., Daughtry, C., Morisette, J., Schaaf, C., and Strahler, A. (2002). Validating MODIS land surface reflectance and albedo products: Methods and preliminary results. *Remote Sensing of Environment*, **83**, 149–162.

Morisette, J. T., Privette, J. L., and Justice, C. O. (2002). A framework for the validation of MODIS land products. *Remote Sensing of Environment*, **83**, 77–96.

Nishihama, M., Wolfe, R. E., Solomon, D., Patt, F., Blanchette, J., Fleig, A., and Masuoka, E. (1997). *MODIS Level 1A Earth Location: Algorithm Theoretical Basis Document. NASA Technical Memorandum*, SDST-092, Version 3. Goddard Space Flight Center, Greenbelt, MD.

Privette, J. L., Myneni, R. B., Knyazikhin, Y., Mukufute, M., Roberts, G., Tian, Y., Wang, Y., and Leblanc, S. G. (2002). Early spatial and temporal validation of MODIS LAI product in Africa. *Remote Sensing of Environment*, **83**, 232–243.

Tan, B., Hu, J., Zhang, P., Huang, D., Shavanov, V. N., Weiss, M., Knyazikhin, Y., and Myneni, R. B. (2005). Validation of MODIS LAI product in croplands of Alpilles, France. *Journal of Geophysical Research*, **110**, D01107, doi:10.1029/2004JD004860.

Tian, Y., Woodcock, C. E., Wang, Y., Privette, J. L., Shabanov, N. V., Zhou, L., Zhang, Y., Buermann, W., Dong, J., Veikkanen, B., Häme, T., Andersson, T. K., Ozdogan, M., Knyazikhin, Y., and Myneni, R. B. (2002). Multiscale analysis and validation of the MODIS LAI product: I. Uncertainty assessment. *Remote Sensing of Environment*, **83**, 414–430.

Wan, Z., Zhang, Y., Zhang, Q., and Li, Z. (2002). Validation of the land-surface temperature products retrieved from Terra Moderate Resolution Imaging Spectroradiometer data. *Remote Sensing of Environment*, **83**, 163–180.

Wang, Y., Woodcock, C. E., Buermann, W., Stenberg, P., Voipio, P., Smolander, H., Hame, T., Tian, Y., Hu, J., Knyazikhin, Y., and Myneni, R. B. (2004). Evaluation of the MODIS LAI algorithm at a coniferous forest site in Finland. *Remote Sensing of Environment*, **91**, 114–127.

Wolfe, R. E., Nishihama, M., Fleig, A. J., Kuyper, J. A., Roy, D. P., Storey, J. C., and Patt, F. S. (2002). Achieving sub-pixel geolocation accuracy in support of MODIS land science. *Remote Sensing of Environment*, **83**, 31–49.

Woodcock, C. E., Collins, J. B., and Jupp, D. L. B. (1997). Scaling remote sensing models. In *Scaling-up: From Cell to Landscape*, P. R. van Gardingen, G. M. Foody, and P. J. Curran, eds. Cambridge: Cambridge University Press, pp. 61–77.

Yang, K., and Wolfe, R. E. (2001). MODIS level 2 grid with the ISIN map projection. In *Proceedings of the IEEE International Geoscience and Remote Sensing Symposium*, Sydney, Australia, Vol. 7, pp. 3291–3293.

3

Survey of image registration methods

ROGER D. EASTMAN, NATHAN S. NETANYAHU,
AND JACQUELINE LE MOIGNE

3.1 Introduction

Automatic image registration, bringing two images into alignment by computing
a moderately small set of transformation parameters, might seem a well-defined,
limited problem that should have a clear, universal solution. Unfortunately, this
is far from the state of the art. With a wide spectrum of applications to diverse
categories of data, image registration has evolved into a complex and challenging
problem that admits many solution strategies. The growing availability of digital
imagery in remote sensing, medicine, and numerous other areas has driven a
substantial increase in research in image registration over the past 20 years. This
growth in research stems from both this increasing diversity in image sources,
as image registration is applied to new instruments like hyperspectral sensors in
remote sensing and medical imaging scanners in medicine, and new algorithmic
principles, as researchers have applied techniques such as wavelet-based features,
information theoretic metrics and stochastic numeric optimization.

This chapter surveys the diversity of image registration strategies applied to
remote sensing. The objectives of the survey are to explain basic concepts used in
the literature, review selected algorithms, give an overall framework to categorize
and compare algorithms, and point the reader to the literature for more detailed
explanations. While manual and semi-manual approaches are still important in
remote sensing, our primary intent is to review research approaches for building
fully automatic and operational registration systems. Following the survey article
by Brown (1992), we review an algorithm by considering the basic principles from
which it is constructed. These principles include, among others, the measure of
similarity used to compare images and the optimization algorithm used to optimize
the measure. Most image registration algorithms in the literature and those used
in practice are based on variations on these basic elements and their combination.
Indeed, a reader familiar with the basic, major principles and their various combinations, can easily gain a good understanding of a new image registration technique

or system. A survey of the scientific literature on image registration specifically for remote sensing is provided by Fonseca and Manjunath (1996). Previous surveys of the general image registration literature include Brown (1992) and Zitová and Flusser (2003), while general books on registration include Modersitzki (2004) and Goshtasby (2005). Surveys limited to medical applications, with specific focus on mutual information, include Maintz and Viergever (1998) and Pluim *et al.* (2003).

Within the wide spectrum of image registration principles and applications, this chapter focuses on techniques relevant to automatic registration of regular two-dimensional image data from Earth satellite instruments used for remote sensing, primarily those instruments sensing in the visible or near-visible spectra. We do not treat extensively methods for instruments that use radar, those that directly produce range information, or sensors borne aloft by airplanes and balloons. These forms of imagery may differ from satellite imagery in perspective and other characteristics. However, since new image registration techniques are often imported into remote sensing from research undertaken in other fields, we also review some methods from those fields as appropriate. Other fields with active research in image registration include medical imaging, video analysis for multimedia, and robotics. These fields differ from remote sensing in their requirements for image registration. In contrast to remote sensing, medical image registration works with a large number of imaging modalities from whole body to retinal scans, and involves tissues that can deform or change drastically. Similarly, image registration for multimedia video analysis and robotics works on short-range imagery of complex 3D scenes, rather than long-range imagery of planetary surfaces. We specifically do not include in our review articles on elastic or nonrigid registration, 3D volumetric registration, and 3D range registration. Readers interested in the latter topics are referred to Maintz and Viergever (1998), Lester and Arridge (1999), Goshtasby (2005), and Salvi *et al.* (2006).

Regardless of the field of application, the articles reviewed in this chapter originate roughly from three overlapping communities. The image processing community focuses on developing methodologies and techniques, and a research contribution presents usually a novel technique or an evaluation method. Such contributions include mostly proof-of-concept demonstrations on limited image sets so the techniques described can be considered promising prototypes. A second set of contributions comes from the community of ground support satellite teams. This community specializes in the design of effective, practical algorithms used to create orthorectified image products for a limited set of instruments. The techniques they describe serve as a basis for successful operational systems run on extensive datasets. Operational satellite teams have an end-to-end understanding of data processing used in their satellite system, with inside knowledge of sensor and satellite engineering. They use image registration to update orbital and navigational models, calibrate sensors, and validate image products. Given the initial satellite position and orientation, their imagery may be closely aligned prior to registration, which reduces the search time during registration. The last set of contributions is made by the remote sensing user community. The articles by this community raise various issues and describe common practice of using image registration in the analysis of imagery for specific applications, such as land use planning or crop yield forecasting. The remote sensing user community has a broad set of interests, including determining the best algorithm to use for remotely sensed datasets, the extent to which misregistration errors impact their analyses, whether their data have been registered properly, and which algorithm best combines heterogeneous satellite image sources with maps, Geographic Information Systems (GIS), airborne instrument imagery, and field data. End user datasets may vary greatly in initial alignment, so fully automatic registration software would need to be more robust to large geometric transformations. In Eastman *et al.* (2007) we briefly surveyed current practice by the operational and user communities, and noted the different requirements for them.

Each article on image registration addresses some variation of the general problem. The variation addressed may vary by community or application, or the research intentions of the authors. An overriding objective is to develop an accurate, fully automatic registration algorithm that is successful without human intervention. Other objectives an article may address include:

(1) Improving robustness and reliability, so a registration algorithm better handles noise, occlusions, and other problems.
(2) Refining the geometric transformation to better model the complex physical imaging process of an instrument and satellite.
(3) Increasing the subpixel accuracy of the transformation computed to meet higher operational requirements.
(4) Speeding up the registration process to handle greater throughput and more complex algorithms.
(5) Handling large and unknown initial transformation estimates, so an algorithm can match across greater displacements.
(6) Managing multimodal registration, so an algorithm can be applicable to images with radiometric, scale, and other differences that might be present across band, instrument, or platform.

As the topic of image registration for remote sensing is broad and can be broken down into many different categorical schemes, this chapter addresses each subtopic from different aspects. In Section 3.2 we define a terminology for the components of image registration. This terminology will allow us to decompose systems into their major components. In Section 3.3 we review major algorithmic categories to examine how these various components are most commonly combined into full systems. In Section 3.4 and Section 3.5 we review, respectively, particular characteristics of geometric and radiometric transformations relevant to remote sensing. In Section 3.6 we review articles that evaluate image registration algorithms and systems. Finally, Section 3.7 contains concluding remarks.

3.2 Elements of image registration algorithms

An instance of a pairwise image registration problem is to align one image, the *sensed image*, onto a second image, the *reference image*, by computing a transformation that is optimal in some sense. We will define an image registration problem by the following five elements:

(1) Reference image, $I_1(x, y)$, that is generally taken to be unchanged.
(2) Sensed image, $I_2(x, y)$, that is transformed to match the reference.
(3) Geometric transformation, f, that maps spatial positions in one image to the other.
(4) Radiometric transformation, g, that transforms intensity values in one image to the other.
(5) Noise term, $n(x, y)$, that models sensor and other imaging noise.

These definitions lead to the following relationship between the reference and sensed images, where (x, y) represent the coordinate system in the reference image, and (u, v) the coordinate system in the sensed image. Two image locations in the reference and sensed image are said to correspond or to be corresponding points if they are mapped into each other under the geometric transformation, i.e.,

$$I_1(x, y) = g(I_2(f_x(u, v), f_y(u, v))) + n(x, y). \tag{3.1}$$

The above informal model is only intended to provide context for our review. Readers interested in a formal mathematical definition of the image registration problem are referred to Modersitzki (2004).

Intuitively, solving an instance of the image registration problem requires computing the geometric and radiometric transformations, so that Eq. (3.1) holds. More generally, the image registration problem is to devise an algorithm that, for all pairs of images from two sets of images in the domain of interest, will compute the optimal transformations for any instance, where the definition of optimal depends on the choice of similarity measure as defined below.

This simplified, general model has three immediate caveats. Requiring strict equality between corresponding image intensities in the reference and sensed images is a strong constraint. This constraint is frequently relaxed to a statistical relationship (Roche *et al.*, 2000) or an information-theoretic one (Maes *et al.*, 1999), which require that corresponding regions show similar distributions under the optimal transformations. Also, the noise term can confound a number of factors, both additive and multiplicative, so a single additive term is only a weak approximation. Some image differences can be the result of atmospheric or sensor noise, while others could stem from temporal ground changes that are the signal of interest. Thus, care must be used in defining image differences as "noise."

From the general model relating the reference and sensed image, we may categorize algorithms according to the widely accepted framework given in Brown (1992). Brown's review described four standard elements in the design of an image registration algorithm:

(1) Search space that defines the possible transformations between the reference and target images, both geometric and radiometric, that will be considered.
(2) Feature space of information content extracted from the images to be used in their comparison.
(3) Similarity metric that defines the merit of matching image features under given transformations.
(4) Search strategy used to find the optimal transformation, i.e., the transformation in the parameter space that maximizes the similarity metric.

Beyond these four standard elements of an image registration scheme, there is a fifth element critical to their understanding, namely the validation of an image registration algorithm as accurate and reliable. While not an integral component of the registration scheme itself, the procedures of testing, evaluating, and validating these schemes comprise an important element in studying various approaches of image registration.

3.3 General approaches to registration

Image registration algorithms are often characterized as *area-based* or *feature-based*. In area-based algorithms, areas or regions of the original image data are matched with minimal preprocessing or with preprocessing that preserves most of the image data. These algorithms may compute the differences of raw image pixel values, or use all pixel values to compute an intermediate full-information representation like the Fourier coefficients. Area-based methods are often labeled correspondence-less matching, as an entire area is matched without constructing an explicit correspondence between points in the two images. In feature-based algorithms, on the other hand, the original images are preprocessed to extract distinctive, highly informative features that are used for matching. These algorithms may extract a few distinctive control points (CPs), or detect dense edges/contour maps for matching. Image registration systems for remote sensing often combine the two approaches at different levels. At one stage the system may use feature-based control points, followed at a later stage by area-based matching. This may be very appropriate when large distortions or displacements make feature-point matching more robust. This is because local regions are warped, so individual pixels align very poorly, but derived features are more invariant to the distortions. After those large transformations are initially accounted for, area-based matching can be effectively performed.

3.3.1 Manual registration

While laborious and prone to error, manual registration, which is based on selection of control points, is still widely used and highly regarded in the remote sensing community. The user visually selects distinctive matching points from two images to be registered, and then uses those points to compute and validate a geometric transformation. Remote sensing software packages, such as ITT's ENVI and PCI Geomatics' Geomatica, typically support manual selection. One advantage to manual registration is that it is easy to understand and implement. Manual selection allows the user to refine the set of control points for a number of reasons, including focusing on a region of interest while ignoring sections of the image that are not under study, fine-tuning the geometric transformation to meet accuracy objectives, and basing the control points on known ground features. And, while some imagery can be difficult for manual control point selection, in general, users can adapt to more data sources than a typical algorithm. In some cases, considerable effort and care goes into the selection of ground control points. Researchers may visit a study site to select and label with Global Positioning System (GPS) coordinates a set of robust ground features like road intersections (Wang and Ellis, 2005b). These georeferenced control points are then located in the imagery and integrated into an image-to-image registration scheme. Manually selected control points have been used to initialize automatic matching steps (Kennedy and Cohen, 2003) and are often used for accuracy studies on automatic systems. The research emphasis with manual registration is typically on the quality of the final geometric transformation, as the software systems support complex empirical and physical models such as rational polynomials. In addition to manual selection of control points, remote sensing software packages usually support the automatic selection and correlation-based registration of control points, although they vary in the nature and extent of the automatic selection.

3.3.2 Correlation-related methods

Correlation-related methods directly compute a similarity measure for corresponding image regions by pixelwise comparisons of intensity values. Often called template matching, a region from one image is translated around the other to find the alignment that optimizes the similarity measure. The similarity measure for the absolute difference of pixel intensities is given by

$$D[\Delta x, \Delta y] = \sum_x \sum_y |I_1(x, y) - I_2(x + \Delta x, y + \Delta y)|, \qquad (3.2)$$

where Δx, Δy denote, respectively, the horizontal and vertical shifts in the sensed image, and the summation is carried out over all x, y locations of an image region. The above similarity measure, expressed by the L_1-norm as the sum of absolute differences, can be expressed alternatively in terms of the L_2-norm as the sum of squared differences. The correlation coefficient itself sums the product of intensity values, in essence maximizing an inner product. The naive brute-force search approach is to evaluate these similarity measures according to Eq. (3.2), that is, by summing over all x, y in the region, and letting Δx, Δy vary over a predefined search window until finding the optimal translations. Although correlation-related methods are fundamental to image registration, they have a number of drawbacks that need to be addressed for practical applications.

First, the brute-force approach is computationally expensive. Specifically, it requires $O(n^2 m^2)$ operations, where n and m denote, respectively, the dimension of the image region and search window. Although this may be acceptable for limited size regions and search windows, faster methods should be used, in general. Indeed, there are a number of methods for speeding up the calculation of the above correlation measures. A fast computation of the correlation coefficient in the frequency domain deserves special consideration. See Subsection 3.3.3 below. Other approaches include partial computations, coarse-to-fine pyramid search, specialized parallel hardware, and numerical optimization.

For partial computation, the full similarity measure is not computed for all locations in the search window. Instead, it is computed at sampled locations in the search window (Althof *et al.*, 1997). Alternatively, the computation can be truncated if it exceeds a certain threshold or a previously computed minimum. The *sequential similarity detection algorithm* (SSDA) accumulates the sum of absolute differences until the measure becomes too large for the current alignment to provide the likely minimum (Barnea and Silverman, 1972). Huseby *et al.* (2005) used the SSDA method to register Advanced Very High Resolution Radiometer (AVHRR) and Moderate Resolution Imaging Spectroradiometer (MODIS) data. Another method, which projects a 2D region into two 1D correlations, lowers the computational cost and adds a noise-smoothing effect (Cain *et al.*, 2001). The 2D to 1D projection is handled as follows. The columns of the subimage being matched are summed to produce a vertical, 1D array. Similarly, the rows are summed to produce a horizontal, 1D array.

For coarse-to-fine pyramid search, the image is blurred and decimated into a sequence of smaller, lower-resolution images of sizes that are a decreasing sequence, typically of powers of two. Translations in a given image are clearly reduced in the smaller images, so the search window for these images becomes smaller accordingly. Once a solution is found for a smaller image, the resulting alignment is extrapolated to the next higher-resolution image. The search result is refined repeatedly in this fashion through the highest-resolution image. Coarse-to-fine search has the added advantage that the blurring and decimation can smooth the objective function (i.e., the similarity measure), thereby reducing the impact of local minima and noise.

Regarding specialized hardware, the inherent parallel nature of the problem can be exploited to parallelize the correlation computation by using parallel processing units (Le Moigne *et al.*, 2002), *application specific integrated circuits* (ASICs) (Gupta, 2007), *field-programmable gate arrays* (FPGAs) (Sen *et al.*, 2008), or *general purpose graphics processing units* (GPUs) in graphics cards (Köhn *et al.*, 2006).

Finally, another alternative to brute-force search is a variant of gradient descent numerical optimization. In this approach the L_2-norm similarity measure is computed by an iterative solution to a least-squares formulation (Dewdney, 1978; Lucas and Kanade, 1981; Irani and Peleg, 1991; Thévenaz *et al.*, 1998; Irani and Anandan, 1999; Baker and Matthews, 2001). Assuming that the similarity measure is smooth without local minima, the iterative process will converge to the minimum cost alignment in $O(n^2)$ time (Dewdney, 1978).

The brute-force approach becomes more expensive when used for subpixel accuracy, as the use of fractional pixel increments increases the computational cost over integral increments. The previously mentioned gradient descent approach can address this, as the iterative process naturally converges to subpixel accuracy (Irani and Peleg, 1991; Baker and Matthews, 2001). An alternative approach is to compute the similarity measure at integral displacements, and then use polynomials computed over a local neighborhood to interpolate a subpixel optimum (Lee *et al.*, 2004). Even when interpolated over small 3×3 neighborhoods, the latter can be effective to a tenth of a pixel.

The computational cost of the brute-force approach can be even more significant with a complex geometric transformation beyond translation. The addition of each parameter, due to rotation, scale, skew, and high-order distortions, multiplies the size of the search space. In practice, correlation is often used on small image regions or chips, where translation is an adequate approximation, even if the entire image is to be registered by a complex physical model (Theiler *et al.*, 2002). When more complex models are required, albeit still relatively low order, such as rotation, scale, and translation (RST), affine, or homography, the extra parameters can be incorporated into the least-squares formulation by a linearized approximation (Irani and Peleg, 1991; Thévenaz *et al.*, 1998).

The basic correlation similarity measures can be weakened by noise, occlusions, temporal changes, radiometric differences in multimodal imagery, and other sources that may influence pixel differences. This is a major reason for the development of more statistically complex similarity measures, or feature-based methods, as discussed later in this section.

Still, correlation measures themselves can be made more robust and more adapted to complex imagery. Images can be preprocessed to enhance image frequencies less susceptible to noise sources, either by similar smoothing due to the coarse-to-fine projection approaches mentioned previously, or by edge enhancement (Andrus *et al.*, 1975). Keller and Averbuch (2006) presented an *implicit similarity measure*, which treats the two images asymmetrically by computing the gradient magnitude in one image, and edges in the other. For any displacement, the measure is computed by summing the gradient magnitude covered by edges in the second image. Kaneko *et al.* (2003) dealt with occlusion by selective masking. Robust statistical measures, such as M-estimators, can be used to reduce the influence of outlying noise values. Arya *et al.* (2007) presented a version of normalized correlation, based on M-estimators, which is robust to occlusions and noise. Kim and Fessler (2004) used M-estimators in intensity correlation for a medical application, and demonstrated better performance than that obtained by using the *mutual information* (MI) measure. Radiometric differences between images can also be handled by a gradient descent approach to the least-squares formulation, provided that they can be modeled by a small set of parameters. Examples include affine radiometric transforms (Gruen, 1985) and gamma correction (Thévenaz *et al.*, 1998). Georgescu and Meer (2004) integrated radiometric correction with robust M-estimators in a gradient descent approach. Statistical alternatives to basic correlation are the correlation ratio and the Woods measure (Roche *et al.*, 2000).

While the deficiencies of correlation methods are often cited in papers on new registration techniques, the approach is important and remains widely used in remote sensing applications (Emery *et al.*, 2003; Lee *et al.*, 2004). In Eastman *et al.* (2007) we reviewed several image registration schemes in major satellite ground systems, six of which used area-based correlation. These schemes are presented in Table 3.1 and described below. All the operational systems were developed by ground support teams, and they share a number of characteristics beyond the use of normalized intensity correlation. They all perform matching in local regions (rather than global matching), deal only with translation (since it dominates in small regions), and use preconstructed databases of carefully selected image regions, topographic relief correction of features before correlation, and cloud masking or thresholds to eliminate cloudy regions. Most systems use subpixel

Table 3.1 *Operational image registration systems based on correlation*

Instrument	Satellite	Resolution	Similarity
ASTER	Terra	15–90 m	Correlation to DEM corrected CPs
MISR	Terra	275 m	Correlation to DEM corrected CPs
MODIS	Terra	250 m–1 km	Correlation to DEM corrected CPs
HRS	SPOT	2.5 m	Correlation to DEM corrected data
ETM+	Landsat-7	15–60 m	Correlation to arid region CPs
VEGETATION	SPOT	1 km	Correlation to DEM corrected CPs

estimation but vary in how the subpixel transformation components are computed. Also, operational groups report the following practical registration issues that need to be addressed: effectiveness of normalized correlation in cross-band registration, adaption to thermal changes in satellite geometry and minor problems in orbit data, inadequate uniform sampling of CPs across the image, and suitability of a specific ground location for correlation for different reasons. A ground location can be unsuitable for correlation because the ground features are uniform and indistinct, because seasonal changes in temperature and vegetation cause the imagery to significantly vary, or because human activity causes the imagery to vary.

Iwasaki and Fujisada (2005) described the image registration system for the Advanced Spaceborne Thermal Emission and Reflection Radiometer (ASTER), a 14-band multispectral imager launched in 1999 on the Terra (EOS-AM1) satellite. Registration was done to a database of about 300–600 CPs which were mapped onto topographic maps. The similarity measure used was normalized correlation with transformation limited to translation. Matches were rejected for correlation less than 0.7 or if clouds were detected. Subpixel estimation was calculated by fitting a second-order polynomial to the correlation values. In a retrospective study the authors cited accuracies of 50 m, 0.2 pixels and 0.1 pixels, respectively, for three bands of different resolution. Jovanovic *et al.* (2002) described the geometric correction system for the Multi-angle Imaging SpectroRadiometer (MISR) instrument on the Terra satellite. Registration was conducted on a database of 120 ground control points (GCPs) represented by nine 64 × 64 image chips from Landsat Thematic Mapper (TM) images, one chip from each spectral band. Each control CP was of an identifiable ground feature that could be located to 30 m, and was selected for seasonally invariant features. The chips were mapped onto terrain-corrected imagery and a ray-casting algorithm was used to warp each chip to the appropriate geometry for the appropriate MISR camera. Chip matching was done to subpixel accuracy for translation, potentially to 1/8 of a pixel, using least-squares optimization (Ackerman, 1984).

Wolfe *et al.* (2002) described a geolocaton system (see also Xiong *et al.*, 2005) for the MODIS instrument on the Terra satellite. Registration was carried out on a database of 605 land CPs known to 15 m in 3D. For each CP, 24-km^2 chips (of 30-m/pixel resolution) were constructed from Landsat TM bands 3 and 4 over cloud-free areas, predominantly along coastlines and waterways. Higher-resolution TM chips were resampled to MODIS resolution, using the MODIS point spread function and nominal MODIS position information. A fractional sampling interval was used to get subpixel accuracy with a threshold of 0.6 for rejection.

Baillarin *et al.* (2005) described the automatic orthoimage production system, ANDORRE, with its algorithmic core TARIFA (French acronym for Automatic Image Rectification and Fusion Processing) for the SPOT-5 High Resolution Stereoscopic (HRS) instrument, launched in 2002. (The French satellite SPOT stands for Earth observation satellite.) ANDORRE uses an extensive database of orthoimagery tiles integrated with Digital Elevation Model (DEM) data to generate, via ray tracing, a simulated image for matching. Matching is done using multiresolution search with the number of levels set to keep a 5 × 5 pixel size search window. CPs are automatically found, matched by correlation, and used to calibrate a parametric model. Geometric outliers and CPs with correlation coefficient below 0.80 are rejected.

Lee *et al.* (2004) described image registration for the geolocation of ETM+, an instrument on the Landsat-7 satellite that was launched in 1999. ETM+ specifications include locating an image pixel to 250-m absolute geodetic accuracy (namely, each pixel is off by at most 250 m from its true ground position), 0.28 pixels band-to-band (that is, bands have a relative misregistration of less than a 1/3 of a pixel at most), and 0.4 pixels temporal registration (i.e., images taken of the same area over time have less than 1/2 pixel misregistration). These three requirements lead to multiple registration approaches for calibration and assessment. Geodetic accuracy is updated by correlating systematically corrected panchromatic band regions against a database of CP image chips extracted from USGS digital orthophotos. A second correlation algorithm evaluates band-to-band alignment by subpixel registration with second-order fit to the correlation surface.

Sylvander *et al.* (2000) described the geolocation system for the VEGETATION instrument on the SPOT satellites, which became operational since the launch of SPOT-4 in 1999. A database of approximately 3500 CPs was built from VEGETATION images. Each distinct ground CP location was represented in the database as four image chips taken from images of different seasons and orientations. Correlation-based matching was done under the control of human operators who ensure that there are ten matched points per orbit to compute satellite position corrections. Multispectral registration to subpixel accuracy of 0.11 pixels was reported.

3.3.3 *Fourier-domain and other transform-based methods*

Frequency domain methods provide primarily a fast alternative for computing the correlation coefficient similarity measure. A frequency transform converts a digital signal, or an image, to a collection of frequency coefficients that represent the strength of each frequency in the original signal. For example, if the original image has a large number of ridges, then the frequency domain will have large coefficients related to the regular distance between the ridges. Transform-based registration methods are based on the premise that the information in the transformed image will make the geometric transformation easier to recover.

Fourier domain methods are based on Fourier's *shift theorem*, which states that if $g(x, y)$ is a translated version of a signal $f(x, y)$ in the spatial xy domain, then the corresponding Fourier transforms $G(u, v)$ and $F(u, v)$ in the frequency uv domain are related by a phase shift that can be recovered efficiently. For image registration, the two images are transformed to the frequency domain by the *fast Fourier transform* (FFT). These transforms are then multiplied efficiently in the frequency domain, and their product is transformed back (via the inverse FFT) to the spatial domain for recovery of the translation. (The desired translation is found at the location of the inverse transform peak.) While the computation itself is equivalent to that of the correlation coefficient in the spatial domain, the computational cost of the above method is $O((n + m)^2 \log(n + m))$, rather than the brute-force $O(n^2 m^2)$ (Dewdney, 1978).

Notwithstanding the relative computational efficiency of standard Fourier-based methods, in comparison to the brute-force computation of the correlation coefficient in the spatial domain, the two approaches share many drawbacks in terms of handling subpixel displacements, geometric transformations beyond translation, and sensitivity to noise. Also, it would be difficult for the Fourier approach to recover large displacements. Thanks to various modifications, however, Fourier-based registration in the frequency domain can be adapted to handle these problems. De Castro and Morandi (1987) extended the method to rotation, while Chen *et al.* (1994) and Reddy and Chatterji (1996) extended it to transformations that consist of rotation and scale, as well as translation. Specifically, Chen *et al.* (1994) combined the shift-invariant Fourier transform and the scale-invariant Mellin transform, leading to the term "Fourier-Mellin transform." Their method is also known in the literature as *symmetric phase-only matched filtering* (SPOMF). Abdelfattah and Nicolas (2005) applied the invariant Fourier-Mellin transform to register Interferometric Synthetic Aperture Radar (In-SAR) images, and demonstrated an improved signal-to-noise ratio (SNR) over standard correlation. Keller *et al.* (2005) used the pseudopolar Fourier transform for computing rotation in a more robust and efficient manner.

Shekarforoush *et al.* (1996) extended the method to subpixel displacements. This was further refined by Stone, Orchard, Chang, and Martucci (2001) and Foroosh, Zerubia, and Berthod (2002) who demonstrated effective registration across spectral bands in SPOT satellite imagery. The high accuracy required of subpixel extensions emphasizes aliasing artifacts from rotation effects at image boundaries. These effects, which come from interpolation of rotated images, were addressed by Stone *et al.* (2003). The same authors also addressed the integration of cloud masks into Fourier methods in McGuire and Stone (2000). Zokai and Wolberg (2005) adapted the method to large displacements with perspective distortions. Orchard (2007) also addressed the recovery of large displacements and rotations, integrating the Fourier transform into a gradient descent framework for an exhaustive global search, while demonstrating robustness to multimodal radiometric differences in remote sensing and medical imagery. An additional Fourier-based method is due to Leprince *et al.* (2007), who combined a Fourier method on local image regions with a global model of the imaging systems to achieve subpixel registration within 1/50 of a pixel.

The Fourier basis functions have infinite support and they do not spatially localize the frequency response. On the other hand, wavelet basis functions, such as the Haar, Gabor, Daubechies, finite Walsh and Simoncelli wavelets used in image registration (Li *et al.*, 1995; Hsieh *et al.*, 1997; Le Moigne *et al.*, 2000; Zavorin and Le Moigne, 2005; Lazaridis and Petrou, 2006), do have finite support and thus do localize the frequency response of an image feature. A Fourier coefficient represents a particular frequency but does not indicate where in the image features with that frequency occurred. Wavelet features, on the other hand, give both frequency and position information. As a result, wavelets are not used in the same way as Fourier coefficients. Rather than apply the shift theorem (as in the Fourier case), wavelets are used to decompose the image into higher-frequency, edge-like features, and lower-frequency features. The higher-frequency features are used typically as dense edge features for correlation- or feature-based matching. This will be further discussed in Subsection 3.3.5.

3.3.4 Mutual information and distribution-based approaches

Requiring equality between image intensities is a strong constraint, which is not valid in many cases. For cross-band and cross-instrument fusion, the image intensities may be related by nonmonotonic or even nonfunctional relationships, requiring powerful similarity measures for effective matching that are based on statistical relationships. In this case, the similarity measure is based on comparing local intensity distributions rather than individual pixel values.

The most effective of these measures has proven to be the *mutual information* (MI) measure and its variants. MI-based registration was first introduced in the medical imaging domain by Maes *et al.* (1997) and Viola and Wells (1997). See also Pluim *et al.* (2000), Thévenaz and Unser (2000) and the review of MI-based methods in medical imaging (Pluim *et al.*, 2003). These methods make very weak assumptions about the relationship between pixel intensities in the two images, and they do not require a monotonic or any functional model. Mutual information is an information-theoretic quantity that measures the spread of the values in a joint histogram represented by a 2D matrix. If two images match perfectly, then the joint histogram of their intensities should cluster around the diagonal of the matrix; otherwise, the values spread off the diagonal. Denoting the image intensities by the random variables X and Y, the mutual information measure is defined by

$$MI(X;Y) = \sum_{x \in X} \sum_{y \in Y} p_{X,Y}(x,y) \log \left(\frac{p_{X,Y}(x,y)}{p_X(x)\, p_Y(y)} \right), \qquad (3.3)$$

where $p_X(x)$ and $p_Y(y)$ denote, respectively, the probability density function (PDF) of X and Y, and $p_{X,Y}(x,y)$ denotes the joint PDF of X and Y.

A number of authors have addressed various issues, such as efficiency, accuracy, and robustness, regarding the application of MI in image registration. Maes *et al.* (1999) looked at ways of obtaining efficient speedups, by investigating a multi-resolution pyramid approach in combination with several (non)gradient-based optimization algorithms. Shams *et al.* (2007) applied MI to gradient values to compute more efficiently an initial scale and translation estimate before refining the registration by a variation of Powell's numeric optimization. The MI measure is subject to scalloping artifacts that stem from interpolation errors that introduce false local optima. The scalloping artifacts show up as ridges in the similarity measure surface. A number of authors have considered refinements to improve the computation of the joint histogram. See, for example, Chen *et al.* (2003), Dowson and Bowden (2006), Ceccarelli *et al.* (2008), Dowson *et al.* (2008), and Rajwade *et al.* (2009). Other authors have looked at improving, generalizing, and unifying MI with other information-theoretic measures. See Bardera *et al.* (2004), Pluim *et al.* (2004), and Knops *et al.* (2006).

Despite its wide use in medical imaging and good potential for multimodal fusion, MI-based methods have yet to be used extensively in remote sensing applications. Still, research-oriented articles have demonstrated promising approaches on limited image sets. Kern and Pattichis (2007) gave a thorough review of MI implementation details, and developed a gradient descent version that was extensively tested on synthetically generated, cross-band image pairs acquired by the Multispectral Thermal Imager (MTI) satellite. Cole-Rhodes *et al.* (2003) integrated MI with a wavelet pyramid, and used a stochastic gradient numeric optimization search approach to register multimodal, multiscale imagery acuired by IKONOS,

Landsat ETM, MODIS, and Sea-viewing Wide Field-of-view Sensor (SeaWiFS). Chen *et al.* (2003) applied MI with different interpolations (nearest neighbor, linear, cubic, and partial volume) to register Landsat TM, Indian Remote Sensing Satellite (IRS), Panchromatic (PAN), and Radarsat Synthetic Aperture Radar (Radarsat/SAR) images over the San Francisco Bay Area, California. (Partial volume interpolation was found to be most effective.) In similar tests, Chen *et al.* (2003) used Landsat TM imagery to establish the ability of generalized partial volume algorithms to reduce interpolation artifacts in the similarity measure. Inglada and Giros (2004) systematically investigated the application of different similarity measures to multisensor satellite registration of SPOT and ERS-2 data, including correlation, correlation ratio, normalized standard deviation, MI and the related Kolmogorov measure. Cariou and Chehdi (2008) used MI with gradient descent to register airborne pushbroom sensor data to a reference orthoimage, computing a geometric transformation based on explicit airplane orientation parameters. A recent report (Nies *et al.*, 2008) has discussed the application of MI to the rather noisy SAR imagery. He *et al.* (2003) applied a generalization of MI, the Jensen-Rènyi divergence measure, to the registration of airplane profiles in Inverse Synthetic Aperture Radar (ISAR) images.

3.3.5 Feature-point methods

A feature-point registration algorithm first extracts a set of distinctive, highly informative feature points from both images, and then matches them based on local image properties. Feature points are referred to by different names in the literature, including control points, ground control points, tie-points, and landmarks. Care must be taken, though, with respect to the terminology used in a paper to distinguish feature points defined only by local, syntactic image calculations from other types of control points, such as those manually selected or on the basis of semantic characteristics. Feature points are determined by the application of an interest operator to the images for finding candidate feature points, and the subsequent extraction of feature descriptors to be used in computing pointwise similarity measures between corresponding features. After individual pairs of feature points are tentatively matched between images, the algorithm often applies a geometric consistency similarity measure to verify individual matches and reject outliers. (Note that this overlaps with area-based methods that register local regions or chips, essentially treating them as control points, so there is no clear distinction between feature- and area-based methods.) Feature-based methods can be quite complex, involving multiple passes of registration, geometric transformations that contain several parameters, various similarity measures, and different search approaches.

We can categorize feature-point methods into two general classes: A class of methods that match dense sets of low-information-content features (typically edge points or wavelets), and a class of methods that match smaller sets of features that contain higher-information content, where the feature extraction and the pointwise similar measure are more complex to evaluate. Methods based on dense feature sets may require less initial work but a more sophisticated geometric consistency measure (for defining a match), as well as a more intensive search effort on the feature sets. In the extreme, each feature point is considered individually without any additional information, and the problem reduces to a purely geometric one of geometric point set matching in a robust and efficient manner.

In dense feature point matching, the similarity measure is based on large sets or neighborhoods of point features. The point features are typically found using a fast, low computational cost edge or interest operator. For the methods considered in this section, the feature points are represented only by their position, and not labeled with intensity or other descriptors computed from their local neighborhood, nor grouped into contours or other structures. One approach is to apply binary or edge correlation to regions of points (Andrus *et al.*, 1975). Both images are reduced to binary images of edge responses, and a correlation coefficient computed. The correlation can be computed with addition rather than multiplication on the binary images. Some methods are asymmetric in their use of the binary edge images. Keller and Averbuch (2006), previously discussed, and Madini *et al.* (2004), discussed later, both used implicit similarity measures that use a binary edge or vector description in one image to sum the magnitude of edge response in the other image. Both binary to binary correlation, and binary to edge response correlation, show robustness to cross-sensor registration.

A large number of papers have been written on the pure point pattern matching problem in the fields of image processing, pattern recognition, and computational geometry. Perhaps the simplest similarity measure among unlabeled point sets involves the Hausdorff distance and its variants (Alt *et al.*, 1994; Chew *et al.*, 1997; Goodrich *et al.*, 1999; Huttenlocher *et al.*, 1993a). The Hausdorff distance is a geometric measure based on the maximum of the minimum distances between points in the two sets. In effect, the measure evaluates a proposed correspondence between two point sets based on the greatest remaining distance among individual point correspondences. Compared to binary edge correlation, the Hausdorff measure is more tolerant of subpixel variations in edge location. The standard notion of the Hausdorff distance, however, is not suitable for typical remote sensing applications, since it requires that every point (from at least one set) have a nearby matching point in the other set. Also, computing the optimal alignment of two-point sets even under the relatively simple Hausdorff distance is computationally intensive. In an attempt to alleviate the high complexity of point pattern matching, some researchers have considered *alignment-based* algorithms. These algorithms use alignments between small subsets of points to generate potential aligning transformations, the best of which are then subjected to more detailed analysis. Examples of these approaches include early work in the field of image processing (Stockman *et al.*, 1982; Goshtasby and Stockman, 1985; Goshtasby *et al.*, 1986) and more recent work in the field of computational geometry (Heffernan and Schirra 1994; Goodrich *et al.*, 1999; Gavrilov *et al.*, 2004; Cho and Mount, 2005; Choi and Goyal, 2006). Alignments can also be part of a more complex algorithm. For example, Kedem and Yarmovski (1996) presented a method for performing stereo matching under translation based on propagation of local matches for computing good global matches.

For remote sensing applications it is important that the distance measure be robust, in the sense that it is insensitive to a significant number of feature points from either set that have no matching point in the other set. Examples of a robust distance measures include the *partial Hausdorff distance* (PHD) introduced by Huttenlocher and Rucklidge (1993) and Huttenlocher *et al.* (1993b), and symmetric and absolute differences discussed in Alt *et al.* (1996) and Hagedoorn and Veltkamp (1999). (See Chapter 8 for specific details.) Both measures were used within a branch-and-bound algorithmic framework for searching in transformation space. Drawing on this framework, Mount *et al.* (1999) proposed two ways of reducing the computational complexity of feature-point matching for image registration. Specifically, they considered an approximation branch-and-bound algorithm and a randomized, Monte Carlo algorithm (called *bounded alignment*) to further accelerate the search. A robust, hierarchical image registration scheme based on this algorithmic approach was applied by Netanyahu *et al.* (2004) for georegistering Landsat data to subpixel accuracy.

Also, Chen and Huang (2007) presented a multilevel algorithm for the subpixel registration of high-resolution SAR data. They employed a Hausdorff metric with bi-tree searching to a pyramid of edge maps to compute a rough alignment. The SSDA area measure (mentioned in the previous subsection) was then employed to acquire tie-points, which were used to refine the alignment to subpixel accuracy.

As noted above, large or moderate sets of feature points can be contaminated with many outliers and spurious points that do not have proper correspondences. Wong and Clausi (2007) described the system, Automatic Registration of Remote Sensing Images (ARRSI), which uses a variation on the *random sample consensus* (RANSAC) algorithm (Fischler and Bolles, 1981) to match the control points. The system works by selecting random subsets of control points from which a tentative geometric transformation is computed. If the transformation consistently extends to a significant portion of the full set, then it is accepted as correct. The authors tested their approach on cross-band and cross-sensor Landsat and X-SAR images.

Similarity, voting or accumulation schemes like the Hough transform (Yam and Davis, 1981; Li *et al.*, 1995) can be used for potential matching of pairs to vote on the most likely transformation. In the Hough transform two feasible feature matches, whether correct or not, are used to compute a set of geometric transformations that map the feature in one image into the feature in the second image. Those sets are combined in a voting scheme, and the transformation chosen that gathers the most votes from the most potential matches. Correct matches are assumed to vote consistently for the correct match, while incorrect matches are assumed to vote for a random collection of inconsistent matches. Yang and Cohen (1999) computed affine invariants and convex hulls from local groups of points to drive the matching process.

Methods based on smaller, sparse sets of feature points require more effort to locate and describe the feature points. As mentioned, the first step is to apply an interest operator to the images for finding distinctive points. A syntactic interest operator is based only on general image properties, such as the magnitude of a local image derivative, or the persistence of the feature across scales. On the other hand, a semantic feature is based on the interpretation of the local image region as a road intersection, building corner, or other real world entity that can be stable over time. The same ground feature may be available for a long period of time for registration applications (Wang *et al.*, 2005). In Subsection 3.3.2 we described six operational systems that used databases of manually selected ground control points (GCPs) represented as image chips. Remote sensing users often give semantic rules for the selection of GCPs (Sester *et al.*, 1998; Wang *et al.*, 2005). In other registration applications, for example, in robotic vision or medical imaging, the constant flow of new images emphasizes the use of real-time syntactic interest operators, while in remote sensing it is appropriate to invest considerable effort in collecting feature points for long-term use.

A preliminary phase in image registration involves the detection and extraction of image features that have suitable content for reliable matching. In other words, the original images are converted into feature sets for computing an optimal match. Syntactic interest operators look for distinctive regions with high-content information that can be localized in two directions simultaneously, such as corners or line intersections. They do so by computing local image properties (such as first and second image derivatives). Commonly used operators are due to Harris and Stephens (1988) and Förstner and Gülch (1987). Schmid *et al.* (2000) systematically reviewed and evaluated a number of interest operators and concluded, based on the criterion they defined, that the Harris operator was the most repeatable and informative. Mikolajczyk and Schmid (2004) extended the Harris detector to an affine invariant, scale space version (Harris-affine), that detects local maxima of the detector at multiple image scales. Mikolajczyk *et al.* (2005) reviewed six different affine invariant operators, including the Harris-affine, and concluded under what conditions each operator was most useful. Kelman *et al.* (2007) evaluated the three interest operators: Laplacian-of-Gaussian, Harris, and *maximally stable extremal regions* (MSERs), for the repeatability of their locations and orientations. The topic is rich enough to support a book surveying interest and feature operators (Tuytelaars and Mikolajczyk, 2008). However, much of the literature does not account for the complexity of remote sensing imagery. Hong *et al.* (2006) emphasized the complexity of extracting GCPs in remote sensing data, as the extraction operator needs to account for the physical imaging models of the sensor, the satellite orbit, and the terrain's DEM to accurately recover invariant feature points in SAR and optical images.

Once feature points have been detected, a feature descriptor operator extracts local information to define a similarity measure for point-to-point matching. By reducing a neighborhood to a smaller set of descriptors relatively invariant to geometric and radiometric transformations, the matching process can be made more efficient and robust. Local descriptors include moments of intensity or gradient information, local histograms of intensity of gradients, or geometric relationships between local edges.

Lowe defined the *scale-invariant feature transform* (SIFT) (Lowe, 1999; Lowe, 2004), which is a weighted, normalized histogram of local gradient edge directions invariant to minor affine transformations. The SIFT operator has been widely used in a number of registration applications. Yang *et al.* (2006) used SIFT in an extension of the *dual bootstrap iterative closest point* (ICP) algorithm, originally developed for retinal image registration. As originally defined by Stewart *et al.* (2003), dual bootstrap is a robust algorithm that proved successful on more than 99.5% of retinal image pairs, and 100% on pairs with an overlap of at least 35%. The dual bootstrap ICP algorithm is a multistage procedure. It starts by matching interest points found in retinal vessel bifurcations. In the retinal version, the invariants used were the ratio of blood vessel widths and their orientations, yielding a five-component feature-point descriptor. The SIFT operator was then employed to obtain an extended version that performed successfully on the challenging image pairs that were experimented with. Li *et al.* (2009) have recently refined the SIFT operator to increase its robustness and applied it to cross-band and cross-sensor satellite images. The reader is referred to the following additional relevant papers: Dufournaud *et al.* (2004); Ke and Sukthankar (2004); Mikolajczyk and Schmid (2004); Mikolajczyk and Schmid (2005); and Mikolajczyk *et al.* (2005).

In remote sensing, there have been a number of applications of feature point detectors and descriptors. Igbokwe (1999) performed subpixel registration of Landsat Multi Spectral Scanner (MSS) and TM data to 0.28 pixel accuracy. They located feature points using the interest operator, *target defined ground operator* (TDGO), described in Chen and Lee (1992), and then applied least-squares iterative subpixel matching in local regions. Grodecki and Lutes (2005) performed registration for redundant validation of IKONOS geometric calibration. They used two test sites with features points selected manually and automatically with the Förstner and Gülch interest operator. Bentoutou *et al.* (2005) registered SPOT and SAR images to subpixel accuracy, with a root mean square error (RMSE) of 0.2 pixels. They located feature points using an enhanced Harris detector, and then matched the points with affine invariant region descriptors with a subpixel interpolation of the similarity measure. Carrion *et al.* (2002) described the GEOREF system that the authors tested with Landsat imagery. In the GEOREF system, features extracted by the Förstner operator are located and matched at multiple levels using an image pyramid. An interesting twist is that once two feature points are matched, the match is refined in a local neighborhood around the feature point by a least-squares, area-based method. Here, feature- and area-based approaches are combined to best match two feature points.

At an extreme for the semantic content of features, Growe and Tönjes (1997) used a semantic, expert system approach to segment and understand image features for matching. Raw, low-level image features like edges were interpreted through a semantic net as meaningful high-level features like roads and houses. They matched the resulting feature with the A* algorithm to control search.

Arevalo and Gonzalez (2008b) presented a complex algorithm for a nonrigid, piecewise local polynomial geometric transformation that integrates a number of the methods explained above. Control points were detected by the Harris operator and then matched to subpixel accuracy using gradient descent. An initial transform was then estimated with RANSAC, followed by a computation of a triangular mesh of the control points. Since some of the triangular regions in the mesh may overlap discontinuities, such as building edges, the mesh was refined through an iterative process to swap edges until regions became more homogeneous. The homogeneity was evaluated by a mutual information measure.

3.3.6 *Contour- and region-based approaches*

Contour- and region-based approaches use extended features such as lines, contours, and regions. These extended features, which can be obtained by connecting edge points, or segmenting the images into compact regions, can prove robust in cross-band and cross-sensor registration. Contour- and region-based approaches are particularly useful for water features, such as lakes, rivers, and coastlines, or for man-made structures like roads. However, these features are all subject to change over time so they cannot be assumed constant.

Line matching, based on linear features extracted in IKONOS and SPOT imagery, was described in Habib and Alruzouq (2004) and Habib and Al-Ruzouq

(2005). Manually outlined linear segments of road networks were provided as input. The collections of these line primitives were then compared by a *modified iterated Hough transform* (MIHT) to compute the desired affine transformation. Wang and Chen (1997) performed line matching by using invariant properties of line segments.

Contour matching uses pixel sequences or polynomial curves fit to edge data, such as *non-uniform rational B-splines* (NURBS), to compute optimal geometric transformations for matching and registration. See, for example, Li *et al.* (1995), Carr *et al.* (1997), Eugenio and Marques (2003), and Pan *et al.* (2008). We briefly review a few related methods below. Dai and Khorram (1999) derived a contour-based matching scheme, and demonstrated its applicability on Landsat TM images. Edge pixels were first extracted by a Laplacian-of-Gaussian (LoG) zero-crossing operator, and then linked/sorted to detect contours. Regions defined by these closed contours were then matched (in feature space), combining invariant moments of the planar regions and chain-code representations of their contours. Finally, the initial registration step was refined (in image space) by matching a set of GCPs defined by centers of gravity of matched regions. Subpixel accuracy of roughly 0.2 pixels (in terms of the RMSE at the GCPs) was reported. Similarly, Eugenio and Marques (2003) registered AVHRR and SeaWiFS images by extracting closed and open-boundary contours of SeaWiFS island targets. They combined invariant region descriptors, individual contour matching, and a final global registration step on all contour points. Govindu and Shekhar (1999) carried out image matching by using contour invariants under affine transformations for both connected and disconnected contours. Xia and Liu (2004) performed efficient matching by using "super-curves," which are formed by superimposing pairs of affine-related curves in one coordinate system, and then finding simultaneously their B-spline approximations and registration.

Madani *et al.* (2004) described the operational AutoLandmark system for registration of the Geostationary Operational Environmental Satellite (GOES) imagery. GOES I-M is a series of 7-band weather satellites, which were launched from 1994 to 2001. Since the satellite images contain considerable ocean regions, the registration could be complicated. On the other hand, this type of image could support a coastline matching strategy. Note that the daily warming/cooling cycle affects the infrared bands, in the sense that land and water can reverse in radiation, causing coastal edges to invert, migrate, or even disappear. To avoid these radiometric effects, the registration was performed on a database of coastlines, which were represented in vector format. Specifically, 24×24 (or 96×96) landmark neighborhoods were extracted from the sensed image, and edge pixels (for each neighborhood) were detected using the Sobel operator. The edge map obtained was matched against a binary image rasterized from the coastline vector database, taking the sum of relevant edge strengths as a measure of correlation, and limiting the transformation to translations only. Cloud regions were masked out along the process. Bisection search was finally used to obtain subpixel accuracy. The overall quality measure of the registration was defined as a combination of the edge correlation, the fraction of cloud contamination, and the contrast (and possibly illumination) in the image.

Registration based on region matching uses local patches or closed boundary curves, found by intensity segmentation or curve extraction. Dare and Dowman (2001) segmented SPOT and SAR images into homogeneous patches, and then used straightforward parameters like area, perimeter, width, height, and overlapping area to compute a similarity measure between regions. Following this, they matched strong edge pixels by dynamic programming. Flusser and Suk (1994) detected close boundaries (from edge pixels obtained by the Sobel operator) and then used affine invariant moments of closed boundaries to register SPOT and Landsat images to subpixel accuracy under an affine transformation. See also Goshtasby *et al.* (1986) and Yuan *et al.* (2006) for additional related papers.

3.4 Geometric transformations

Image registration assumes a coherent geometric transformation between the sensed and reference images. If sufficient knowledge is available about the imaging model of the sensor, geometric distortions from satellite orbit and attitude variations, atmospheric effects, and topographic relief, then a physically accurate model of the transformation can be constructed, and an appropriate algorithm for the model can be chosen (Huseby *et al.*, 2005). If the above information is not available or is too complex, an approximate empirical model can be used. A physically accurate model is of course most suitable for refining a raw remote sensing image into an orthoimage, that is, a distortion-free version of the image that represents an ideal projection onto geodetic coordinates of the ground plane. Fully corrected orthoimages are best suited for integration into geographic information systems and fusion with all types of geographic data. Toutin (2004) noted several factors that drive the use of physical models of higher accuracy for orthorectification. These factors include higher-resolution sensors, additional imagery taken from off-nadir viewing angles, greater precision needed for digital processing, and the fusion of data from multiple image and vector sources. Armston *et al.* (2002) compared the use of physical and empirical geometric models in landcover mapping in Australia, using models available in commercial software packages. They evaluated the models used on field-collected GCPs and manually selected tie-points initially found using syntactic and semantic properties and then refining the GCP matches by image chip correlation. Using the RMSE of the GCPs as a quality metric, they found that empirical rational polynomial models gave higher errors than the physical models.

Still, the transformations used in the matching phase of image registration are usually based on empirical models that offer sufficient accuracy for the task but are not physically accurate. In particular, when an image registration algorithm is based on matching small image regions as chips or control points, empirical models of a simple geometric transformation as translation can be sufficiently accurate even without taking into account additional elements such as perspective. Thus, many image registration algorithms use empirical, low-order geometric models over small regions as chips or control points, and then use the control points to compute more accurate, parametric geometric models of higher order. Many algorithmic techniques, such as those based on numerical optimization or the Fourier transform, are primarily designed for the explicit recovery of the parameters of a low-order model, such as translation, rotation, and scale. Empirical models include:

(1) Rotation, scale, and translation (RST), i.e., transformations with four parameters. The RST model is a useful subset of the affine transformations.
(2) Affine transformation of six parameters.
(3) Projective transformation of eight parameters.
(4) Homography, which consists of eight degrees of freedom.
(5) Higher order 2D and 3D polynomial functions.
(6) Rational polynomials.

There are many sources of information on the mathematical details of geometric transformations, and we do not pursue the details here. For a general treatment, the reader is referred to Goshtasby (2005), and for details on projective and homography transformations the reader is referred to Faugeras (1993). Toutin (2004) is a review article on empirical and physical geometric models, including rational polynomials.

While the geometric transformations used for local matching are usually global and rigid, nonrigid or locally rigid transforms are important to account for distortions that vary across an image pair. Here, global means a single transformation is used across the entire image. Sharp topography, such as tall buildings and steep natural features, can introduce significant local perspective effects that are not accounted for in global, low-order transformations. Goshtasby (1988) introduced methods for piecewise local nonrigid methods, over which the images are segmented into regions in which local polynomial functions are used. In Zagorchev and Goshtasby (2006), the authors reviewed and compared these methods, including thin-plate spline, multiquadric, piecewise linear and weighted mean transformations. Arevalo and Gonzalez (2008a) compared nonrigid geometric transformations on QuickBird imagery. These transformations included piecewise (linear or cubic) functions, weighted mean functions, radial basis functions, and B-spline functions.

Approximate image alignment was first done manually. This was then refined using feature points that were obtained due to the Harris detector. Finally, gradient descent was used to achieve subpixel accuracy. The authors found the piecewise polynomial model superior on nonorthorectified imagery, and global fourth-order polynomials most adequate for orthorectified images.

3.5 Radiometric transformations and resampling

Under the correct geometric transformation, two corresponding image points should have related intensity values. If we assume that the corresponding image points represent the same ground feature, then the radiometric readings of that feature should have some consistency between two sensing events. That consistency or relationship confounds a number of factors, for example, the radiometric responses of the two sensors, the time of day and other environmental factors, the angle of view, the specular response of the ground feature, etc. All this is further confounded by multiband sensors, so in registering two images we may be relating two spectral bands which may or may not overlap. If we can account for all these factors, we can build a radiometric model (or function g) that maps intensities in one image to another, and use this model in designing a registration algorithm. However, few image registration algorithms use explicit radiometric relationships during the matching process. This is because the relationship may be difficult to calibrate, the algorithm may be used on many sensors (so a single relationship is not useful or appropriate), and the confounding environmental factors may overwhelm the relationship. Instead, most registration algorithms base their similarity measures on general assumptions made about the relationship, either explicitly or implicitly. Thus, choosing the appropriate assumptions is important in algorithm design and selection. The assumptions made about the radiometric model were classified in Roche *et al.* (2000) as follows:

(1) Identity relationship, i.e., assuming that the image intensity is invariant. This assumption is explicit in methods based on similarity measures that sum directly the square of absolute values of intensity differences.
(2) Affine relationship, i.e., assuming that the intensities differ by a linear gain and bias. This assumption is implicit in methods based on the correlation coefficient or least-squares minimization.
(3) Functional relationship, i.e., assuming that the intensities differ by a general function. This assumption is implicit in a few measures, including the correlation and Woods measures.
(4) Statistical relationship, i.e., assuming that while individual corresponding points may have differing intensities, in local neighborhoods they are drawn from the same statistical distribution. This is implicit in mutual information and similar approaches.

There are algorithms that integrate an explicit radiometric term during the matching process. Georgescu and Meer (2004) integrated a radiometric correction with robust M-estimators in a gradient descent approach. Thévenaz *et al.* (1998) integrated an exponential gamma model in a least-squares gradient descent approach, while Bartoli (2008) used a general multiband affine relationship to combine different color channels dynamically during gradient descent registration. Orchard (2007) used linear regression to estimate a piecewise linear relationship between semi-registered images. Hong and Zhang (2008) considered different radiometric normalization methods for high-resolution satellite images, such as scattergrams and histogram matching, to bring different sensors into the same color metric. This approach might be appropriate to preprocess images before matching.

Since digital images are represented by values on a discrete grid, typically a uniformly spaced rectangular grid, for one image to be geometrically transformed to match another, the values have to be resampled to the new grid locations. This can be done for two purposes: (1) Obtain an end product of image registration for further use in remote sensing applications, and (2) produce intermediate images (during the registration process) for incremental use, for example, an iterative computation of the geometric transformation. Greater care must be taken in the former case than in the latter, since values in the final image product may be used in further image analysis steps, while the intermediate resampled images are only a means to registration and will be deleted after use. As a result, an image registration system may use multiple resampling techniques. In both cases, it is important to avoid resampling artifacts, as this may degrade the data itself or the accuracy of the image registration.

Resampling is commonly done by reverse sampling of image values to remap the sensed image to a new image. If $f(u, v)$ is the geometric transformation that maps the sensed image into the reference image, then the inverse transformation $f^{-1}(x, y)$ is applied to map a pixel in the new sensed image to a subpixel location (u', v') between four surrounding input pixels in the grid of the old sensed image. (See Fig. 3.1.) The values of these four pixels (or pixels in a larger surrounding neighborhood) are then interpolated to compute the new pixel value. Increasing fidelity of the interpolation usually incurs greater computational cost.

The basic interpolation methods below are well documented in Goshtasby (2005). These methods include, in order of computational effort:

(1) *Nearest neighbor*, where the output pixel is given the value of the input pixel whose location is closest to the reverse sampled position (x, y). The advantage of nearest neighbor resampling is that the output image only contains intensity values present in the original image. However, it can produce aliasing "jaggies," particularly with rotation.

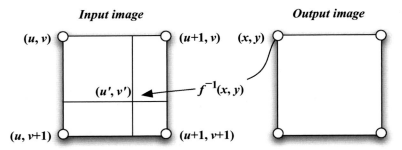

Figure 3.1. Resampling neighborhood.

(2) *Bilinear*, where the output pixel value is a linear interpolation of the local neighborhood, usually the four surrounding input pixels. The advantage of bilinear interpolation is that it is fast. Also, its results are visually similar to those obtained by more complex interpolators, although it is not as accurate as the bicubic interpolator or the other higher-order methods.

(3) *Bicubic*, where the output pixel value is obtained by a cubic polynomial interpolation of the values in a local neighborhood.

(4) *Spline*, where the output pixel value is computed by a polynomial spline interpolation (e.g., B-spline) of the local neighborhood.

(5) *Sinc function*, where the output pixel value is obtained by an interpolation based on the sinc function, $\sin(x, y)/r$ (where $r = \sqrt{x^2 + y^2}$) over a local neighborhood.

Each method can result in resampling (or interpolation) artifacts in the resulting image and in similarity measures derived from the image values. Inglada *et al.* (2007) reported on artifacts from bilinear, bicubic, and sinc interpolators, and noted that the interpolators had the effect of blurring the sensed image, although the magnitude of the blurring depended on the subpixel value of the shift. For example, integral shifts land exactly on a pixel and result in little blurring, while shifts of half a pixel result in maximum blurring. This can result in a "scalloping" effect that introduces artificial local maxima at subpixel shifts in similarity measures, such as mutual information (Pluim *et al.*, 2000; Tsao, 2003; Thévenaz *et al.*, 2008) and correlation measures (Salvado and Wilson, 2007; Rohde *et al.*, 2008). These artifacts can be addressed by careful prefiltering to reduce noise (Salvado and Wilson, 2007) or by proper randomized sampling (Tsao, 2003; Thévenaz *et al.*, 2008). However, this must be done carefully, as noise blurring contributes to the scalloping effect.

3.6 Evaluation of image registration algorithms

A typical satellite is very expensive to develop and operate, and the data produced can influence major governmental and industrial decisions on land use, crop production and other economic or political issues. Given the commercial and social impacts of research and development in remote sensing, proper evaluation of image registration reliability and accuracy is thus an essential component.

In general, performance evaluation may be associated with different aspects of the image analysis process. The image registration research community, by and large, is concerned with the derivation of proper metrics for technique evaluation and ranking. This community works to evaluate and rank basic registration algorithms, with the objective of choosing the best registration algorithms for general classes of data. On the other hand, the operational community focuses on validating the performance specifications of its image products, and end users focus on whether the image products meet their needs for commercial and scientific analysis. They are more concerned with the quality of the ultimate data of interest than with the best registration algorithm. Ideally, metrics for image registration evaluation would articulate and incorporate end-user requirements so the imaging community could better serve and communicate with end users.

The evaluation of image registration is relatively well established within the medical imaging community, which has access to techniques that are not as easily available in remote sensing. For example, in medical image registration, ground truth (or "gold standards") are produced by introducing fiducial marks into images or using phantom targets (Penney *et al.*, 1998), which facilitates algorithm comparison. In contrast, it is harder in remote sensing to produce ground truth for naturally collected data.

In this section we look into the issue of registration evaluation in two respects. First, we review evaluation aspects concerning the operational and user communities, for example, the impact of registration errors on operational teams and practitioners. Secondly, we discuss evaluation issues concerning the image registration algorithms themselves.

3.6.1 Operational and user oriented evaluation

How good must an image registration algorithm have to be? To answer this, one should consider the application in question. If used to mosaic multiple images for a visual display, the requirements from an algorithm may be less rigorous than for aligning two images to detect small changes in vegetative cover. Operational satellite teams have performance requirements set during the system design, which they work to meet. Users set requirements that come from the needs of their analytic techniques. Both sets of requirements can be informative in the design and evaluation of registration algorithms.

Accuracy needs of image registration for change detection has been one of the most studied application needs. Townshend *et al.* (1992) studied four regions in

Landsat TM imagery, and found that to keep the Normalized Difference Vegetation Index (NDVI) pixel-based change estimations down to an error of 10%, the registration subpixel accuracy had to be within 0.2 pixels. Studying, however, three additional homogeneous, arid regions, they found that the registration accuracy required for the above 10% error was within 0.5 to 1.0 pixels. For detection of changes over time, Dai and Khorram (1998) showed that a registration accuracy of less than 1/5 of a pixel was required to achieve a change detection error of less than 10%. Other articles on the impact of misregistration on change measurements include, for example, Roy (2000), Salas *et al.* (2003), and Wang and Ellis (2005a). Bruzzone and Cossu (2003) formulated an approach to evaluate change detection that adapts to and mitigates misregistration errors. Other issues that influence the impact of misregistration on change detection include the number of classes, region heterogeneity (Verbyla and Boles, 2000), and the resolution at which change detection is computed (Wang and Ellis, 2005b). In the latter paper the authors considered how measures of region heterogeneity in high-resolution imagery interact with change resolution and misregistration errors. Specifically, these authors have considered how to adapt to misregistration errors either by estimating them or lowering the resolution of change detection (Verbyla and Boles, 2000; Bruzzone and Cossu, 2003; Wang and Ellis, 2005b).

Another well-studied area is the evaluation of errors in orthorectified imagery, which is critical for operational teams. Zhou and Li (2000), Toutin (2003), Grodecki and Lutes (2005), and Wang *et al.* (2005) all addressed the geometric accuracy of IKONOS images. Looking at IKONOS panchromatic and multiband images over seven study sites in Canada, Toutin investigated the relationship between the accuracy of GCPs and its effect on precision correction of IKONOS for orthorectification. He found that the accuracy of GCPs affected how many of them were required to compute accurate geometric models for orthorectification, but that the error sensitivity of the model was low and good results could be obtained despite inaccurate GCPs. Zhou and Li (2000) performed a similar analysis to compare the influence of the number of GCPs, their distribution, and their measurement error, as well as the order of the geometric correction polynomial on the error obtained. Grodecki and Lutes (2005) performed backup validation of IKONOS geometric calibration through manually and automatically selected GCPs, and found that it met the operational specifications. Wang *et al.* (2005) looked specifically at four geometric models (translation, translation and scale, affine, and second-order polynomial models) to further reduce ground point errors in IKONOS, with the objective of determining if lower cost IKONOS geosatellite imagery could be adequately refined for practical use.

Ideally, there should be mechanisms for the propagation analysis of image registration errors through the processing of remotely sensed images. Krupnik (2003) developed a mathematical formulation for error propagation in orthorectified image production for predicting the accuracy of the orthoimage obtained. The formulation combined several parameters, such as the accuracy of the digital elevation model, control points, and image measurements, as well as the configuration of fiducial, tie, and control points, and the ground slope. Poe *et al.* (2008) developed a similar formulation for geolocation errors observed with data from the Special Sensor Microwave Imager/Sounder (F-16 SSMIS) radiometer observations.

3.6.2 Algorithm evaluation

A traditional means of evaluating registration and geolocation accuracy is through the residual error of manually selected control points, that is, their root mean square error (RMSE). The assessment of transformations produced by manually selected CPs is often judged by repeatability and precision rather than formal accuracy, as the RMSE and variance evaluate the tightness of the CP residuals in the absence of ground truth data. This is often done by computing residuals under a transformation in question with respect to a separate set of independent check points using *hold-out validation* (HOV), or by recomputing the residual of one control point via *leave-one-out cross-validation* (LOOCV) (Brovelli *et al.* 2006). (In each iteration, a different control point is left out.) In general, the accuracy increases with the number of CPs up to a diminishing return (Toutin, 2004; Wang and Ellis, 2005a). Spatial and orientation distributions of residuals have also been considered (Buiten and van Putten, 1997; Fitzpatrick *et al.*, 1998; Armston *et al.*, 2002). Felicsimo *et al.* (2006) verified registration by careful analysis of residuals, looking at homogeneity and isotropy of spatial uncertainty by means of GPS kinematic check lines and circular statistics.

In addition to evaluating the end result of registration algorithms, image registration research can concentrate on evaluating alternative elements of registration algorithms. Maes *et al.* (1999) evaluated alternative optimization methods for use with mutual information. Recently, Klein *et al.* have also evaluated optimization methods (Klein *et al.*, 2007; 2009). Jenkinson and Smith (2001) looked at the efficiency and robustness of global optimization schemes, observing that local optimization schemes often did not find the optimum in their test medical imagery. They also tested alternative statistical similarity measures. A number of other articles focused on surveying, comparing, and unifying similarity measures. Kirby *et al.* (2006) implemented and compared several measures on images from the airborne Variable Interference Filter Imaging Spectrometer (VIFIS). The measures included normalized correlation, ordinal correlation, and invariant intensity moments. They found normalized correlation the most effective. In a medical application, Skerl *et al.* (2006) proposed an evaluation protocol for evaluation of similarity measures independent of the optimization procedure used. They implemented nine different multimodal statistical measures, including mutual information, normalized mutual information, correlation ratio and Woods criterion, and applied them to different image types. The authors compared the performance obtained with these measures with respect to several parameters, such as accuracy, capture, distinctiveness of the global optimum, and robustness. Their results varied for the image types studied, so their protocol can be seen as a way of recommending a similarity measure according to the relevant application. Inglada and Giros (2004) systematically investigated the application of different similarity measures to multisensor satellite registration of SPOT and ERS-2 data, including correlation, correlation ratio, normalized standard deviation, mutual information, and the related Kolmogorov measure. Pluim *et al.* (2004) compared the performance with respect to seven different information-theoretic similarity measures (including mutual information) on medical images. They considered specifically issues such as smoothness of the optimization surface, accuracy, and optimization difficulty.

One problem in evaluating image registration algorithms is the comparison of apples to oranges, that is, the comparison of algorithms implemented by different research groups using software components that may not be directly equivalent. It can be hard to compare search strategies, for example, while holding all other elements of the algorithm fixed, as two implementations may differ slightly in edge detection or other processing details. In Le Moigne *et al.* (2003), the authors presented a modular framework for implementing several algorithms in a common software system so as to hold processing details fixed, and to allow systematic variation on elements such as feature extraction, similarity measure and search strategy. This approach is often taken for comparison of a single element, such as variations in search or similarity measure, but rarely as to allow variation of several elements. A second problem for evaluation of registration algorithms is the development of useful metrics for algorithm performance. In Le Moigne *et al.* (2004), the authors extended the previously described general framework to studies of the sensitivity of registration algorithms to initial estimates of transformation parameters. The study found that the domains of parameter convergence for different algorithms can be complex in shape and extent.

Change detection, that is, the recovery of significant temporal or modal changes between two images, is an important analytic tool in remote sensing, which is closely related to similarity measure. An underlying assumption associated with similarity measure is that two corresponding image regions are related. This assumption is invalid, however, under various circumstances, for example, a forest has been cut down or a field has been converted to provide housing. In other words, whereas most of the image may match perfectly, certain isolated regions do not. If these regions can be detected as outliers, then the registration algorithm can ignore them. For example, cloud detection and masking are essential elements. Radke *et al.* (2005) surveyed change detection algorithms, while Hasler *et al.* (2003) modeled outliers by finding regions with differing statistics. Im *et al.* (2007) demonstrated change detection using correlation image analysis and image segmentation.

3.7 Summary

Image registration is a key, essential element in analysis of Earth remote sensing imagery. Registration is critical both for initial processing of raw satellite data for validation, preparation, and distribution of image products, and for end-user processing of those image products for data fusion, change detection, cartography, and other analyses. Given the increasing, immense rate of image capture by satellite systems and the growing complexity of end-user analyses, there is a stronger need for accurate, validated, fully automatic, and efficient registration algorithms and software systems. This chapter has surveyed the recent image registration literature to provide an overview of the current best approaches to this broad, multifaceted problem.

The image registration literature continues to grow, and the reader may wish to pursue more recent publications in specialized venues. General surveys and books on image registration were previously mentioned in the introduction. Examples of publication venues that frequently include articles on image registration for remote sensing are the *IEEE Transactions on Geoscience and Remote Sensing* (*TGRS*) and the *IEEE Transactions on Image Processing* (*TIP*), which deal, respectively, with remote sensing issues and image processing and mathematical models. Other venues where readers can keep their knowledge up-to-date include the yearly *IEEE International Geoscience and Remote Sensing Symposium* (*IGARSS*), the *American Geophysical Union* (*AGU*) Fall and Spring meetings, the annual *IEEE Conference on Computer Vision and Pattern Recognition* (*CVPR*), the *IEEE International Conference on Image Processing* (*ICIP*), and the conferences on *Neural Information Processing Systems* (*NIPS*). This short list of venues provides merely a starting point for further exploration of numerous relevant papers in the area of image registration.

References

Abdelfattah, R. and Nicolas, J. M. (2005). InSAR image co-registration using the Fourier-Mellin transform. *International Journal of Remote Sensing*, **26**(13), 2865–2876.

Ackerman, F. (1984). Digital image correlation: Performance and potential application in photogrammetry. *The Photogrammetric Record*, **11**(64), 429–439.

Alt, H., Aichholzer, O., and Rote, G. (1994). Matching shapes with a reference point. In *Proceedings of the Tenth Annual ACM Symposium on Computational Geometry*, Stony Brook, NY, pp. 85–91.

Alt, H., Fuchs, U., Rote, G., and Weber, G. (1996). Matching convex shapes with respect to the symmetric difference. In *Proceedings of the Fourth Annual European Symposium on Algorithms*. London: Springer-Verlag, pp. 320–333.

Althof, R. J., Wind, M. G., and Dobbins, J. T. I. (1997). A rapid and automatic image registration algorithm with subpixel accuracy. *IEEE Transactions on Medical Imaging*, **16**(2), 308–316.

Andrus, J., Campbell, C., and Jayroe, R. (1975). Digital image registration method using boundary maps. *IEEE Transactions on Computers*, **24**(9), 935–940.

Arevalo, V. and Gonzalez, J. (2008a). An experimental evaluation of non-rigid registration techniques on QuickBird satellite imagery. *International Journal of Remote Sensing*, **29**(2), 513–527.

Arevalo, V. and Gonzalez, J. (2008b). Improving piecewise linear registration of high-resolution satellite images through mesh optimization. *IEEE Transactions on Geoscience and Remote Sensing*, **46**(11), 3792–3803.

Armston, J., Danaher, T., Goulevitch, B., and Byrne, M. (2002). Geometric correction of Landsat MSS, TM, and ETM+ imagery for mapping of woody vegetation cover and change detection in Queensland. In *Proceedings of the Eleventh Australasian Remote Sensing and Photogrammetry Conference*, Brisbane, Queensland, Australia, pp. 1–23.

Arya, K., Gupta, P., Kalra, P., and Mitra, P. (2007). Image registration using robust M-estimators. *Pattern Recognition Letters*, **28**, 1957–1968.

Baillarin, S., Bouillon, A., Bernard, M., and Chikhi, M. (2005). Using a three dimensional spatial database to orthorectify automatically remote sensing images. In *Proceedings of the ISPRS Hangzhou 2005 Workshop on Service and Application of Spatial Data Infrastructure*, Hangzhou, China, pp. 89–93.

Baker, S. and Matthews, I. (2001). Equivalence and efficiency of image alignment algorithms. In *Proceedings of the IEEE Conference on Computer Vision and Pattern Recognition*, Kauai, Hawaii, pp. 1090–1097.

Bardera, A., Feixas, M., and Boada, I. (2004). Normalized similarity measures for medical image registration. In *Proceedings of the SPIE International Symposium on Medical Imaging*, San Diego, CA, pp. 108–118.

Barnea, D. J. and Silverman, H. F. (1972). A class of algorithms for fast digital image registration. *IEEE Transactions on Computing*, **21**, 179–186.

Bartoli, A. (2008). Groupwise geometric and photometric direct image registration. *IEEE Transactions on Pattern Analysis and Machine Intelligence*, **30**(12), 2098–2108.

Bentoutou, Y., Taleb, N., Kpalma, K., and Ronsin, J. (2005). An automatic image registration for applications in remote sensing. *IEEE Transactions on Geoscience and Remote Sensing*, **43**(9), 2127–2137.

Brovelli, M. A., Crespi, M., Fratarcangeli, F., Giannone, F., and Realin, E. (2006). Accuracy assessment of high resolution satellite imagery by leave-one-out method. In *Proceedings of the Seventh International Symposium on Spatial Accuracy Assessment in Natural Resources and Environmental Science*, Lisbon, Portugal, pp. 533–542.

Brown, L. G. (1992). A survey of image registration techniques. *ACM Computing Surveys*, **24**(4), 325–376.

Bruzzone, L. and Cossu, R. (2003). An adaptive approach to reducing registration noise effects in unsupervised change detection. *IEEE Transactions on Geoscience and Remote Sensing*, **41**(11), 2455–2465.

Buiten, H. J. and van Putten, B. (1997). Quality assessment of remote sensing image registration: Analysis and testing of control point residuals. *ISPRS Journal of Photogrammetry and Remote Sensing*, **52**, 57–73.

Cain, S., Hayat, M., and Armstrong, E. (2001). Projection-based image registration in the presence of fixed-pattern noise. *IEEE Transactions on Image Processing*, **10**(12), 1860–1872.

Cariou, C. and Chehdi, K. (2008). Automatic georeferencing of airborne pushbroom scanner images with missing ancillary data using mutual information. *IEEE Transactions on Geoscience and Remote Sensing*, **46**(5), 1290–1300.

Carr, J. L., Mangolini, M., Pourcelot, B., and Baucom, A. W. (1997). Automated and robust image geometry measurement techniques with application to meteorological satellite imaging. In *Proceedings of the CESDIS Image Registration Workshop*, NASA Goddard Space Flight Center, Greenbelt, MD, pp. 89–99.

Carrion, D., Gianinetto, M., and Scaioni, M. (2002). GEOREF: A software for improving the use of remote sensing images in environmental applications. In *Proceedings of the First Biennial Meeting of the International Environmental Modelling and Software Society*, Vol. 3, Lugano, Switzerland, pp. 360–365.

Ceccarelli, M., di Bisceglie, M., Galdi, C., Giangregorio, G., and Liberata, U. S. (2008). Image registration using non-linear diffusion. In *Proceedings of the IEEE International Geoscience and Remote Sensing Symposium*, Boston, MA, pp. 220–223.

Chen, H. M., Arora, M. K., and Varshney, P. K. (2003). Mutual information-based image registration for remote sensing data. *International Journal of Remote Sensing*, **24**(18), 3701–3706.

Chen, H., Varshney, P. K., and Arora, M. K. (2003). Performance of mutual information similarity measure for registration of multitemporal remote sensing images. *IEEE Transactions on Geoscience and Remote Sensing*, **41**(11), 2445–2454.

Chen, L.-C. and Lee, L.-H. (1992). Progressive generation of control frame works for image registration. *Photogrammetric Engineering & Remote Sensing*, **58**(9), 1321–1328.

Chen, Q., Defrise, M., and Deconinck, F. (1994). Symmetric phase-only matched filtering of Fourier-Mellin transforms for images registration and recognition. *IEEE Transactions on Pattern Analysis and Machine Intelligence*, **16**(12), 1156–1168.

Chen, T. and Huang, T. S. (2007). Optimizing image registration by mutually exclusive scale components. In *Proceedings of the IEEE International Conference on Computer Vision*, Rio de Janeiro, Brazil, pp. 1–8.

Chew, L. P., Goodrich, M. T., Huttenlocher, D. P., Kedem, K., Kleinberg, J. M., and Kravets, D. (1997). Geometric pattern matching under Euclidean motion. *Computational Geometry Theory and Applications*, **7**, 113–124.

Cho, M. and Mount, D. M. (2005). Improved approximation bounds for planar point pattern matching. In *Proceedings of the Ninth Workshop on Algorithms and Data Structures*, Waterloo, Canada; *Lecture Notes in Computer Science*, Springer-Verlag, Vol. 3608, pp. 432–443.

Choi, V. and Goyal, N. (2006). An efficient approximation algorithm for point pattern matching under noise. In *Proceedings of the Seventh Latin American Theoretical Informatics Symposium*, Valdivia, Chile; *Lecture Notes in Computer Science*, Springer-Verlag, Vol. 3887, pp. 298–310.

Cole-Rhodes, A., Johnson, K., Le Moigne, J., and Zavorin, I. (2003). Multiresolution registration of remote sensing images using stochastic gradient. *IEEE Transactions on Image Processing*, **12**(12), 1495–1511.

Dai, X. L. and Khorram, S. (1998). The effects of image misregistration on the accuracy of remotely sensed change detection. *IEEE Transactions on Geoscience and Remote Sensing*, **36**(5), 1566–1577.

Dai, X. L. and Khorram, S. (1999). A feature-based image registration algorithm using improved chain-code representation combined with invariant moments. *IEEE Transactions on Geoscience and Remote Sensing*, **37**(5), 2351–2362.

Dare, P. and Dowman, I. (2001). An improved model for automatic feature-based registration of SAR and SPOT images. *ISPRS Journal of Photogrammetry and Remote Sensing*, **56**(1), 13–28.

De Castro, E. and Morandi, C. (1987). Registration of translated and rotated images using finite Fourier transforms. *IEEE Transactions on Pattern Analysis and Machine Intelligence*, **9**(5), 700–703.

Dewdney, A. K. (1978). Analysis of a steepest-descent image-matching algorithm. *Pattern Recognition*, **10**(1), 31–39.

Dowson, N. and Bowden, R. (2006). A unifying framework for mutual information methods for use in non-linear optimisation. In *Proceedings of the Ninth European Conference on Computer Vision*, Graz, Austria; *Lecture Notes in Computer Science*, Springer, Vol. 3951, pp. 365–378.

Dowson, N., Kadir, T., and Bowden, R. (2008). Estimating the joint statistics of images using nonparametric windows with application to registration using mutual information. *IEEE Transactions on Pattern Analysis and Machine Intelligence*, **30**(10), 1841–1857.

Dufournaud, Y., Schmid, C., and Horaud, R. (2004). Image matching with scale adjustment. *Computer Vision and Image Understanding*, **93**(2), 175–194.

Eastman, R. D., Le Moigne, J., and Netanyahu, N. S. (2007). Research issues in image registration for remote sensing. In *Proceedings of the IEEE Computer Vision and Pattern Recognition Workshop on Image Registration and Fusion*, Minneapolis, MN, pp. 3233–3240.

Emery, W., Baldwin, D., and Matthews, D. (2003). Maximum cross correlation automatic satellite image navigation and attitude corrections for open-ocean image navigation. *IEEE Transactions on Geoscience and Remote Sensing*, **41**(1), 33–42.

Eugenio, F. and Marques, F. (2003). Automatic satellite image georeferencing using a contour-matching approach. *IEEE Transactions on Geoscience and Remote Sensing*, **41**(12), 2869–2880.

Faugeras, O. (1993). *Three-Dimensional Computer Vision*. Boston, MA: MIT Press.

Felicsimo, A. M., Cuartero, A., and Polo, M. E. (2006). Analysis of homogeneity and isotropy of spatial uncertainty by means of GPS kinematic check lines and circular statistics. In *Proceedings of the Seventh International Symposium on Spatial Accuracy Assessment in Natural Resources and Environmental Science*, Lisbon, Portugal, pp. 85–90.

Fischler, M. A. and Bolles, R. C. (1981). Random sample consensus: A paradigm for model fitting with applications to image analysis and automated cartography. *Communications of the ACM*, **24**(6), 381–395.

Fitzpatrick, J., West, J. B., and Maurer, C. R. (1998). Predicting error in rigid-body point-based registration. *IEEE Transactions on Medical Imaging*, **17**(5), 694–702.

Flusser, J. and Suk, T. (1994). A moment-based approach to registration of images with affine geometric distortion. *IEEE Transactions on Geoscience and Remote Sensing*, **32**(2), 382–387.

Fonseca, L. M. G. and Manjunath, B. S. (1996). Registration techniques for multisensor remotely sensed imagery. *Photogrammetric Engineering & Remote Sensing*, **62**(9), 1049–1056.

Foroosh, H., Zerubia, J. B., and Berthod, M. (2002). Extension of phase correlation to subpixel registration. *IEEE Transactions on Image Processing*, **11**(3), 188–200.

Förstner, W. and Gülch, E. (1987). A fast operator for detection and precise location of distinct points, corners, and centers of circular features. In *Proceedings of the ISPRS Intercommission Workshop on Fast Processing of Photogrammetric Data*, Interlaken, Switzerland, pp. 281–385.

Gavrilov, M., Indyk, P., Motwani, R., and Venkatasubramanian, S. (2004). Combinatorial and experimental methods for approximate point pattern matching. *Algorithmica*, **38**(1), 59–90.

Georgescu, B. and Meer, P. (2004). Point matching under large image deformations and illumination changes. *IEEE Transactions on Pattern Analysis and Machine Intelligence*, **26**(6), 674–688.

Goodrich, M. T., Mitchell, J. S. B., and Orletsky, M. W. (1999). Approximate geometric pattern matching under rigid motions. *IEEE Transactions on Pattern Analysis and Machine Intelligence*, **21**(4), 371–379.

Goshtasby, A. (1988). Registration of images with geometric distortions. *IEEE Transactions on Geoscience and Remote Sensing*, **26**(1), 60–64.

Goshtasby, A. (2005). *2-D and 3-D Image Registration for Medical, Remote Sensing, and Industrial Applications*. Hoboken, NJ: John Wiley and Sons.

Goshtasby, A. and Stockman, G. C. (1985). Point pattern matching using convex hull edges. *IEEE Transactions on Systems, Man, and Cybernetics*, **15**(5), 631–637.

Goshtasby, A., Stockman, G., and Page, C. (1986). A region-based approach to digital image registration with subpixel accuracy. *IEEE Transactions on Geoscience and Remote Sensing*, **24**(3), 330–399.

Govindu, V. and Shekhar, C. (1999). Alignment using distributions of local geometric properties. *IEEE Transactions on Pattern Analysis and Machine Intelligence*, **21**(10), 1031–1043.

Grodecki, J. and Lutes, J. (2005). IKONOS geometric calibrations. Technical report, Space Imaging.

Growe, S. and Tönjes, R. (1997). A knowledge based approach to automatic image registration. In *Proceedings of the International Conference on Image Processing*, Washington, DC, pp. 228–231.

Gruen, A. (1985). Adaptive least squares correlation: A powerful image matching technique. *South African Journal of Photogrammetry, Remote Sensing and Cartography*, **14**, 175–187.

Gupta, N. (2007). A VLSI architecture for image registration in real time. *IEEE Transactions on Very Large Scale Integration (VLSI) Systems*, **15**(9), 981–989.

Habib, A. F. and Alruzouq, R. I. (2004). Line-based modified iterated Hough transform for automatic registration of multi-source imagery. *The Photogrammetric Record*, **19**(105), 5–21.

Habib, A. F. and Al-Ruzouq, R. (2005). Semi-automatic registration of multi-source satellite imagery with varying geometric resolutions. *Photogrammetric Engineering & Remote Sensing*, **71**(3), 325–332.

Hagedoorn, M. and Veltkamp, R. C. (1999). Reliable and efficient pattern matching using an affine invariant metric. *International Journal of Computer Vision*, **31**(2–3), 103–115.

Harris, C. and Stephens, M. (1988). A combined corner and edge detector. In *Proceedings of the Fourth Alvey Vision Conference*, Manchester, UK, pp. 147–151.

Hasler, D., Sbaiz, L., Susstrunk, S., and Vetterli, M. (2003). Outlier modeling in image matching. *IEEE Transactions on Pattern Analysis and Machine Intelligence*, **25**(3), 301–315.

He, Y., Hamza, A. B. and Krim, H. (2003). A generalized divergence measure for robust image registration. *IEEE Transactions on Signal Processing*, **51**(5), 1211–1220.

Heffernan, P. J. and Schirra, S. (1994). Approximate decision algorithms for point set congruence. *Computational Geometry Theory and Applications*, **4**(3), 137–156.

Hong, G. and Zhang, Y. (2008). A comparative study on radiometric normalization using high resolution satellite images. *International Journal of Remote Sensing*, **29**(2), 425–438.

Hong, S. H., Jung, H. S., and Won, J. S. (2006). Extraction of ground control points (GCPs) from synthetic aperture radar images and SRTM DEM. *International Journal of Remote Sensing*, **27**(18), 3813–3829.

Hsieh, J.-W., Liao, H.-Y. M., Fan, K.-C., Ko, M.-T., and Hung, Y.-P. (1997). Image registration using a new edge-based approach. *Computer Vision and Image Understanding*, **67**(2), 112–130.

Huseby, R. B., Halck, O. M., and Solberg, R. (2005). A model-based approach for geometrical correction of optical satellite images. *International Journal of Remote Sensing*, **26**(15), 3205–3223.

Huttenlocher, D. P. and Rucklidge, W. J. (1993). A multi-resolution technique for comparing images using the Hausdorff distance. In *Proceedings of the IEEE Conference on Computer Vision and Pattern Recognition*, New York, NY, pp. 705–706.

Huttenlocher, D. P., Kedem, K., and Sharir, M. (1993a). The upper envelope of Voronoi surfaces and its applications. *Discrete and Computational Geometry*, **9**, 267–291.

Huttenlocher, D. P., Klanderman, G. A., and Rucklidge, W. J. (1993b). Comparing images using the Hausdorff distance. *IEEE Transactions on Pattern Analysis and Machine Intelligence*, **15**(9), 850–863.

Igbokwe, J. I. (1999). Geometrical processing of multi-sensoral multi-temporal satellite images for change detection studies. *International Journal of Remote Sensing*, **20**(6), 1141–1148.

Im, J., Rhee, J., Jensen, J. R., and Hodgson, M. E. (2007). An automated binary change detection model using a calibration approach. *Remote Sensing of Environment*, **106**(1), 89–105.

Inglada, J. and Giros, A. (2004). On the possibility of automatic multisensor image registration. *IEEE Transactions on Geoscience and Remote Sensing*, **41**(10), 2104–2120.

Inglada, J., Muron, V., Pichard, D., and Feuvrier, T. (2007). Analysis of artifacts in subpixel remote sensing image registration. *IEEE Transactions on Geoscience and Remote Sensing*, **45**(1), 254–264.

Irani, M. and Anandan, P. (1999). About direct methods. In *Proceedings of the IEEE ICCV Workshop on Vision Algorithms*, Corfu, Greece, pp. 267–277.

Irani, M. and Peleg, S. (1991). Improving resolution by image registration. *CVGIP: Graphical Models and Image Processing*, **53**(3), 231–239.

Iwasaki, A. and Fujisada, H. (2005). ASTER geometric performance. *IEEE Transactions on Geoscience and Remote Sensing*, **43**(12), 2700–2706.

Jenkinson, M. and Smith, S. (2001). A global optimisation method for robust affine registration of brain images. *Medical Image Analysis*, **5**, 142–156.

Jovanovic, V. M., Bull, M. A., Smyth, M. M., and Zong, J. (2002). MISR in-flight camera geometric model calibration and georectification performance. *IEEE Transactions on Geoscience and Remote Sensing*, **40**(7), 1512–1519.

Kaneko, S., Satohb, Y., and Igarashia, S. (2003). Using selective correlation coefficient for robust image registration. *Pattern Recognition*, **36**(5), 1165–1173.

Ke, Y. and Sukthankar, R. (2004). PCA-SIFT: A more distinctive representation for local image descriptors. In *Proceedings of the IEEE Conference on Computer Vision and Pattern Recognition*, Washington, DC, pp. 506–513.

Kedem, K. and Yarmovski, Y. (1996). Curve based stereo matching using the minimum Hausdorff distance. In *Proceedings of the Twelfth Annual ACM Symposium on Computational Geometry*, Philadelphia, PA, pp. C15–C18.

Keller, Y. and Averbuch, A. (2006). Multisensor image registration via implicit similarity. *IEEE Transactions on Pattern Analysis and Machine Intelligence*, **28**(5), 794–801.

Keller, Y., Shkolnisky, Y., and Averbuch, A. (2005). The angular difference function and its application to image registration. *IEEE Transactions on Pattern Analysis and Machine Intelligence*, **27**(9), 969–976.

Kelman, A., Sofka, M., and Stewart, C. V. (2007). Keypoint descriptors for matching across multiple image modalities and non-linear intensity variations. In *Proceedings of the IEEE Computer Vision and Pattern Recognition Workshop on Image Registration and Fusion*, Minneapolis, MN, pp. 1–7.

Kennedy, R. E. and Cohen, W. B. (2003). Automated designation of tie-points for image-to-image coregistration. *International Journal of Remote Sensing*, **24**(17), 3467–3490.

Kern, J. and Pattichis, M. (2007). Robust multispectral image registration using mutual-information models. *IEEE Transactions on Geoscience and Remote Sensing*, **45**(5), 1494–1505.

Kim, J. and Fessler, J. A. (2004). Intensity-based image registration using robust correlation coefficients. *IEEE Transactions on Medical Imaging*, **23**(11), 1430–1444.

Kirby, N. E., Cracknell, A. P., Monk, J. G. C., and Anderson, J. A. (2006). The automatic alignment and mosaic of video frames from the variable interference filter imaging spectrometer. *International Journal of Remote Sensing*, **27**(21), 4885–4898.

Klein, S., Pluim, J. P. W., Staring, M., and Viergever, M. A. (2009). Adaptive stochastic gradient descent optimisation for image registration. *International Journal of Computer Vision*, **81**(3), 227–239.

Klein, S., Staring, M., and Pluim, J. P. W. (2007). Evaluation of optimization methods for nonrigid medical image registration using mutual information and B-splines. *IEEE Transactions on Image Processing*, **16**(12), 2879–2890.

Knops, Z. F., Maintz, J. B. A., Viergever, M. A., and Pluim, J. P. W. (2006). Normalized mutual information based registration using k-means clustering and shading correction. *Medical Image Analysis*, **10**(3), 432–439.

Köhn, A., Drexl, J., Ritter, F., König, M., and Peitgen, H. O. (2006). GPU accelerated image registration in two and three dimensions. In *Proceedings of Bildverarbeitung für die Medizin 2006*. Hamburg: Springer Verlag, pp. 261–265.

Krupnik, A. (2003). Accuracy prediction for ortho-image generation. *The Photogrammetric Record*, **18**(101), 41–58.

Lazaridis, G. and Petrou, M. (2006). Image registration using the Walsh transform. *IEEE Transactions on Image Processing*, **15**(8), 2343–2357.

Le Moigne, J., Campbell, W., Cromp, R., Branch, A., and Center, N. (2002). An automated parallel image registration technique based on the correlation of wavelet features. *IEEE Transactions on Geoscience and Remote Sensing*, **40**(8), 1849–1864.

Le Moigne, J., Cole-Rhodes, A., Johnson, K., Morisette, J., Netanyahu, N. S., Eastman, R., Stone, H., and Zavorin, I. (2003). Earth science imagery registration. In *Proceedings of the IEEE International Geoscience and Remote Sensing Symposium*, Toulouse, France, pp. 161–163.

Le Moigne, J., Morisette, J., Cole-Rhodes, A., Johnson, K., Netanyahu, N. S., Eastman, R., Stone, H., Zavorin, I., and Jain, P. (2004). A study of the sensitivity of automatic image registration algorithms to initial conditions. In *Proceedings of the IEEE International Geoscience and Remote Sensing Symposium*, Anchorage, Alaska, pp. 1390–1393.

Le Moigne, J., Netanyahu, N. S., Masek, J. G., Mount, D., Goward, S., and Honzak, M. (2000). Geo-registration of Landsat data by robust matching of wavelet features. In *Proceedings of the IEEE International Geoscience and Remote Sensing Symposium*, Honolulu, Hawaii, pp. 1610–1612.

Lee, D. S., Storey, J. C., Choate, M. J., and Hayes, R. W. (2004). Four years of Landsat-7 on-orbit geometric calibration and performance. *IEEE Transactions on Geoscience and Remote Sensing*, **42**(12), 2786–2795.

Leprince, S., Barbot, S., Ayoub, F., and Avouac, J. (2007). Automatic and precise orthorectification, coregistration, and subpixel correlation of satellite images, application to ground deformation measurements. *IEEE Transactions on Geoscience and Remote Sensing*, **45**(6), 1529–1558.

Lester, H. and Arridge, S. R. (1999). A survey of hierarchical non-linear medical image registration. *Pattern Recognition*, **32**(1), 129–149.

Li, H., Manjunath, B., and Mitra, S. (1995). Multisensor image fusion using the wavelet transform. *CVGIP: Graphical Models and Image Processing*, **57**(3), 235–245.

Li, Q., Wang, G., Liu, J., and Chen, S. (2009). Robust scale-invariant feature matching for remote sensing image registration. *IEEE Geoscience and Remote Sensing Letters*, **6**(2), 287–291.

Lowe, D. (1999). Object recognition from local scale-invariant features. In *Proceedings of the IEEE International Conference on Computer Vision*, Corfu, Greece, Vol. 2, pp. 1150–1157.

Lowe, D. G. (2004). Distinctive image features from scale-invariant keypoints. *International Journal of Computer Vision*, **20**(2), 91–110.

Lucas, B. D. and Kanade, T. (1981). An iterative image registration technique with an application to stereo vision. In *Proceedings of the International Joint Conference on Artificial Intelligence*, Vancouver, British Columbia, Canada, pp. 674–679.

Madani, H., Carr, J. L., and Schoeser, C. (2004). Image registration using AutoLand mark. In *Proceedings of the IEEE International Geoscience and Remote Sensing Symposium*, Anchorage, Alaska, pp. 3778–3781.

Maes, F., Collignon, A., Vandermeulen, D., Marchal, G., and Suetens, P. (1997). Multimodality image registration by maximization of mutual information. *IEEE Transactions on Medical Imaging*, **16**(2), 187–198.

Maes, F., Vandermeulen, D., and Suetens, P. (1999). Comparative evaluation of multiresolution optimization strategies for multimodality image registration by maximization of mutual information. *Medical Image Analysis*, **3**(4), 373–386.

Maintz, J. and Viergever, M. (1998). A survey of medical image registration. *Medical Image Analysis*, **2**(1), 1–36.

McGuire, M. and Stone, H. S. (2000). Techniques for multiresolution image registration in the presence of occlusions. *IEEE Transactions on Geoscience and Remote Sensing*, **38**(3), 1476–1479.

Mikolajczyk, K. and Schmid, C. (2004). Scale & affine invariant interest point detectors. *International Journal of Computer Vision*, **60**(1), 63–86.

Mikolajczyk, K. and Schmid, C. (2005). A performance evaluation of local descriptors. *IEEE Transactions on Pattern Analysis and Machine Intelligence*, **27**(10), 1615–1630.

Mikolajczyk, K., Tuytelaars, T., Schmid, C., and Zisserman, A. (2005). A comparison of affine region detectors. *International Journal of Computer Vision*, **65**(1–2), 43–72.

Modersitzki, J. (2004). *Numerical Methods for Image Registration*. Oxford: Oxford University Press.

Mount, D. M., Netanyahu, N. S., and Le Moigne, J. (1999). Efficient algorithms for robust feature matching. *Pattern Recognition*, **32**(1), 17–38.

Netanyahu, N. S., Le Moigne, J., and Masek, J. G. (2004). Georegistration of Landsat data via robust matching of multiresolution features. *IEEE Transactions on Geoscience and Remote Sensing*, **42**(7), 1586–1600.

Nies, H., Loffeld, O., Dönmez, B., Hammadi, A. B., and Wang, R. (2008). Image registration of TerraSAR-X data using different information measures. In *Proceedings of the IEEE International Geoscience and Remote Sensing Symposium*, Boston, MA, pp. 419–422.

Orchard, J. (2007). Efficient least squares multimodal registration with a globally exhaustive alignment search. *IEEE Transactions on Image Processing*, **16**(10), 2526–2534.

Pan, C., Zhang, Z., Yan, H., Wu, G., and Ma, S. (2008). Multisource data registration based on NURBS description of contours. *International Journal of Remote Sensing*, **29**(2), 569–591.

Penney, G. P., Weese, J., Little, J. A., Desmedt, P., Hill, D. L. G., and Hawkes, D. J. (1998). A comparison of similarity measures for use in 2-D-3-D medical image registration. *IEEE Transactions on Medical Imaging*, **17**(4), 586–595.

Pluim, J. P. W., Maintz, J. B. A., and Viergever, M. A. (2000). Interpolation artefacts in mutual information based image registration. *Computer Vision and Image Understanding*, **77**(2), 211–232.

Pluim, J., Maintz, J., and Viergever, M. (2003). Mutual-information-based registration of medical images: A survey. *IEEE Transactions on Medical Imaging*, **22**(8), 986–1004.

Pluim, J. P., Maintz, J. B. A., and Viergever, M. A. (2004). *f*-Information measures in medical image registration. *IEEE Transactions on Medical Imaging*, **23**(12), 1506–1518.

Poe, G. A., Uliana, E. A., Gardiner, B. A., von Rentzell, T. E., and Kunkee, D. B. (2008). Geolocation error analysis of the special sensor microwave imager/sounder. *IEEE Transactions on Geoscience and Remote Sensing*, **46**(4), 913–922.

Radke, R. J., Andra, S., Al-Kofahi, O., and Roysam, B. (2005). Image change detection algorithms: A systematic survey. *IEEE Transactions on Image Processing*, **14**(3), 294–307.

Rajwade, A., Banerjee, A., and Rangarajan, A. (2009). Probability density estimation using isocontours and isosurfaces: Applications to information-theoretic image registration. *IEEE Transactions on Pattern Analysis and Machine Intelligence*, **31**(3), 475–491.

Reddy, B. and Chatterji, B. (1996). An FFT-based technique for translation, rotation and scale-invariant image registration. *IEEE Transactions on Image Processing*, **5**(8), 1266–1271.

Roche, A., Malandain, G., and Ayache, N. (2000). Unifying maximum likelihood approaches in medical image registration. Special Issue on 3D Imaging. *International Journal of Imaging Systems and Technology*, **11**(1), 71–80.

Rohde, G., Aldroubi, A., and Healy, D. (2008). Interpolation artifacts in sub-pixel image registration. *IEEE Transactions on Image Processing*, **18**(2), 333–345.

Roy, D. (2000). The impact of misregistration upon composited wide field of view satellite data and implications for change detection. *IEEE Transactions on Geoscience and Remote Sensing*, **38**(4), 2017–2032.

Salas, W. A., Boles, S. H., Frolking, S., Xiao, X., and Li, C. (2003). The perimeter/area ratio as an index of misregistration bias in land cover change estimates. *International Journal of Remote Sensing*, **24**(5), 1165–1170.

Salvado, O. and Wilson, D. L. (2007). Removal of local and biased global maxima in intensity-based registration. *Medical Image Analysis*, **11**(2), 183–196.

Salvi, J., Matabosch, C., Fofi, D., and Forest, J. (2006). A review of recent range image registration methods with accuracy evaluation. *Image and Vision Computing*, **25**(5), 578–596.

Schmid, C., Mohr, R., and Bauckhage, C. (2000). Evaluation of interest point detectors. *International Journal of Computer Vision*, **37**(2), 151–172.

Sen, M., Hemaraj, Y., Plishker, W., Shekhar, R., and Bhattacharyya, S. S. (2008). Model-based mapping of reconfigurable image registration on FPGA platforms. *Journal of Real-Time Image Processing*, **3**(3), 149–162.

Sester, M., Hild, H., and Fritsch, D. (1998). Definition of ground-control features for image registration using GIS-data. *International Archives of Photogrammetry and Remote Sensing*, **32/4**, 537–543.

Shams, R., Sadeghi, P., and Kennedy, R. A. (2007). Gradient intensity: A new mutual information based registration method. In *Proceedings of the IEEE Computer Vision and Pattern Recognition Workshop on Image Registration and Fusion*, Minneapolis, MN, pp. 3249–3256.

Shekarforoush, H., Berthod, M., and Zerubia, J. (1996). Subpixel image registration by estimating the polyphase decomposition of cross power spectrum. In *Proceedings of the IEEE Conference on Computer Vision and Pattern Recognition*, San Francisco, CA, pp. 532–537.

Skerl, D., Likar, B., and Pernus, F. (2006). A protocol for evaluation of similarity measures for rigid registration. *IEEE Transactions on Medical Imaging*, **25**(6), 779–791.

Stewart, C., Tsai, C., and Roysam, B. (2003). The dual-bootstrap iterative closest point algorithm with application to retinal image registration. *IEEE Transactions on Medical Imaging*, **22**(11), 1379–1394.

Stockman, G. C., Kopstein, S., and Benett, S. (1982). Matching images to models for registration and object detection via clustering. *IEEE Transactions on Pattern Analysis and Machine Intelligence*, **4**(3), 229–241.

Stone, H. S., Orchard, M., Chang, E. C., and Martucci, S. (2001). A fast direct Fourier-based algorithm for subpixel segistration of images. *IEEE Transactions on Geoscience and Remote Sensing*, **39**(10), 2235–2243.

Stone, H. S., Tao, B., and McGuire, M. (2003). Analysis of image registration noise due to rotationally dependent aliasing. *Journal of Visual Communication and Image Representation*, **14**, 114–135.

Sylvander, S., Henry, P., Bastien-Thiry, C., Meunier, F., and Fuster, D. (2000). VEGETATION geometrical image quality. *Société Française de Photogrammétrie et de Télédétection*, **159**, 59–65.

Theiler, J. P., Galbraith, A. E., Pope, P. A., Ramsey, K. A., and Szymanski, J. J. (2002). Automated coregistration of MTI spectral bands. In *Proceedings of the SPIE Conference on Algorithms and Technologies for Multispectral, Hyperspectral, and Ultraspectral Imagery VIII*, Orlando, FL, Vol. 4725, pp. 314–327.

Thévenaz, P. and Unser, M. (2000). Optimization of mutual information for multiresolution image registration. *IEEE Transactions on Image Processing*, **9**(12), 2083–2099.

Thévenaz, P., Bierlaire, M., and Unser, M. (2008). Halton sampling for image registration based on mutual information. *Sampling Theory in Signal and Image Processing*, **7**(2), 141–171.

Thévenaz, P., Ruttimann, U., and Unser, M. (1998). A pyramid approach to subpixel registration based on intensity. *IEEE Transactions on Image Processing*, **7**(1), 27–41.

Toutin, T. (2003). Error tracking in IKONOS geometric processing. *Photogrammetric Engineering & Remote Sensing*, **69**(1), 43–51.

Toutin, T. (2004). Review article: Geometric processing of remote sensing images: Models, algorithms and methods. *International Journal of Remote Sensing*, **25**(10), 1893–1924.

Townshend, J., Justice, C., Gurney, C., and McManus, J. (1992). The impact of misregistration on change detection. *IEEE Transactions on Geoscience and Remote Sensing*, **30**(5), 1054–1060.

Tsao, J. (2003). Interpolation artifacts in multimodality image registration based on maximization of mutual information. *IEEE Transactions on Medical Imaging*, **22**(7), 854–864.

Tuytelaars, T. and Mikolajczyk, K. (2008). *Local Invariant Feature Detectors: A Survey*. Hanover, MA: Now Publishers, Inc.

Verbyla, D. L. and Boles, S. H. (2000). Bias in land cover change estimates due to misregistration. *International Journal of Remote Sensing*, **21**(18), 3553–3560.

Viola, P. A. and Wells, W. M. I. (1997). Alignment by maximization of mutual information. *International Journal of Computer Vision*, **24**(2), 137–154.

Wang, G., Gertner, G. Z., Shoufan, F., and Anderson, A. B. (2005). A methodology for spatial uncertainty analysis of remote sensing and GIS products. *Photogrammetric Engineering & Remote Sensing*, **71**(12), 1423–1432.

Wang, H. and Ellis, E. C. (2005a). Image misregistration error in change measurements. *Photogrammetric Engineering & Remote Sensing*, **71**(9), 1037–1044.

Wang, H. and Ellis, E. C. (2005b). Spatial accuracy of orthorectified IKONOS imagery and historical aerial photographs across five sites in China. *International Journal of Remote Sensing*, **26**(9), 1893–1911.

Wang, J., Di, K., and Li, R. (2005). Evaluation and improvement of geopositioning accuracy of IKONOS stereo imagery. *Journal of Surveying Engineering*, **131**(2), 35–42.

Wang, W.-H. and Chen, Y.-C. (1997). Image registration by control points pairing using the invariant properties of line segments. *Pattern Recognition Letters*, **18**(3), 269–281.

Wolfe, R. E., Nishihama, M., Fleiga, A. J., Kuyper, J. A., Roy, D. P., Storey, J. C., and Patt, F. S. (2002). Achieving sub-pixel geolocation accuracy in support of MODIS land science. *Remote Sensing of Environment*, **83**(1–2), 31–49.

Wong, A. and Clausi, D. A. (2007). ARRSI: Automatic registration of remote-sensing images. *IEEE Transactions on Geoscience and Remote Sensing*, **45**(5), 1483–1493.

Xia, M. and Liu, B. (2004). Image registration by "super-curves". *IEEE Transactions on Image Processing*, **13**(5), 720–732.

Xiong, X., Che, N., and Barnes, W. (2005). Terra MODIS on-orbit spatial characterization and performance. *IEEE Transactions on Geoscience and Remote Sensing*, **43**(2), 355–365.

Yam, S. and Davis, L. (1981). Image registration using generalized Hough transform. In *Proceedings of the IEEE Conference on Pattern Recognition and Image Processing*, Dallas, Texas, pp. 525–533.

Yang, G., Stewart, C. V., Sofka, M., and Tsai, C. L. (2006). Automatic robust image registration system: Initialization, estimation, and decision. In *Proceedings of the Fourth IEEE International Conference on Computer Vision Systems*, New York, NY, pp. 23–31.

Yang, Z. and Cohen, F. S. (1999). Image registration and object recognition using affine invariants and convex hulls. *IEEE Transactions on Image Processing*, **8**(7), 934–947.

Yuan, Z.-M., Wu, F., and Zhuang, Y.-T. (2006). Multi-sensor image registration using multi-resolution shape analysis. *Journal of Zhejiang University SCIENCE A*, **7**(4), 549–555.

Zagorchev, L. and Goshtasby, A. (2006). A comparative study of transformation functions for nonrigid image registration. *IEEE Transactions on Image Processing*, **15**(3), 529–538.

Zavorin, I. and Le Moigne, J. (2005). Use of multiresolution wavelet feature pyramids for automatic registration of multisensor imagery. *IEEE Transactions on Image Processing*, **14**(6), 770–782.

Zhou, G. and Li, R. (2000). Accuracy evaluation of ground points from IKONOS high-resolution satellite imagery. *Photogrammetric Engineering & Remote Sensing*, **66**(3), 1103–1112.

Zitová, B. and Flusser, J. (2003). Image registration methods: A survey. *Image and Vision Computing*, **21**(11), 977–1000.

Zokai, S. and Wolberg, G. (2005). Image registration using log-polar mappings for recovery of large-scale similarity and projective transformations. *IEEE Transactions on Image Processing*, **14**(10), 1422–1434.

Part II

Similarity Metrics for Image Registration

4

Fast correlation and phase correlation

HAROLD S. STONE

Abstract

Correlation is an extremely powerful technique for finding similarities between two images. This chapter describes why correlation has proved to be a valuable tool, how to implement correlation to achieve extremely high performance processing, and indicates the limits of correlation so that it can be used where it is appropriate. Section 4.1 gives the underlying theory for fast correlation, which is the well-known *convolution theorem*. It is this theory that gives correlation a huge processing advantage in many applications. Also covered is normalized correlation, which is a form of correlation that allows images to be matched in spite of differences in the images due to uniform changes of intensity. Section 4.2 treats the practical implementation of correlation, including the use of masks to eliminate irrelevant or obscured portions of images. The implementations described in this section treat images that differ only by translation, and otherwise have the same orientation and scale. Section 4.3 introduces an extension of the basic algorithm to allow for small differences of scale and orientation as well as translation. Section 4.4 presents very high precision registration. This section shows how to make use of Fourier phase to determine translational differences down to a few hundredths of a pixel. The images must be oriented and scaled identically and have a translational difference that does not exceed half of a pixel. Section 4.5 deals with fast rotational registration that uses phase correlation to discover the rotational difference between two images. The two images need not be scaled identically nor registered with respect to translation, but the images must have substantial overlap in order to be registered successfully. The same technique also produces an estimate of the scale difference between two images.

4.1 Underlying theory

The development is described in this section for one-dimensional data, but it extends in a straightforward way to two dimensions. The general problem in one dimension with discrete coordinates is to locate the best possible match for pattern vector **y** of length M within a reference vector **x** of length N. For example, let **y** be [1, −1, −1, 1] and **x** be [1, 1, −1, 1, −1, −1, 1, 1, −1, −1, −1, 1]. The problem is to find the position at which **y** best matches a subvector of **x**. Since **x** has length 12 and **y** has length 4, we can lay **y** on **x** in each of 9 distinct positions. It is obvious that we should lay **y** on **x** in each of those 9 positions and compare the corresponding components. Let us do that and count the number of corresponding components that agree (A) and the number that disagree (D), and then compute $A - D$. We get the following result: [2, 0, −2, 4, 0, −4, 0, 2, 2].

The maximum of the result is 4, which occurs at index position 3 (with the first index having a value of 0). This indicates that the best match occurs when **y** is shifted to the right 3 positions before it is placed on **x** and compared. Indeed, at that position the 4 components of **x** are [1, −1, −1, 1] which are identical to those of **y**. Note also that the minimum value of the comparison vector is −4, which occurs where the components of **x** are [−1, 1, 1, −1]. In this position the components are the negatives of the corresponding **y** components. This also could be a perfect match if the pattern vector **y** could be produced by a natural process that transforms an input reference **x** by negating its components, among other things.

The process of comparing vectors component-by-component and noting where they are the same is a type of process that is often called *correlation*. In this particular example, the correlation is done by subtracting the number of disagreements from the number of agreements, which is an appropriate measure when input data are binary valued. However, when input data can be multivalued, for example, 8-bit pixel data, simple counts of agreements and disagreements are not appropriate. Two images of the same scene are likely to differ in ways that make some discrepancies more likely than others. One image may be captured by an instrument that is uniformly more sensitive than the instrument that captured the second image, or the natural illumination may be greater for one image than the other. Consequently, if a pattern vector **y** is similar to a reference vector **x** in a particular position except for a scaling of the pixel values, then **y** is likely to be an image of same feature as **x**, whereas if **y** differs from **x** in a specific position by much more than a rescaling of the pixel values, then **y** does not likely depict the same feature.

The world of statistical mathematics has an appropriate mathematical measure that expresses the likeness of two numerical vectors when the two could differ by a transformation that maps pixel values y_i of the first vector into pixel values x_i of the second by a mapping such that $x_i = ay_i + b$, where a and b are constants. The function is known as the *correlation coefficient* of **x** and **y** (Edwards, 1976). The correlation coefficient of a pattern $\mathbf{y} = [y_0, y_1, \ldots, y_{M-1}]$ with a reference vector $\mathbf{x} = [x_0, x_1, \ldots, x_{N-1}]$ is a vector **corrcoef(x, y)** of length $N - M + 1$ whose jth

$(j = 0, 1, \ldots, N - M + 1)$ component is given by

$$
\begin{aligned}
&\mathbf{corrcoef}(\mathbf{x}, \mathbf{y})_j \\
&= (\mathbf{x} \bullet \mathbf{y})_j \\
&= \frac{\left(\sum_{i=0}^{M-1} x_{i+j} y_i - (1/M) \sum_{i=0}^{M-1} x_{i+j} \sum_{i=0}^{M-1} y_i \right)}{\sqrt{\left(\sum_{i=0}^{M-1} x_{i+j}^2 - (1/M) \left(\sum_{i=0}^{M-1} x_{i+j} \right)^2 \right) \left(\sum_{i=0}^{M-1} y_i^2 - (1/M) \left(\sum_{i=0}^{M-1} y_i \right)^2 \right)}}.
\end{aligned}
$$
(4.1)

This function is normalized so that its possible values lie in the range $[-1, +1]$. These values are attained when there is a map between the pixel values of $\mathbf{y}$ and $\mathbf{x}$ such that $x_{i+j} = ay_i + b$ for all subscripts i, $i = 0, 1, \ldots, M - 1$. If $a > 0$, the correlation coefficient is 1, and if $a < 0$ the correlation coefficient is -1. As an example, consider the vectors $\mathbf{y} = [1, 2, 2, 1]$ and $\mathbf{x} = [3, 4, 7, 4, 1, 1, 4, 6, 6, 4, 7, 2]$. The function $\mathbf{corrcoef}(\mathbf{x}, \mathbf{y})$ is $[0.6667, 0.7071, -0.3015, -1.0000, -0.2357, 0.3665, 1.0000, -0.6882, 0.3906]$. Note the correlation coefficient is 1.000 where $\mathbf{y}$ matches the subvector $[4, 6, 6, 4]$ because the mapping between $\mathbf{y}$ and $\mathbf{x}$ uses $a = 2$ and $b = 2$. Also note the correlation of -1.000 where $\mathbf{y}$ matches the subvector $[4, 1, 1, 4]$ using $a = -3$ and $b = 7$.

Although Eq. (4.1) looks formidable to calculate, actually it can be calculated extremely quickly for large images by using some mathematical legerdemain. The cost of calculation is on the order of 100 floating-point operations per reference pixel, growing very slightly faster than linearly in the size of the reference. The trick is to make use of the well-known convolution theorem. To explain the convolution theorem, we first define the convolution of two vectors.

Let $\mathbf{x}$ and $\mathbf{y}$ be vectors of length N with subscripts running from 0 to $N - 1$. Let $\mathbf{x} \otimes \mathbf{y}$ denote the *circular convolution* of $\mathbf{x}$ and $\mathbf{y}$ defined by

$$
\mathbf{circconv}(\mathbf{x}, \mathbf{y})_j = (\mathbf{x} \otimes \mathbf{y})_j = \sum_{i=0}^{N-1} x_{(j-i) \bmod N} y_i.
$$
(4.2)

The use of the term *circular* is a consequence of the arithmetic on the subscripts being done modulo N, so that when $i = 4$, $j = 2$, and $N = 10$, the calculation of $j - i$ is done modulo 10, and the result is 8, not -2. As an example of the computation of a circular convolution, let $\mathbf{x}$ be the vector $[0, 1, 1, 2, 2, 0, 0, 0, 0, 0]$ and $\mathbf{y}$ be the vector $[0, 0, 0, 0, 1, 1, 2, 2, 0, 0]$. Then $\mathbf{x} \otimes \mathbf{y} = [8, 8, 4, 0, 0, 0, 1, 2, 5, 8]$. A visual explanation of the circular convolution is that the vector $\mathbf{x}$ is flipped from left to right, and then correlated to the vector $\mathbf{y}$ in each of N different circular phases.

Convolution theorem: Let *Fourier*(**x**)$_\omega$ denote frequency component ω of the Fourier transform of **x**. Then

$$Fourier(\mathbf{x} \otimes \mathbf{y})_\omega = Fourier(\mathbf{x})_\omega \cdot Fourier(\mathbf{y})_\omega. \tag{4.3}$$

(For a proof of the convolution theorem, see Strang, 1986, p. 297*ff*; Stone, 1998.)

The image registration operations of interest to us depend on correlations, rather than on convolutions. Correlations have the same structure as convolutions, except that when correlating a vector **x** to a vector **y**, the vector **x** is aligned to the vector **y** without first reversing **x**, as is done for convolutions. Following the steps above for convolutions, we define a circular correlation, and state a *correlation theorem* for circular correlations that corresponds to the convolution theorem for circular convolutions.

Let **x** and **y** be vectors of length N with subscripts running from 0 to $N - 1$. Let **x** ∘ **y** denote the *circular* correlation of **x** and **y** defined by

$$\mathbf{circcorr}(\mathbf{x}, \mathbf{y})_j = (\mathbf{x} \circ \mathbf{y})_j = \sum_{i=0}^{N-1} x_{(j+i) \bmod N} y_i. \tag{4.4}$$

Note that the difference between a circular correlation and a circular convolution is in the subscript of the element of **x** in the summation. For convolutions, the subscript is the difference between j and i, whereas for correlations, the subscript is the sum of j and i. As an example of the computation of a circular correlation, let **x** be the vector [0, 1, 1, 2, 2, 0, 0, 0, 0, 0] and **y** be the vector [0, 0, 0, 0, 1, 1, 2, 2, 0, 0], as used in an earlier example. Then **x** ∘ **y** = [2, 0, 0, 0, 2, 4, 7, 10, 7, 4]. A visual explanation of the circular correlation is that the vector **x** is correlated to the vector **y** in each of N different circular phases. The circulation correlation vector is sometimes called the *unnormalized correlation vector* of **x** and **y** because the values of the correlations can take on any real numbered values.

Because circular correlations are very similar to circular convolutions, there is a correlation theorem that corresponds to the convolution theorem. The difference between circular correlations and circular convolutions is the sign of a subscript, and that causes a dependence on the complex conjugate of a Fourier transform instead of a dependence on the Fourier transform itself.

Correlation theorem: Let *Fourier*(**x**)$_\omega$ denote frequency component ω of the Fourier transform of **x** and *conj*(*Fourier*(**x**)$_\omega$) denote the complex conjugate of ω. Then

$$Fourier(\mathbf{x} \circ \mathbf{y})_\omega = Fourier(\mathbf{x})_\omega \cdot conj(Fourier(\mathbf{y})_\omega). \tag{4.5}$$

(For a proof of the correlation theorem, see Stone, 1998.)

The importance of the correlation theorem lies in its use in correlating one image to another image as the central step of registering the two images. To correlate two images, each with N pixels, in each of N relative positions requires N^2 operations. The reason is that N operations are required to produce one correlation value because this step multiplies and sums the N pairs of corresponding pixels. There are N distinct relative positions of two images with N pixels. So there are on the order of N^2 operations in the entire process. The correlation theorem enables the computations to be done in the Fourier domain instead of in the pixel domain. In the Fourier domain, only N operations are required to find the circular correlation of two images instead of N^2 operations. Therefore, if the cost of transforming images into the Fourier domain and transforming the correlation back into the real number domain is much less than N^2 operations, the cost of the correlation calculation is greatly reduced by doing it in the Fourier domain. In fact, well-known methods exist for calculating the Fourier transform of an image with N pixels using a total number of operations that grows as $N\log N$ rather than as N^2. The original idea for the fast Fourier transform is attributed to I. J. Good (1958). Cooley and Tukey (1965) developed computer algorithms that significantly influenced subsequent developments and they are credited with making possible the widespread use of the algorithm. For a thorough discussion of fast Fourier transform techniques, see Agarwal and Cooley (1977). Anuta (1970) is one of the earliest references that describe the use of fast Fourier transforms for image registration.

At this point, we have shown that Fourier transforms enable the fast computation of unnormalized correlations of pairs of images. The following section shows how these techniques are easily extended to deal with normalized correlations as described in Eq. (4.1), which are somewhat more complex than the unnormalized circular correlations of Eq. (4.4).

4.2 Practical implementations of the theory

This section discusses the practical implementation of Fourier-based image correlation techniques for image correlation. The series of steps explained here are:

(1) Extension of the method to compute normalized correlations.
(2) Incorporating masks for invalid pixels.
(3) An optimization technique that reduces the number of computations.

The extension of Eq. (4.1) to a Fourier-based calculation is quite straightforward. When **x** and **y** are both vectors of length M, Eq. (4.1) breaks down into a collection of circular correlations. Consider the first term in the numerator of the equation, which is:

$$\sum_{i=0}^{M-1} x_{i+j} y_i.$$

This is exactly equal to $(\mathbf{x} \circ \mathbf{y})_j$. Because of our assumption that both vectors are of length M, the summation of the elements in the numerator of Eq. (4.1) of $\mathbf{x}$ does not depend on j, nor do the summations of x_{i+j}^2 in the denominator. Therefore, Eq. (4.1) can be written to be

$$\mathbf{corrcoef}(\mathbf{x}, \mathbf{y})_j$$
$$= (\mathbf{x} \bullet \mathbf{y})_j$$
$$= \frac{\left((\mathbf{x} \circ \mathbf{y})_j - (1/M) \sum_{i=0}^{M-1} x_i \sum_{i=0}^{M-1} y_i \right)}{\sqrt{\left(\sum_{i=0}^{M-1} x_i^2 - (1/M) \left(\sum_{i=0}^{M-1} x_i \right)^2 \right) \left(\sum_{i=0}^{M-1} y_i^2 - (1/M) \left(\sum_{i=0}^{M-1} y_i \right)^2 \right)}}. \tag{4.6}$$

In this form, it is clear that the correlations can be calculated quickly in the Fourier domain, and the remaining elements of the correlation coefficient are independent of the relative positions of $\mathbf{x}$ and $\mathbf{y}$. Thus they can be computed one time and used repeatedly in Eq. (4.6) when computing normalized values.

The hypothetical conditions for the use of Eq. (4.6) rarely apply because almost inevitably there will be invalid pixels created when an image $\mathbf{x}$ is shifted relative to image $\mathbf{y}$ to determine the quality of the registration in that relative position, or they will be present because of occlusions due to clouds and other factors. When considering prospective alignments, a registration algorithm should also identify boundary regions of one image for which the second has no corresponding pixels, and prevent such pixels from taking part in the computation. The image masks used to identify invalid pixels due to cloud occlusions can also be used for identifying the invalid pixels at image borders that arise during the registration process, so one mechanism suffices to deal with all invalid pixels.

To derive the form of the correlation equation to apply when one or both registration images contain invalid pixels, assume that $\mathbf{x}$ is a one-dimensional vector with N elements, and that $\mathbf{y}$ is a one-dimensional vector with M elements, $M < N$. (Although the discussion assumes that the images are one-dimensional, the discussion applies equally well to two or more dimensions. The extension to two and higher dimensions is straightforward.)

Because the correlation theorem relies on Fourier transforms of vectors of equal lengths, instead of treating **y** as a vector of length M, let it be represented instead as a vector of length N together with a mask of length of N whose 1 bits identify valid pixels of **y** and whose 0 bits identify invalid or unknown pixel values of **y**. For example, let $N = 8$, and $M = 4$. Then the image **x** may be the vector [1, 4, 5, 1, 2, 6, 6, 3] and the image **y** may be the vector [2, 6, 6, 1]. For the purposes of the image registration algorithm, **y** will be extended to a vector of length 8 whose values are [2, 6, 6, 1, 0, 0, 0, 0] together with a corresponding mask **m**, called a *pattern mask*, whose values are [1, 1, 1, 1, 0, 0, 0, 0]. The first four elements of **m** are 1's, which signify that the corresponding elements of **y** are valid. The last four elements of **m** are 0's, which signify that the last four elements of **y** are invalid. These are the four elements that have been appended to **y** to make its size equal to that of **x**. The mask **m** can also indicate values of **y** that are deemed to be invalid for reasons such as being occluded by clouds, or due to other factors that could create pixel invalidity. For example, a separate program could determine that the second element value of **y** could be that of a cloud rather than of the Earth's surface. To identify that element as invalid, its value in **y** would be set to the value 0 and the corresponding bit in the mask **m** would also be set to 0 to distinguish that specific pixel value from a different valid pixel whose intensity happens to be 0. The values of **y** and **m** in the case when the second pixel is determined to be a cloud pixel would then be [2, 0, 6, 1, 0, 0, 0, 0] and [1, 0, 1, 1, 0, 0, 0, 0], respectively.

Similarly, the image **x** can have invalid pixels, because they may be pixels occluded by clouds or are invalid for other similar reasons. Consequently, we associate a separate mask **h** with **x**, and use the values of **h** to identify which pixels of **x** are valid, just as the elements of **m** identify which pixels of **y** are valid. Invalid pixels of **x** must also be forced to the value 0 in the remainder of the algorithm, even if they have nonzero values in the unprocessed images. The use of masks together with Fourier transforms for fast registrations was first described in the context of least squares and average intensity measures (see Stone and Li, 1996, and Stone and Shamoon, 1997).

Now we can examine each of the terms of Eq. (4.1) and determine how they break down into circular correlations, each of which can be done in the Fourier domain. In the numerator of Eq. (4.1), the term $\sum_{i=0}^{M-1} y_i$ is a summation of the elements of **y**. However, these elements are further qualified to be elements that correspond to positions in which the vector **x** has valid elements. That is, the summation should properly be changed to $\sum_{i=0}^{M-1} h_{j+i} y_i$. This term can be calculated in the Fourier domain because it is equal to $(\mathbf{h} \circ \mathbf{y})_j$. Likewise, the term $\sum_{i=0}^{M-1} x_{j+i}$ should sum only those values of **x** for which the corresponding values of **y** are valid. That term should properly be written as $\sum_{i=0}^{M-1} x_{j+i} m_i = (\mathbf{x} \circ \mathbf{m})_j$. Finally, the factor $1/M$ in the numerator of Eq. (4.1) needs to be altered to incorporate the effect of invalid pixels. In the absence of invalid pixels, the value of M is equal to the number of multiplication products that are summed in $(\mathbf{x} \circ \mathbf{y})_j$. When invalid pixels are present, then the value of M needs to be replaced by the actual number of valid terms that are summed by the expression $\sum_{i=0}^{M-1} x_{i+j} y_i$. Observe that the number of terms for any relative position of $\mathbf{x}$ and $\mathbf{y}$ is equal to the inner product of the image masks $\mathbf{m}$ and $\mathbf{h}$ with the same relative alignment. (A product $x_{i+j} y_i$ has a valid value if the corresponding image masks for those positions are both valid, that is, if $h_{i+j} m_i = 1$. Therefore, the sum $\sum_{i=0}^{M-1} h_{i+j} m_i$ counts the number of pairs that actually contribute to the correlation sum.) Then the value $\sum_{i=0}^{M-1} h_{i+j} m_i$ must be used in place of the value M in the numerator of Eq. (4.1). But this product is equal to $(\mathbf{h} \circ \mathbf{m})_j$. Similar changes apply to the denominator of Eq. (4.1). The result of these observations is that Eq. (4.1) becomes the following:

$$\mathbf{corrcoef}(\mathbf{x}, \mathbf{y})_j$$

$$= (\mathbf{x} \bullet \mathbf{y})_j$$

$$= \frac{((\mathbf{x} \circ \mathbf{y})_j - (1/(\mathbf{h} \circ \mathbf{m})_j)(\mathbf{x} \circ \mathbf{m})_j(\mathbf{h} \circ \mathbf{y})_j)}{\sqrt{((\mathbf{x}^2 \circ \mathbf{m})_j - (1/(\mathbf{h} \circ \mathbf{m})_j)((\mathbf{x} \circ \mathbf{m})_j)^2)((\mathbf{h} \circ \mathbf{y}^2)_j - (1/(\mathbf{h} \circ \mathbf{m})_j)((\mathbf{h} \circ \mathbf{y})_j)^2)}}.$$

$$(4.7)$$

Note that the term $\mathbf{x}^2$, which is a factor of the term $(\mathbf{x}^2 \circ \mathbf{m})$, is shorthand for a vector, each of whose elements is the square of the corresponding element of $\mathbf{x}$. Thus, if $\mathbf{x} = [1, 2, 3]$, then $\mathbf{x}^2 = [1, 4, 9]$.

In this form, it is clear that the normalized correlation of $\mathbf{x}$ and $\mathbf{y}$ depends on the following six circular correlations, each of which can be computed in the Fourier domain.

(1) $(\mathbf{h} \circ \mathbf{m})$
(2) $(\mathbf{x} \circ \mathbf{m})$
(3) $(\mathbf{h} \circ \mathbf{y})$
(4) $(\mathbf{x} \circ \mathbf{y})$
(5) $(\mathbf{x} \circ \mathbf{y}^2)$
(6) $(\mathbf{x}^2 \circ \mathbf{y})$

To compute the normalized correlations of $\mathbf{x}$ and $\mathbf{y}$ for all relative positions j, calculate the six circular correlations in the Fourier domain, and transform back to the pixel domain. Then combine the jth components of the respective transforms according to Eq. (4.7). As expressed here, the computation of the algorithm grows as $N \log N$ for images with N pixels. The algorithm for normalized correlation based on masks and Fourier transforms first appeared in Stone (1999).

In practice, it is necessary to expend the computations required for only three of the six circular correlations in the Fourier domain. The reason is that all of the vectors whose transforms are being computed are real valued. Their Fourier transforms take on a special form. There is a well-known trick that allows us to reduce the algorithm complexity by transforming two real vectors together in one operation (Cooley, 1966; Singleton, 1967).

Consider two vectors, $\mathbf{x}$ and $\mathbf{y}$, both of length N and both with real values. The Fourier transform of a real valued vector is conjugate antisymmetric. That is, the Fourier transform of $\mathbf{x}$ is a complex valued vector whose Fourier components satisfy the equation

$$F(\mathbf{x})_{\omega} = conj(F(\mathbf{x})_{N-\omega}) \text{ for } \omega = 1, 2, \ldots, N/2 - 1, \tag{4.8}$$

where $conj(F(\mathbf{x}))$ denotes the conjugate of the complex Fourier transform of $\mathbf{x}$, and the Fourier transform indices are $\omega = 0, 1, \ldots, N-1$. Similarly, consider the Fourier transform of the purely imaginary valued vector $i \cdot \mathbf{y}$ obtained by multiplying $\mathbf{y}$ by i. $F(i \cdot \mathbf{y})_{\omega} = conj(F(i \cdot \mathbf{y})_{N-\omega})$ for $\omega = 1, 2, \ldots, N/2 - 1$, which is equivalent to the following statements:

$$\text{Real}\{F(i \cdot \mathbf{y})_{\omega}\} = -\text{Real}\{F(i \cdot \mathbf{y})_{N-\omega}\} \quad \text{for } \omega = 1, 2, \ldots, N/2 - 1.$$
$$\text{Imag}\{F(i \cdot \mathbf{y})_{\omega}\} = \text{Imag}\{F(i \cdot \mathbf{y})_{N-\omega}\} \quad \text{for } \omega = 1, 2, \ldots, N/2 - 1. \tag{4.9}$$

From Eqs. (4.8) and (4.9), one finds that

$$F(\mathbf{x})_{\omega} = \left[\text{Real}\{F(\mathbf{x}+i\mathbf{y})_{\omega} + F(\mathbf{x}+i\mathbf{y})_{N-\omega}\} + i \cdot \text{Imag}\{F(\mathbf{x}+i\mathbf{y})_{\omega}\right.$$
$$\left. - F(\mathbf{x}+i\mathbf{y})_{N-\omega}\}\right]/2,$$
$$F(\mathbf{y})_{\omega} = \left[\text{Imag}\{F(\mathbf{x}+i\mathbf{y})_{\omega} + F(\mathbf{x}+i\mathbf{y})_{N-\omega}\} - i \cdot \text{Real}\{F(\mathbf{x}+i\mathbf{y})_{\omega}\right.$$
$$\left. - F(\mathbf{x}+i\mathbf{y})_{N-\omega}\}\right]/2. \tag{4.10}$$

Therefore, to compute the Fourier transforms of both $F(\mathbf{x})$ and $F(\mathbf{y})$, the recipe is to compute the transform $F(\mathbf{x}+i\mathbf{y})$ and to partition this into the separate transforms of $F(\mathbf{x})$ and $F(\mathbf{y})$ by using Eq. (4.10).

The method is even easier to apply in the inverse direction. Assume that $F(\mathbf{x})$ and $F(\mathbf{y})$ are Fourier transforms of real valued vectors $\mathbf{x}$ and $\mathbf{y}$. Then the inverse transform of $F(\mathbf{x}) + i \cdot F(\mathbf{y})$ is $\mathbf{x} + i\mathbf{y}$. So the real part of the inverse transform is $\mathbf{x}$ and the imaginary part of the inverse transform is $\mathbf{y}$. Using this trick, the normalized correlation can be computed by forming three direct Fourier transforms and three inverse Fourier transforms.

When applying Eq. (4.7) to perform image registration, the normalized correlation function is computed for all integer translations (corresponding to all values of j in the one-dimensional form shown in Eq. (4.7)). The position with the highest correlation is deemed to be the best match. Since the correlation function is normalized, its value at the position of the best match gives a measure of quality of the match. If the correlation function is greater than 0.9, almost surely the position of the peak is the best registration position for translation, to within a precision of the nearest integer valued translation position. Lower values of the peak correlation correspond to a lower confidence that the peak is at the true registration position. Correlation values of 0.7 or higher are usually accurate. Below 0.7, the probability of the true registration position being at a different point than at the position of the correlation peak becomes significantly higher. Correlation values of 0.3 and below usually indicate that the correlation function has not found the true registration position.

4.3 Finding rotation, scale, and translation using Fourier techniques

The main advantage of Fourier-based correlation is speed. However, the direct use of the method is fast only with respect to finding translation offsets. It is not directly useful for finding rotational differences or scale changes between images. This section shows how to use fast correlation *indirectly* to find rotational and scale change parameters. The idea is to use a grid of small chips from one image to match chips of a second image. The resulting registrations give a corresponding grid pattern on the second image. We can then find a matrix of coefficients that best maps the known grid of the first image to the discovered grid of the second image. Using the mapping, we can then find and eliminate chips that are poorly matched by that mapping, eliminate those chips, and repeat the process. After one, possibly two iterations, the process converges to a matrix that explains the coefficients well, or indicates that the mapping is unsuccessful. The process gives a mean-square estimate of the residual registration error, and it also shows explicitly how each chip maps with respect to the original grid. The examples in this section use real images of the Mars surface in which the translational displacement was a few hundred pixels.

Figure 4.1 shows two images of the Martian surface that were registered by using the fast correlation described in the previous section using small chips from the Martian images.

Note the second image has black bands that correspond to dropouts during image capture or transmission back to Earth. The second image is processed in conjunction with a mask that indicates the valid pixels, as discussed above. The comparable mask for the first image is all 1's, because all pixels are valid.

Under the assumption that the two images differ by a rigid rotation, the problem is to find the rotation angle θ, the scale s, and the translations Δx and Δy by which

(a)

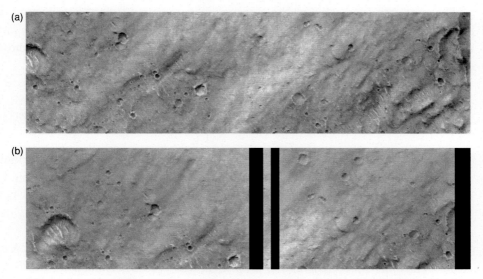

(b)

Figure 4.1. (a), (b) Two images of the Martian surface. (Courtesy: NASA Goddard Space Flight Center.)

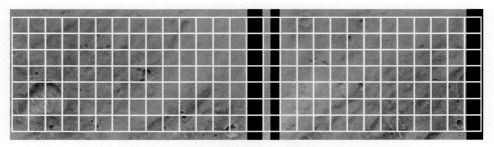

Figure 4.2. Image chips derived from the second Mars image.

they differ. The first step of the process is to produce a collection of "chips" from the second image. The chips are shown in Fig. 4.2.

Each of the image chips is registered separately to the first image using normalized correlation. Some of the registrations fail because the chips of the second image lie outside the boundaries of the first image or because they contain a substantial number of invalid pixels. The correlation values indicate the quality of the individual registrations, and allow the algorithm to retain only the registrations that are likely to have succeeded. For the processing in this example, the correlation threshold was set to 0.6. This passes all correct registrations as well as some false ones. Subsequent processing described below eliminates the false registrations.

The next step is to estimate the registration parameters from the collection of chip registrations. The method for performing this operation is based on the fact that the registration parameters can be encapsulated into a linear matrix $\mathbf{A}$ whose components are given in Eq. (4.11):

$$\mathbf{A} = \begin{bmatrix} 1 & 0 & -\Delta x \\ 0 & 1 & -\Delta y \\ 0 & 0 & 1 \end{bmatrix} \begin{bmatrix} s & 0 & 0 \\ 0 & s & 0 \\ 0 & 0 & 1 \end{bmatrix} \begin{bmatrix} \cos\theta & -\sin\theta & 0 \\ \sin\theta & \cos\theta & 0 \\ 0 & 0 & 1 \end{bmatrix}. \qquad (4.11)$$

The matrix $\mathbf{A}$ maps a point (x_2, y_2) from the second image into a corresponding point (x_1, y_1) in the first image by applying a rotation, scale change, and translation, in that order, as described by three component matrices on the right side of Eq. (4.11). For the coordinate system used by matrix $\mathbf{A}$, the point (x_2, y_2) is represented by the column vector $[x_2, y_2, 1]^T$.

Let $\mathbf{c}_1$ denote the center of chip 1 in the vector coordinate system. Then $\mathbf{d}_1 = \mathbf{A}\,\mathbf{c}_1$ is the center of the corresponding chip in Image 1 as determined by the rigid mapping $\mathbf{A}$. The registration algorithm, however, gives a second determination of the position of chip $\mathbf{c}_1$ in Image 1. Let this determination be denoted as $reg(\mathbf{c}_1)$. The objective is to find a set of parameters for the matrix $\mathbf{A}$ such that the discrepancy between $\mathbf{d}_i$ and $reg(\mathbf{c}_i)$ is minimized over all chips $\mathbf{c}_i$. The most used method in practice is to minimize the sum of squares of the discrepancies. There are several popular ways to do this. The data in this section were prepared using the Levenberg-Marquardt methods (Levenberg, 1944; Marquardt, 1963) as embodied in the Matlab function *fminsearch*.

For the images above, the parameters obtained were as follows:

$$\begin{aligned} \Delta x &= 308.7557 \text{ pixels}, \\ \Delta y &= -277.0189 \text{ pixels}, \\ s &= 1.0191, \\ \theta &= -0.4458°. \end{aligned} \qquad (4.12)$$

The estimate was based on a two-step process. All chips whose registrations exceeded the threshold of 0.6 were used to compute a starting set of parameters. Using these parameters, the positions of each of the chips was determined and compared to the positions determined from the registration process. If the distance was greater than a second threshold, the chip was discarded as an outlier, and the parameters were re-estimated based on the chips remaining. In this case, the second iteration used 60 chips to produce the parameter set in Eq. (4.12).

Figure 4.3 shows the original grid and a reconstructed grid from the chip registration to give a further indication of the quality of the registration. For this image pair, one can see a grid distortion that is not captured by the rigid-rotation model.

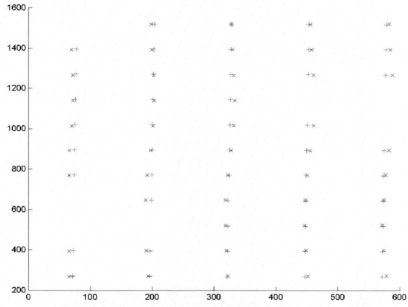

Figure 4.3. An original grid ("+") and a reconstructed grid from registered chips ("x").

It may be due to distortion caused by nonuniform elevation and by the observation angle not being normal to the surface. Figure 4.4 shows a similar registration summary for a different pair of images. This figure shows that the registration is very good throughout the grid.

There are a number of important characteristics of this registration process.

(1) This process does not resample and interpolate images during the registration process so that it is immune to errors caused by image interpolation. All operations are equivalent to translations of images by whole pixels, although the translations do not actually occur because the operations are done in the Fourier domain.

(2) The process produces subpixel precision for translation, and produces as well the angular change and scale change, yet it does not rotate or change scale, or translate images by amounts other than integral amounts. The process works as long as the scale change and rotational changes are relatively small so that chips are transformed into similarly appearing chips with approximately the same orientation and scale as the original chip. If either the rotational angle or scale change is large, then chip registration will not work because the transformed chip will not be recognized to be the same as the original chip through comparisons based solely on translations.

(3) The registration process operates globally because the correlations are done for all possible positions of chip overlap. This allows the algorithm to discover large translations (over 200 pixels in the Mars example). The algorithm is extremely useful for

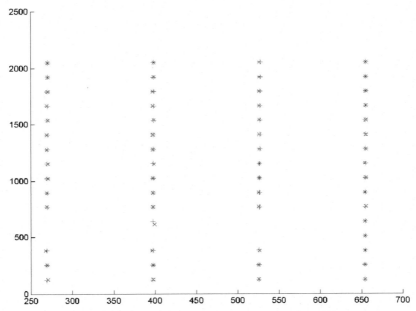

Figure 4.4. A registration summary similar to Fig. 4.3 for a different image pair.

registering images whose translational differences can be very large. In the case of satellite images, often there exist metadata that enable the translation parameters to be known to relatively coarse precision. However, this is not always possible. (There is no GPS on Mars today that can supply such information.)

The computations described for the chip registration process are very fast for each chip because of the Fourier techniques. The large number of chips tends to lengthen the process, but it is still relatively fast. The 60 chips used in the Mars example lengthened the computation by a factor of about 60, but the individual correlations are faster by a factor of about 1000 over pixel-based correlations. So there is a net computational benefit by using the Fourier techniques. The Mars registration required just a few minutes of time on a 2-GHz desktop processor for images of size 3616×1024.

The registration process can be optimized in part by computing the Fourier transforms of the Image 1 data only once, rather than repeat it for each chip registration.

For certain sizes of images, patterns, and chips, there may be no computational speed benefit by using the Fourier techniques. Direct correlations may be faster. An algorithm can determine if Fourier techniques are faster in advance of applying them by estimating the time to register images of given sizes by using Fourier techniques as well as by using pixel-based techniques. If the algorithm determines that pixel-based techniques run faster, then it can apply them instead of the Fourier techniques. The results will be identical because the Fourier techniques do exactly what the corresponding pixel-based techniques do, but it performs them in the Fourier domain. This, indeed, has been used in the implementation of the registration code described in this chapter.

4.4 Very high precision translational registration

In this section, we show a method for determining the translation difference between two images to within a tenth of a pixel. We assume that the images have identical orientations and scales. They differ only by translation. Moreover, we also assume that the translational difference is known to be less than 0.5 pixels. To obtain these input conditions, we assume that a preprocessing algorithm has been applied to create the image pair that satisfies our assumptions.

The method used in this section is based on phase differences in the Fourier transforms of similar images. For translations that are an integral number of pixels, the phase information is independent of whether the images are aliased or not. However, for translations that are a fractional number of pixels, the phase of the Fourier transforms exhibit a complex dependence on the aliasing in the images. The method in this chapter seeks to eliminate the effects of aliasing by using only the portions of the Fourier transform that are uncorrupted by aliasing. Using these data, the algorithm produces subpixel estimates for the translation that have been observed in practice to be accurate to within a tenth of a pixel. The discussion in this section is based on Stone *et al.* (2001).

It is well known that the Fourier transforms of two images that differ only by translation differ by a phase shift. For this discussion, $f_c(x, y)$ is a continuous two-dimensional image, $g_c(x, y)$ is a continuous image that differs from $f_c(x, y)$ by a translation Δx and Δy where both Δx and Δy are fractions less than 0.5, $f(x, y)$ is a discrete image in two dimensions obtained by sampling $f_c(x, y)$, and $g(x, y)$ is the discrete image obtained by sampling $g_c(x, y)$.

In the remainder of this background discussion, we simplify the mathematics by using one-dimensional representations. In this notation, $f_c(x)$ is a continuous image with a Fourier transform $F_c(\Omega)$. The continuous $g_c(x) = f_c(x - x_0)$ is the continuous $f_c(x)$ shifted by an amount x_0. It is well known that the Fourier transform $G_c(\Omega)$ of $g_c(x)$ is given by the formula

$$G_c(\Omega) = F_c(\Omega)e^{-j\Omega x_0}. \tag{4.13}$$

Our objective is to discover x_0. We can do so by using Eq. (4.13) and the transforms $F_c(\Omega)$ and $G_c(\Omega)$. By comparing the phases of the transforms, we can extract the value of x_0, which is the displacement we seek. If the images in question were continuous, the phase of the transforms is all we need to examine. However, the images available to us are sampled images, from which we are not able to construct the continuous transforms $F_c(\Omega)$ and $G_c(\Omega)$. This leads to some additional complexity in the phase estimation process.

We actually observe sampled images $f(x)$ and $g(x) = f(x - x_0)$ where x_0 is a real valued number less than half of the sampling interval. The values of $f(x)$ at the sample points nT are equal to the values of the continuous image $f_c(x)$ at those points, that is, $f(nT) = f_c(nT)$, but the value of $f(x)$ between the sampling values is indeterminate. The usual assumption is that the sampling interval T is sufficiently small to perform sampling at the Nyquist rate, that is, at a rate at least as high as twice the highest frequency present in the image. If the discrete image $f(x)$ is sampled at or above the Nyquist rate, then the continuous image can be reconstructed from the sampled image by using the well-known *sampling theorem* (Shannon, 1949; Unser, 2000). The sampling theorem states that $f(x)$ at any point x is given by

$$f_c(x) = \sum_{n=-\infty}^{+\infty} f(n)\operatorname{sinc}(x/T - n). \tag{4.14}$$

Since $g(x) = f(x - x_0)$, then $g(x)$ is also sampled at the Nyquist rate if it uses the same sample rate as $f(x)$. Then the value of $g_c(x)$ can be approximated by a finite sum derived from Eq. (4.14) as follows:

$$g_c(x) = f_c(x - x_0)$$
$$\approx \sum_{n=0}^{N-1} f(n) \cdot Dirichlet_N((2\pi/N)((x - x_0)/T - n)), \tag{4.15}$$

where the Dirichlet function is defined by the equation

$$Dirichlet_N(x) = \sin(Nx/2)/(N\sin(x/2)). \tag{4.16}$$

Shekarforoush *et al.* (1996) used this as their starting point to derive the registration offset x_0 through the use of phase correlation. Equation (4.15) can be rewritten in the Fourier domain as $G(\omega) = F(\omega)D(\omega)$. One can compute $D(\omega) = G(\omega)/F(\omega)$ in the Fourier domain, then do the inverse transform of $D(\omega)$ to obtain $Dirichlet_N((2\pi/N)((x - x_0)/T))$ which has a very sharp peak at the translational offset x_0. This process is called *phase correlation* because it estimates the relative phase of the images $f(x)$ and $g(x)$ through a Fourier domain process that corresponds to a form of correlation/convolution in the pixel domain.

The theory underlying this phase correlation process relies strongly on the band-limited properties of the sampled images. If the images are sampled below the Nyquist rate, then the continuous forms of the images cannot be reconstructed by using the interpolation of the sampling theorem in Eq. (4.14), nor can $g_c(x)$ be approximated by Eq. (4.15).

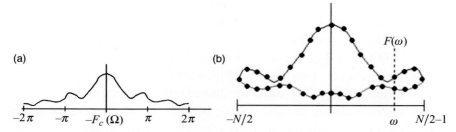

Figure 4.5. (a) The frequency spectrum of a continuous image, and (b) the frequency spectrum of the same image after sampling below the Nyquist rate. (Source: Stone *et al.*, 2001, © IEEE, reprinted with permission.)

Remotely sensed images of the Earth are typically sampled below their Nyquist rates. The Earth views have sharp edges, which lead to high frequencies in the underlying continuous image. To sample at the Nyquist rate would require a low-pass prefilter in the optics system to remove frequencies above the Nyquist rate before sampling. The general trend is to build satellite optics without such a low-pass filter to avoid degradation from this filter, and instead, accept the distortion produced by the aliasing from high frequencies that pass through the optical system.

The remainder of this section describes a process due to Stone *et al.* (2001) which is capable of very precise registrations for images with aliasing. Those images cannot be processed with the same precision by using the technique of Shekarforoush *et al.* (1996).

When continuous scenes are sampled below the Nyquist rate, the resulting discrete image contains aliasing artifacts. The basic idea of the registration algorithm is to observe how aliasing impacts the Fourier spectrum of an image. By isolating the frequencies that are likely to be free from aliasing from those that are likely to be highly aliased, the algorithm then uses just the frequencies that are less corrupted to estimate the relative phase differences in the discrete Fourier transforms of a pair of images $f(x)$ and $g(x)$.

Figure 4.5, reproduced from Stone *et al.* (2001), shows the assumption regarding the frequency spectra of sampled images. Figure 4.5(a) shows the spectrum of a continuous image. The assumption for this section is that the power in the spectrum diminishes rapidly in the higher-frequency regions. In this case, the sampling rate for the image will capture the frequencies that lie between $-\pi$ and π, but will fail to capture correctly the power in the frequencies outside this range. Figure 4.5(a) also shows that almost all of the power in the spectrum lies in the interval -2π and 2π, so that the sampling rate is approximately half of what it should be to meet the conditions of the sampling theorem. Figure 4.5(b) shows the power spectrum of the sampled image when the sampling rate is below the Nyquist rate. The figure shows that the power in the left tail of Fig. 4.5(a) is combined with power in the corresponding frequencies of the right tail of the truncated frequency spectrum shown in Fig. 4.5(b). Similarly the power in the right-hand tail of Fig. 4.5(a) is combined with the power in the left-hand tail of the truncated spectrum shown in Fig. 4.5(b).

The spectrum in Fig. 4.5(a) can be partitioned into two spectra, one part being the spectrum centered around the origin lying between $-\pi$ and π, and the remainder of the spectrum being the part that lies outside the interval between $-\pi$ and π. The sampling process essentially adds the two parts together as expressed by the formula

$$F(\omega) = F_c(2\pi\omega/N) + F_c(2\pi(\omega+N)/N). \qquad (4.17)$$

The formula holds for ω in the interval $-N/2 \le \omega < 0$. For ω in the interval $0 < \omega \le N/2$, replace $\omega + N$ by $\omega - N$. Similarly, the spectrum for the image $G(\omega)$ is expressed as

$$G(\omega) = G_c(2\pi\omega/N)e^{-j2\pi\omega x_0/N} + G_c(2\pi(\omega+N)/N)e^{-j2\pi(\omega+N)x_0/N}. \qquad (4.18)$$

As in Eq. (4.17), Eq. (4.18) holds for ω in the interval $-N/2 \le \omega < 0$. Replace $\omega + N$ by $\omega - N$ to find a formula for ω in the interval $0 < \omega \le N/2$.

To find a precise registration of a pair of images, it is sufficient to determine the phase factor $e^{-2\pi j\omega x_0/N}$ in Eq. (4.18). In the absence of aliasing (as indicated by the second terms in Eqs. (4.17) and (4.18)), the phase factor can be obtained by calculating $G(\omega)/F(\omega)$. In fact, the value of x_0 is the same at each of the values of c in the absence of aliasing, so that for each value of ω, the phase of $G(\omega)/F(\omega)$ gives an estimate of x_0. With multiple estimates of the value of x_0, one can produce a composite which should be very accurate. This is essentially what Shekarforoush *et al.* have proposed in their use of the Dirichlet function of Eq. (4.16). The presence of the second terms in Eq. (4.18) complicates the process because it can have a large impact on the relative phase of $G(\omega)$ and $F(\omega)$. There is little effect on the phase when the magnitude of the second term of Eq. (4.18) is very small compared with the magnitude of the first term. But the effect can be very large if the magnitude of the second term is comparable to or larger than the magnitude of the first term.

Stone *et al.* (2001) proposes to use only a subset of frequencies in the spectra $F(\omega)$ and $G(\omega)$ for estimating the relative phase of the spectra. A brief observation of the spectra in Fig. 4.5(a) indicates that the highest magnitudes in the spectrum lie close to the origin, so that their algorithm tends to select frequencies near the origin. Nevertheless, some frequency components near the origin may have small magnitudes in spite of being close to the origin. The effect of aliasing may be relatively high for such frequencies because the magnitude of the first terms may be small since the total magnitudes are small. So a second component of the algorithm of Stone *et al.* is to discard frequency components near the origin if either the magnitude of $F(\omega)$ or $G(\omega)$ is relatively small compared with the magnitudes of other frequency components near the origin.

The final component of the algorithm deals with the method by which the frequency spectrum of an image is calculated. When calculating the discrete Fourier coefficients of a digital image, the mathematic formula treats the sampled image as being replicated in space so that the left boundary of one image tile is adjacent to the right boundary of the tile on its left, and the right boundary of an image tile is adjacent to the left boundary of the image tile on its right. The "virtual" adjacencies create discontinuities at the boundaries, which create artificially high magnitudes among the high-frequency components in the discrete Fourier transform. These artifacts tend to corrupt the phase extraction process because they tend to create energy in the high-frequency portion of the spectrum that is not present in the original image. Consequently, an essential part of the registration algorithm proposed by Stone *et al.* is the use of a "window" to apply to an image before computing the discrete Fourier transform of the image. The window proposed is a Blackman or Blackman-Harris window whose coefficients multiply the corresponding pixels of the image. The effect of the windowing operation is to produce a new image whose pixel values are zero at the boundary, and whose envelope of pixel magnitudes goes to zero at the boundary in a way that introduces essentially no spurious power in the high frequencies of the image spectrum.

The algorithm as stated in Stone *et al.* (2001) is reproduced here. For details of its derivation, see the full paper.

(1) Register the images $f(x, y)$ and $g(x, y)$ to the nearest integral pixel coordinates. Use, for example, the Fourier-based high-speed correlation method described earlier in this chapter.
(2) Apply a Blackman or Blackman-Harris window to each image to eliminate the boundary effects in the Fourier domain.
(3) Calculate the discrete Fourier transforms of $f(x, y)$ and $g(x, y)$. Let these be $F(\omega, v)$ and $G(\omega, v)$, respectively.
(4) Remove the frequencies that lie outside a radius R of the central peak of the Fourier spectrum. A suitable value of R is $0.6 \cdot N/2$, where N is the minimum number of samples that lie between the central peak and the edge of the frequency spectrum.
(5) Of the remaining frequency components, remove the frequency components at a frequency (ω, v) if either $F(\omega, v)$ or $G(\omega, v)$ have a magnitude less than a threshold α.
(6) Using the frequencies that remain, find a least-squares estimate of the offsets (x_0, y_0) from a least-squares estimate of the phase.

The threshold α in Step 5 can be determined by sorting the frequencies considered by Step 5 by their magnitude. The objective is to determine a value of a constant K, such that only the first K frequencies are used and the remainder is discarded. The idea is to let K vary, and for each value of K, to find an estimate for the offsets according to Step 6 above. As K varies, there should be a range of K for which the estimates of (x_0, y_0) are relatively constant. Use any value of K in this range.

The results of using the algorithm are very good. Stone *et al.* (2001) reports reaching a precision of a few hundreths of a pixel when applying the algorithm to synthetic data with aliasing effects introduced artificially. On real data, the algorithm has been used to register images to a precision of about 0.07 pixels. Because real data does not usually have a ground truth by which precision can be measured exactly, the measurement of precision is performed by coregistering three or more images. For example, with three images, the algorithm gives the offset from A to B, from B to C, and from A to C. In an ideal world the distance from A to C should equal the distance from A to B added to the distance from B to C. The observed precision of 0.07 pixels is the observed discrepancy in the error of the coregistration.

4.5 Estimating rotation and scale

This section shows how to manipulate the Fourier transforms of images to extract the scale and rotational differences of images. The theory on which it is based is that the Fourier transform of an image contains both amplitude and phase information. The translation position of an image is encoded only in the phase, not in the amplitude. Therefore, when you remove the phase information from a Fourier transform, you retain the scale and rotational position information for an image, but you have eliminated dependency on its translational position. You thus can use the amplitude of the Fourier transforms of two images to determine relative rotational position and scaling of the images. The remainder of the discussion in this section follows the presentation of Stone *et al.* (2003).

To find the relative rotation and scaling, recast the Fourier transform amplitude in log polar coordinates in which the transform is expressed as a function of polar angle and the log of the radius of a point (w, v) in Fourier orthogonal frequency components. In theory, if two images differ by a rotation of θ, then their transforms are rotated by θ, which corresponds to a translation by an amount θ in the log polar coordinate system. Hence, the log polar transforms can be treated as if they themselves were real images. They can be registered to find the translation by θ for which they best match. A translation of θ in the log polar coordinate system corresponds to a rotation by θ in the (w, v) domain of the Fourier frequencies, and by θ as well in the (x, y) domain of the pixels. The fast correlation techniques that find translations of images can be applied directly to the log polar Fourier magnitudes to find the angle θ.

A method that has potentially higher precision for finding θ than fast correlation is known as *phase correlation* (Alliney and Morandi, 1986; De Castro and Morandi, 1987; Chen *et al.*, 1994; Alliney *et al.*, 1996; Dasgupta and Chatterji, 1996; Chang *et al.*, 1997). It has the advantage that the phase correlation peak, in theory, is much sharper than a correlation peak, so that it potentially can pin down the value of an offset to a much greater precision than can fast correlation, yet the processing required is about equal to that of fast correlation. To use phase correlation, note that the magnitudes of the Fourier transforms of a pair of images that differ only by a rotational angle θ, also differ (in theory) by a rotation by the same angle θ. The trick to discovering that rotational angle is to remap the magnitudes of those Fourier transforms into a new coordinate system such that rotations in the original coordinate system correspond to translations in the second coordinate system. In the second coordinate system, the Fourier magnitudes act like "images" instead of like Fourier transforms. The problem then reduces to finding the relative translation of these "images" in the second coordinate system. Because the Fourier magnitudes in the second coordinate system differ only by a translation, the respective Fourier transforms of these images are identical in magnitude (to within a constant multiplier) at every point in transform space, but they differ in phase, and the phase is a function of the relative translation in the second coordinate system, which in turn is equal to the rotational difference of the images in the first coordinate system. To find the phase difference of two Fourier transforms, one can phase correlate them, and then take an inverse transform. The inverse transform, in theory, produces a sharp peak centered at a position that corresponds to the difference in translational position of the Fourier magnitudes in the second coordinate system, or equivalently corresponds to the rotational difference of the magnitudes in the first coordinate system.

Phase correlation works well for band-limited images, and has been used successfully to find translational differences between images. The application in this section is for rotational differences between images, not translational differences, and we must rely on additional theory underlying image rotations and their transforms.

Unfortunately, the theory was developed for continuous images. Discrete sampled images do not obey the fundamental property that a rotation of an image by θ in the pixel domain rotates the Fourier transform by an amount θ. Aliasing effects in the transform of a discrete sampled image do not rotate with the transform as the image is rotated. The aliasing effects remain in place. Hence, the Fourier transform of a rotated image has two components. One component is the aliased artifacts which are not rotated. The second component comprises the frequencies of the transform of the unrotated image that are free of aliasing artifacts. These frequencies rotate by an amount θ if the original image is rotated by an amount θ. The result of aliasing is that there is a false rotation peak at $0°$ because the aliasing in the original image correlates with the aliasing of any rotation of the original image at the 0 angle since aliasing artifacts do not rotate. Consequently, a direct application of phase correlation for discovering rotational differences is likely to fail to achieve useful results. However, by noting how aliasing arises and attenuating those effects, we can obtain a useful phase correlation algorithm for deriving rotational differences of images.

The false peak at 0 produced by phase correlation makes it virtually impossible to discover small rotations because the peak of the small rotation is superimposed on and slightly displaced from the false peak. The false peak has a second, devastating effect on the ability to find the real peak. A characteristic of phase correlation is that the Fourier magnitude at the (0, 0) position in the transform domain is unity, so that the real values of the inverse transform must sum to 1. Consequently, the real value of a false peak diminishes the size of the peak at the correct position because the sum of the false peak and the true peak together with any other peaks that exist must sum to unity. Since the false peak can be substantial, say as high as 0.5, the real peak may be quite low, and it is typically smaller than the false peak. In the literature, the size of the real peak has been reported to be as low as 0.1 to 0.3 as compared to an ideal height of 1.0. The result is that the false peak greatly diminishes the signal-to-noise ratio of the real peak no matter where it lies.

We show in this section how to eliminate the false peak at zero to the extent that this is possible. The method requires elimination or attenuation of aliasing effects. A major tool to do this is the use of a Blackman window to window the original image so as to eliminate aliasing attributed to sharp boundaries. The Blackman window has a slope whose shape attenuates high frequencies associated with the edges of images. Another method is to eliminate the Fourier components in a small circle around the (0, 0) coordinate position in Fourier frequency space. Aliasing effects tend to be greatest in this region. The effects of the processing substantially diminish the false peak and substantially increase the true peak.

A critical step in the process described in this chapter is the mapping of the magnitude of a Fourier transform spectrum from the (ω, v) coordinate system into a log polar coordinate sytem (log r, θ) known as *log polar space*. In log polar space, the x coordinate corresponds to the rotational differences in transforms and the y coordinate corresponds to the log of the difference in scale of images. The discussion indicates that phase correlation of Fourier magnitudes can be done along the y-axis as well as along the x-axis, and thereby give an estimate of both the relative angle of rotation and of the relative scales of two images. The log factor inherently leads to less precision in the estimate of the scale because errors in the estimate of the log of the scale factor are amplified when the log is exponentiated. A more fundamental problem is that when the scales of two images differ greatly, there tends to be too few pixel regions in common to perform a precise registration. Consequently, this technique is useful only when the scale factors are fairly close so that the great majority of the pixels in the areas processed is common to both images.

To begin the discussion of the details of rotational registration techniques, historically the use of Fourier techniques for rotational registration has been based on the following principles:

Fourier rotation theorem: When an infinite continuous two-dimensional image is rotated by an angle θ, the Fourier transform $F(\omega, \nu)$ of the image is rotated by the angle θ.

Fourier translation theorem: When an infinite continuous two-dimensional image is translated by an amount (x_0, y_0) the phase of each Fourier transform frequency component $F(\omega, \nu)$ of the image is multiplied by the phase factor $e^{-j2\pi(\omega x_0 + \nu y_0)}$.

Fourier scale theorem: When an infinite continuous two-dimensional image is scaled by an amount s, the Fourier transform $F(\omega, \nu)$ of the image is scaled by the factor $1/s$.

The importance of the first two of these three theorems is that the operations of rotation and translation cause separable actions to be applied to Fourier transforms, whereas they cause interacting actions to be applied to the images. For example, when both a translation and a rotation are applied to an image, the pixels change in a way that depends on both operations. However, the Fourier transforms are affected differently. The translation of an image does not alter the magnitude of a frequency component. Consequently, one can discover the rotation that has been applied by examining only the magnitudes. It is much more efficient to seek a rotation in isolation, and thereby ignore any translation that has been applied, than it is to seek both a rotation and a translation jointly.

Nevertheless, severe problems arise when applying the theorems in practice. The key notion is that the theorems are stated in the context of infinite continuous images, and they do not apply in general to finite discrete images. A simple example of this point is shown in Fig. 4.6, the discrete Fourier transform of a disk.

If the Fourier rotation theorem were true for the discrete Fourier transform of a circular disk, then the transform shown in Fig. 4.6 would have circular symmetry. The rotation of a disk leaves the disk unchanged so the Fourier transform of the disk would have to be unchanged by the rotation of the disk. Clearly, Fig. 4.6 does not have circular symmetry, so that the Fourier rotation theorem cannot hold for this example. The reason that the theorem fails in this case is that the example is finite and discrete. The circular disk whose transform is shown lies at the center of

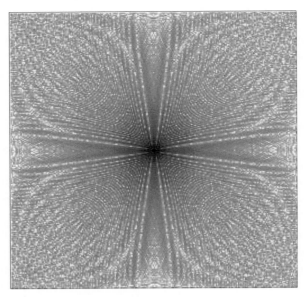

Figure 4.6. The discrete Fourier transform of a circular disk. (Source: Stone *et al.*, 2003, © Elsevier Science, reprinted with permission.)

square tile. The tiled image does not have circular symmetry, even though the disk within the tile does have circular symmetry. The structure of the spectrum in the Fig. 4.6 shows aliasing artifacts mainly due to sampling of the image and to the square boundaries of the image.

In the discussion that follows, the algorithm for finding rotation and scale will be presented in the form that it has generally been proposed, and then it will be modified to remove the aliasing artifacts, and thereby improve its utility.

Here is the basic algorithm, whose utility is impaired because of aliasing artifacts such as those illustrated in Fig. 4.6:

(1) Let $f(x, y)$ and $g(x, y)$ be two images that differ from each other by a rotation, translation, and scale change. To apply the algorithm, we assume that the two images have a substantial overlap. That is, the magnitudes of the rotation, translation, and scale change in combination do not cause large regions of one image to lack corresponding regions in the second image.

(2) Calculate the discrete Fourier transforms of $f(x, y)$ and $g(x, y)$. Let these be $F(\omega, v)$ and $G(\omega, v)$, respectively.

(3) Construct two new images from the magnitudes of the discrete Fourier transforms $F(\omega, v)$ and $G(\omega, v)$. The pixel intensities $\mathbf{F}(\omega, v)$ and $\mathbf{G}(\omega, v)$ at respective locations correspond to the magnitudes of the Fourier transforms $F(\omega, v)$ and $G(\omega, v)$ in these locations.

(4) Map the images $\mathbf{F}(\omega, v)$ and $\mathbf{G}(\omega, v)$ into a new coordinate system where the new coordinates are $(\log r, \theta)$, where r is the radial distance from the point (ω, v) to the center of the image, and θ is the positive angular rotation of the radius from (ω, v) to the image center with respect to a radius pointing straight up. Let these two images be denoted $\mathbf{F}(\log r, \theta)$ and $\mathbf{G}(\log r, \theta)$.

(5) If $f(x, y)$ and $g(x, y)$ differ by a rotation θ, and possibly by a translation, then $\mathbf{F}(\log r, \theta)$ and $\mathbf{G}(\log r, \theta)$ will be circularly shifted with respect to each other in the θ dimension by the amount θ_0, but will not differ because of translational differences between $f(x, y)$ and $g(x, y)$. Apply phase correlation to $\mathbf{F}(\log r, \theta)$ and $\mathbf{G}(\log r, \theta)$ to determine the θ_0 shift amount in the θ dimension by which they differ. The value of θ_0 is the angle of rotation by which $f(x, y)$ and $g(x, y)$ differ.

(6) If $f(x, y)$ and $g(x, y)$ differ by a scale change s, then $\mathbf{F}(\log r, \theta)$ and $\mathbf{G}(\log r, \theta)$ will be circularly shifted with respect to each other in the $\log r$ dimension by the amount $-\log s$, but will not differ because of translational differences between $f(x, y)$ and $g(x, y)$. From the value of $-\log s$, calculate s, which is the scale factor by which $f(x, y)$ and $g(x, y)$ differ.

A practical implementation of these steps must remove or reduce rotationally dependent aliasing. There are at least two methods to reduce the rotationally dependent aliasing present in the Fourier spectra $F(\omega, v)$ and $G(\omega, v)$. Stone *et al.* (2003) suggests the following operations be applied to the spectra:

(1) Apply a Blackman (or Blackman-Harris) window operation to the images $f(x, y)$ and $g(x, y)$ before taking their transforms. The effect of this operation is to remove aliasing due to spurious high frequencies introduced at the boundaries of the images. The windowing operation is applied prior to forming the transforms of the images in Step 2 above.

(2) In Step 4 above, when converting the frequencies of $\mathbf{F}(\omega, v)$ and $\mathbf{G}(\omega, v)$ into coordinates $(\log r, \theta)$, convert only the frequencies that lie within a radius $N/2$ from the center of the frequency plane (the frequency plane is of size N by N) and outside a circle of radius $N/4$ from the center of the frequency plane. Convert only the points that lie with the radial angle θ for $0 \le \theta < \pi$. (This is one half of the rotational range of θ.)

Stone *et al.* (2003) explains the reasons for the advice above. The Blackman (or Blackman-Harris) window eliminates aliasing that would otherwise be introduced at the boundaries of the image. This aliasing is visible in Fig. 4.6. The reason for excluding the frequencies outside of a radius $N/2$ from the origin is that this is the largest circular region that fits within the frequency plane. If frequencies outside the radius $N/2$ were mapped into the $(\log r, \theta)$ representation, then for some values of θ the representation would be deficient, and would not have data in the higher-frequency regions. If frequencies inside the radius $N/4$ were mapped into the $(\log r, \theta)$ representation, then these frequencies would introduce rotationally dependent aliasing artifacts as explained by Stone *et al.* (2003). The removal of these frequencies greatly improves the signal-to-noise ratio in the remainder of processing because it substantially reduces spurious effects caused by aliasing, particular with respect to the spectral values around the zero value of θ. The use of only one half of the spectrum instead of the full spectrum effectively doubles the phase correlation signal-to-noise ratio for reasons explained later in this section.

The first four steps of the algorithm are straightforward, and can be followed from the description in this chapter, possibly with some consultation to the cited literature. Steps 5 and 6 involve a process called *phase correlation*, which has not been discussed in detail in this chapter up to this point. The following discussion of phase correlation should be adequate for practical implementation of the registration algorithm.

The process of phase correlation was first used to find relative shifts of signal waveforms in time. It is very useful for such purposes because it gives an extremely sharp correlation peak. The idea behind phase correlation is that two images can be processed in the frequency domain in such a way that the translational difference of the images is obtained by an inverse transform to the space domain, and the result is a δ function whose peak is at the position that corresponds to the relative displacement of the images.

Here is the idea. Let $f(x, y)$ and $g(x, y)$ be real valued discrete images, and let their images differ by the translation (x_0, y_0). Let $F(\omega, v)$ and $G(\omega, v)$ be the respective discrete Fourier transforms of $f(x, y)$ and $g(x, y)$. Then $F(\omega, v)$ and $G(\omega, v)$ ideally differ by the phase factor $e^{-j2\pi(\omega x_0 + v y_0)}$. Let $\hat{G}(\omega, v)$ denote the complex conjugate of $G(\omega, v)$. The phase correlation of $F(\omega, v)$ and $G(\omega, v)$ is defined to be the inverse Fourier transform of the following function:

$$\frac{F(\omega, v) \cdot \hat{G}(\omega, v)}{\|F(\omega, v)\| \cdot \|G(\omega, v)\|}. \tag{4.19}$$

Since $f(x, y)$ and $g(x, y)$ differ by the translation (x_0, y_0), their transforms satisfy the equation $G(\omega, v) = F(\omega, v) \cdot e^{-j2\pi(\omega x_0 + v y_0)}$. From this it follows that the phase correlation of $F(\omega, v)$ and $G(\omega, v)$ is ideally equal to the inverse Fourier transform of

$$\frac{F(\omega, v) \cdot \hat{G}(\omega, v)}{\|F(\omega, v)\| \cdot \|G(\omega, v)\|} = e^{j2\pi(\omega x_0 + v y_0)}. \tag{4.20}$$

The inverse transform (4.20) is the function $\delta(x - x_0, y - y_0)$. This function is 0 everywhere except at the point (x_0, y_0) where it has the value 1. A very important aspect of Eq. (4.20) noted by Stone *et al.* (2003), is that the value of function $e^{j2\pi(\omega x_0 + v y_0)}$ at the frequency point $(0, 0)$ is 1. This value is equal to the sum of the real values of its inverse Fourier transform. In ideal circumstances, the inverse Fourier transform is exactly $\delta(x - x_0, y - y_0)$, but in less than ideal circumstances

(a)   (b)

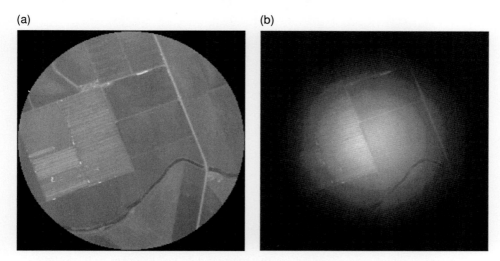

Figure 4.7. (a) An aerial image cropped for rotational registration, and (b) the same image after applying a Blackman window operation. (Source: Stone *et al.*, 2003, © Elsevier Science, reprinted with permission.)

there are noise peaks scattered in the plane of the inverse Fourier transform. Since the sum of the real values of the inverse Fourier transform must sum to 1, the effect of the noise must be to reduce the value of peak at (x_0, y_0).

This explains why the effect of aliasing is very pronounced when using phase correlation to extract rotational position and relative scale. Both of these are done with images derived from the magnitudes of Fourier transforms rather than from the image pixels themselves. The operations of phase correlations on the images derived from Fourier magnitudes incorporates all of the aliasing artifacts that are in those Fourier transforms.

As an example of what those artifacts look like, consider the two images in Fig. 4.7. The image in Fig. 4.7(a) is a typical aerial image of an agricultural area. It has been cropped because it is to be registered with rotated images of itself for varying rotational angles θ. The cropping eliminates effects of pixels outside the uncropped portion of the image. The image in Fig. 4.7(b) has been windowed to eliminate additional spurious elements caused by aliasing. The discrete Fourier transforms of these images are shown in Fig. 4.8.

The Fourier transforms in Fig. 4.8 are derived from the same images but they have vastly different characteristics. The transform in Fig. 4.8(b) is virtually free from aliasing. The strong linear character of the transform is due to the road structure of the original image. If the windowed original image in Fig. 4.7(b) were to be rotated by an angle θ, the strong lines evident in Fig. 4.8(b) would rotate by the same angle θ. On the other hand, Fig. 4.8(a), which is the discrete Fourier transform

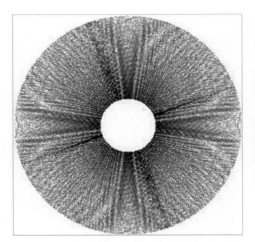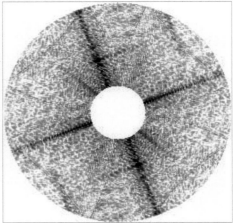

Figure 4.8. Discrete Fourier transforms of the images in Fig. 4.9. Regions that appear in white are regions where the spectral data has been set to 0. (Source: Stone *et al.*, 2003, © Elsevier Science, reprinted with permission.)

of Fig. 4.7(a), exhibits strong aliasing artifacts. The part of the transform due to the road structure of the image, which is the true image content, is barely visible because of the presence of the strong artifacts. The artifacts visible in Fig. 4.8(a) are similar to those in Fig. 4.6, the Fourier transform of a disk, and are present because of the disk-like outline of the image in Fig. 4.7(a). If the original image in Fig. 4.7(a) were to be rotated by an angle θ, then its transform would not be a rotated version of the transform in Fig. 4.8(a). The components of Fig. 4.8(a) due to aliasing would stay in place, because they represent the transform of a disk in square which is unchanged by a rotation of Fig. 4.7(a). On the other hand, the lines of Fig. 4.8(a) that are also present in Fig. 4.8(b) would rotate by the angle θ. As a consequence of the aliasing, phase correlations of transforms of the type shown in Fig. 4.8(a) will exhibit sharp peaks at a rotational angle of $0°$ because the aliasing components in a discrete Fourier transform remain in place when an image is rotated, so there is a nontrivial contribution to the phase correlation function by those components in the position that corresponds to 0 rotation. However, in Fig. 4.8(b), those aliasing components have largely been removed. When the original rotates, virtually the entire structure of its discrete Fourier transform rotates by the same amount. This effectively removes the false peak at $0°$ rotation.

Figure 4.9 illustrates the phase correlation functions obtained from the two examples in Fig. 4.7. For this plot, only one dimension of the phase correlation function is plotted, and it is plotted through the position $\log r = 0$. By limiting the plot to the line through $\log r = 0$, the phase correlations shown correspond to only

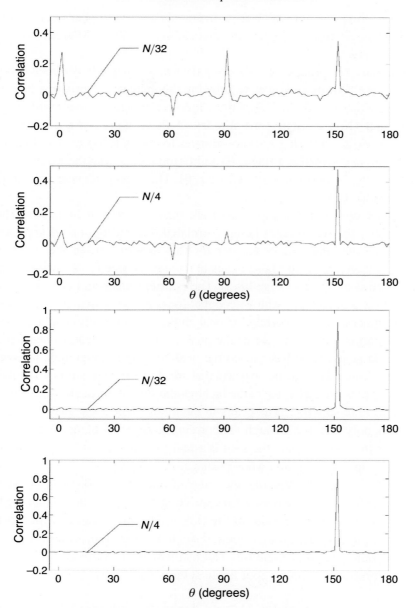

Figure 4.9. (a) Top two graphs, and (b) bottom two graphs. These are the phase correlation functions that derive from the images in Figs. 4.7(a) and 4.7(b), respectively. The upper graph in each of Figs. 4.9(a) and 4.9(b) eliminates frequencies within $N/32$ of the center frequency of the Fourier transform in Fig. 4.8. The lower figure in each of Figs. 4.9(a) and 4.9(b) eliminates frequencies within $N/4$ of the center frequency. (Source: Stone *et al.*, 2003, © Elsevier Science, reprinted with permission.)

those transformations for which scale factor s is unity. The peaks in each plot near 150° show the position of the best estimate of the rotational difference between the original images.

In summary, the process of estimating the rotation angle θ by which two images $f(x, y)$ and $g(x, y)$ differ involves windowing the images, constructing their Fourier transforms $F(\omega, v)$ and $G(\omega, v)$ of the windowed images, constructing the images $\mathbf{F}(\omega, v)$ and $\mathbf{G}(\omega, v)$ of the magnitudes of those transforms, converting those magnitude images into $(\log r, \theta)$ coordinates to obtain $\mathbf{F}(\log r, \theta)$ and $\mathbf{G}(\log r, \theta)$, and then phase correlating $\mathbf{F}(\log r, \theta)$ and $\mathbf{G}(\log r, \theta)$ to find the value θ_0 such that $\mathbf{F}(\log r, \theta - \theta_0)$ is most similar to $\mathbf{G}(\log r, \theta)$. The large peak near 150° in Fig. 4.9 gives the value of θ_0.

Note that both plots in Fig. 4.9(b) are nearly perfect δ functions. The peak indicates that the two images being correlated differed by a θ of approximately 150°, which was the correct answer for this experiment. In the upper plot of Fig. 4.9(a) there are strong false peaks at 0° and 90° due to the artifacts apparent in the spectral structure of Fig. 4.8(a). The Fourier transform in that image clearly shows a four-pointed star, which accounts for correlation matches at multiples of 90°. The upper plot has a correlation peak in the correct position, at approximately 150° of rotation, but the height of the peak is only 0.3 instead of near unity. The reduction in peak height has caused the peak to drop in magnitude to where it is equal approximately to the noise peaks that are at incorrect positions. The reduction in peak height is primarily due to the fact that the sum of the peaks in the correlation function have to sum to 1. The peaks in the figure sum to less than 1 because this plot shows only one line through the correlation function plane. When the sum is taken over the whole plane, the sum is equal to 1 as it must be to be the Fourier inverse of a spectral function whose value at the origin is 1.

The plots in Fig. 4.9 show that the range of rotation is 180° (π radians), instead of the full 360° circle. The reason for restricting rotation to 180° is that the Fourier spectrum magnitude is periodic in the $(\log r, \theta)$ representation with a period of 180°. Recall that the two-dimensional discrete Fourier transform of a real valued image is conjugate antisymmetric. The mathematical expression of this relation when expressed in the (ω, v) coordinate system is given by

$$F(\omega, v) = \hat{F}(N - \omega, N - v). \tag{4.21}$$

In the $(\log r, \theta)$ representation, this equation becomes

$$F(\log r, \theta) = \hat{F}(\log r, \theta + \pi). \tag{4.22}$$

The images used to estimate rotation by means of phase correlation are the images of the magnitudes of the Fourier spectra in the $(\log r, \theta)$ representation. Taking magnitudes of Eq. (4.22) yields the equality shown in Eq. (4.23), i.e.,

$$\| F (\log r, \theta) \| = \| \hat{F} (\log r, \theta + \pi) \|. \tag{4.23}$$

Therefore, the phase correlation function depicted in Fig. 4.9 is periodic in θ, with a period of 180°, and would repeat a second time if the plot were extended through the full 360°. However, the constraint still holds that the sum of the real values is unity so that the peaks shown in Fig. 4.9 would be reduced to half their heights shown if the correlation is taken over 360° instead of over 180°. By allowing θ to range over only half of the unit circle, the phase correlation algorithm enables correlation peaks to attain the maximum value of 1.00 in the ideal case instead of limiting their value to 0.50. This fact was first noted in Stone *et al.* (2003) and provided a partial explanation of the low values of phase correlation peaks in the earlier literature.

The phase correlation algorithm described here applies to the estimation of scale changes as well to estimation of rotational changes. The scale factor s can be estimated by examining the phase correlation function in the full ($\log r, \theta$) plane. Assume that the correlation peak lies at the point ($\log s_0, \theta_0$). Then θ_0 is the best estimate of rotational difference and s_0 is the best estimate of the scale change. That is, $\mathbf{F}(\log r - \log s_0, \theta - \theta_0)$ is most similar to $\mathbf{G}(\log r, \theta)$. Because this formulation uses the logarithm of the scale change instead the scale change itself, the estimate of the value of the logarithm of the scale factor s_0 is inherently less precise than a direct measurement of value of the scale factor.

4.6 Summary and conclusions

The main theme of this section is the use of Fourier techniques in image registration. The primary advantage of Fourier techniques is their raw speed in calculating the normalized correlation of image pairs for all relative translations. They enable global comparisons of relative translations to be done in reasonable time. Techniques that are not Fourier-based are forced to search specific regions of parameter space because a comprehensive search would be too costly.

Another facet of computations in the Fourier domain is that they enable algorithms to derive rotations, scale changes, and translations separately instead of jointly. Separating these factors further reduces the size of the search space because instead of searching over the product of the sizes of the rotation space, scale space, and translation space, the searches are conducted over the sums of the sizes of these spaces.

The chip-based registration technique has a unique advantage of being able to discover translations to subpixel precision as well as small rotations and small scale changes without resorting to image interpolation. All computations are done on the original images, so that the computation does not introduce errors due to image interpolation, nor is there a cost of computation incurred for doing such interpolations.

The Fourier-based method for calculating translational offsets to high precision has proved to be very fast, and has achieved a precision better than 0.1 pixels on real images.

Finally, the phase-correlation technique described for calculating rotation and scale parameters have far greater signal-to-noise ratios than its predecessors. The most important facet of the technique described in this chapter is the elimination of the false peak near 0. The presence of a false peak at 0 destroys the ability to detect and measure small rotations and small changes of scale because they are completely hidden within the false peak.

In spite of the positive characteristics of the algorithms discussed in this chapter, the algorithms are not universally applicable to all registration problems. The success in their use depends very much on the data to which they are applied. Normalized correlations are very useful measures when images differ by changes of intensity that follow a linear relation. But they are not useful when images differ in ways that are not properly modeled by a linear change of intensity. Also, the techniques discussed here tend to be most useful for rigid image transformations, which reduce to rotations, translations, and changes of scale. If the transformations are more complex, then the algorithms may yield less precise answers, or worse yet, may not yield any useful information.

Nevertheless, when the algorithms are applied to images that fall within their scope of utility, they are extremely effective for image registration.

References

Agarwal, R. C. and Cooley, J. W. (1977). New algorithms for digital convolution. *IEEE Transactions on Acoustics, Speech, and Signal Processing*, **25**, 392–410.

Alliney, S. and Morandi C. (1986). Digital image registration using projections. *IEEE Pattern Analysis and Machine Intelligence*, **8**(2), 222–233.

Alliney, S., Cortelazzo, G., and Mian, G. A. (1996). On the registrations of an object translating on a static background. *Pattern Recognition*, **29**(1), 131–141.

Anuta, P. (1970). Spatial registration of multispectral and multitemporal digital imager using fast Fourier transform techniques. *IEEE Transactions on Geoscience Electronics*, **8**, 353–368.

Chang, S. H., Chen, F. H., Hsu, W. H., and Wu, G. Z. (1997). Fast algorithm for point pattern matching: Invariant to translations, rotations and scale changes. *Pattern Recognition*, **30**(2), 311–320.

Chen, Q.-S., Defrise M., and Deconinck, F. (1994). Symmetric phase-only matched filtering of Fourier-Mellin transforms for image registration and recognition. *IEEE Pattern Analysis and Machine Intelligence*, **16**(12), 1156–1168.

Cooley, J. W. (1966). *Harmonic Analysis Complex Fourier Series*. SHARE Program Library No. SDA 3425.

Cooley, J. W. and Tukey, J. W. (1965). An algorithm for the machine calculation of complex Fourier series. *Mathematics of Computation*, **19**(90), 297–301.

Dasgupta, B. and Chatterji, B. N. (1996). Fourier-Mellin transform based image matching algorithm. *Journal of the Institution of Electronics and Telecommunication Engineers*, **42**(1), 3–9.

De Castro, E. and Morandi, C. (1987). Registration of translated and rotated images using finite Fourier transforms. *IEEE Pattern Analysis and Machine Intelligence*, **9**(52), 700–703.

Edwards, A. L. (1976). The correlation coefficient. In *An Introduction to Linear Regression and Correlation*. San Francisco, CA: W. H. Freeman, pp. 33–46.

Good, I. J. (1958). The interaction algorithm and practical Fourier analysis. *Journal of the Royal Statistical Society, Series B (Methodological)*, **20**(2), 361–372.

Levenberg, K. (1944). A method for the solution of certain problems in least squares. *Quarterly of Applied Mathematics*, **2**, 164–168.

Marquardt, D. (1963). An algorithm for least-squares estimation of nonlinear parameters. *Journal of Applied Mathematics*, **11**, 431–441.

Shannon, C. E. (1949). Communication in the presence of noise. *Professional Institute of Radio Engineers*, **37**(1), 10–21; reprinted in *Proceedings of the IEEE*, **86**(2), 447–457.

Shekarforoush, H., Berthod, M., and Zerubia, J. (1996). Subpixel image registration by estimating the polyphase decomposition of cross power spectrum. In *Proceedings of the IEEE Conference on Computer Vision and Pattern Recognition*, pp. 532–537.

Singleton, R. C. (1967). On computing the fast Fourier transform. *Communications of the ACM*, **10**(10), 647–654.

Stone, H. S. (1998). Convolution theorems for linear transforms. *IEEE Transactions on Signal Processing*, **46**(10), 2819–2821.

Stone, H. S. (1999). Progressive wavelet correlation using Fourier techniques. *IEEE Transactions on Signal Processing*, **47**(1), 97–107.

Stone, H. S. and Li, C.-S. (1996). In Image matching by means of intensity and texture matching in the Fourier domain. In *Proceedings of the SPIE Conference on Image and Video Databases, Storage and Retrieval for Still Image and Video Databases IV*, San Jose, CA, Vol. 2670, pp. 337–344.

Stone, H. S. and Shamoon, T. (1997). Search by content of partially occluded images. *International Journal on Digital Libraries*, **1**(4), 329–343.

Stone, H. S., Orchard, M., Chang, E.-C., and Martucci, S. (2001). A fast direct Fourier-based algorithm for subpixel registration of images. *IEEE Transactions on Geoscience and Remote Sensing*, **39**(10), 2235–2243.

Stone, H. S., Tao, B., and McGuire, M. (2003). Analysis of image registration noise due to rotationally dependent aliasing. *Journal of Visual Communication and Image Representation*, **14**(2), 114–135.

Strang, G. (1986). *Introduction to Applied Mathematics*. Wellesley, MA: Wellesley-Cambridge Press.

Unser, M. (2000). Sampling – 50 years after Shannon. *Proceedings of the IEEE*, **88**(4), 569–587.

5

Matched filtering techniques

QIN-SHENG CHEN

Abstract

A matched filter is a linear filtering device developed for signal detection in a noisy environment. The matched filtering technique can be directly employed in image registration for the estimation of image translation. The transfer function of a classic matched filter is the Fourier conjugate of the reference image, divided by the power spectrum of the system noise. If only the phase term of the reference image is used in the construction of the transfer function, the so-called phase-only matched filter generates a sharper peak at the maximum output than that obtained by a classic matched filter. Thus, the image translation can be detected more reliably. If rotation and scale change are also involved, the images to be registered can first be transformed into the Fourier-Mellin domain. The Fourier-Mellin transform of an image is translation invariant and represents rotation and scale changes as translations in the angular and radial coordinates. Employing matched filtering or phase-only matched filtering on the Fourier-Mellin transforms of the images yields the estimation of the image rotation and scaling. After correcting for rotation and scaling, the image translation can be determined by using the matched filtering or phase-only matched filtering method.

5.1 Introduction

Matched filtering is a classic signal processing technique widely used in signal detection (Turin, 1960; Vanderluht, 1969; Whalen, 1971; Kumar and Pochapsky, 1986). A matched filter is a linear filter with a transfer function that maximizes the output signal-to-noise ratio (SNR) for an input signal with known properties. Typically, the transfer function is the Fourier conjugate of the known signal divided by the power spectrum of the system noise. The matched filter is used to determine in an optimal manner the presence of a signal in question from a set of unknown input signals corrupted with noise. The displacement of the maximum output away from a reference time point indicates the shift of the detected signal, such as the delay of a radar signal.

In image registration, it is assumed that the images are acquired from the same object or the same scene. The difference between a reference image and the image to be registered is a global geometric transformation, for example, translation, rotation, and isotropic scaling, possibly with noise contamination. An inverse geometric transformation can realign the images if the transform parameters are known. This chapter discusses how matched filtering techniques can be used to estimate these parameters. For example, if only translation is involved, one can construct a transfer function of the matched filter from the reference image and filter the image to be registered. The location of the maximum output corresponds to the image translation. If rotation and scaling are also present, one can first solve the rotation and scaling problem using a Fourier-Mellin matched filtering technique and then determine the image translation using directly the classic matched filtering method.

This chapter is organized as follows. In Section 5.2 the basic concept of matched filtering is introduced and its application to image registration is described. Section 5.3 discusses the importance of spectral phase in image representation, followed by the introduction in Section 5.4 of phase-only and symmetric phase-only matched filtering, and the comparison of various characteristics of these matched filtering techniques. In Section 5.5 we introduce the Fourier-Mellin transformation and its invariant property. A registration algorithm based on the Fourier-Mellin transformation and matched filtering is provided in Section 5.6. Finally, some experimental results are reported in Section 5.7.

5.2 Matched filter

Consider a signal $b(t)$ which is a replica of a prototype signal $a(t)$ contaminated by additive noise $n(t)$,

$$b(t) = a(t) + n(t). \tag{5.1}$$

If the signal passes through a system with an impulse response $h(t)$, the output is

$$m(t) = h(t) * b(t) = h(t) * a(t) + h(t) * n(t), \tag{5.2}$$

where $*$ denotes convolution. $h(t) * a(t)$ and $h(t) * n(t)$ are the output signal and the output noise, respectively. The average power ratio of these two components defines the output power signal-to-noise ratio (SNR_p), i.e.,

$$\text{SNR}_p = \frac{E\left[|h(t) * a(t)|^2\right]}{E\left[|h(t) * n(t)|^2\right]}. \tag{5.3}$$

The classic matched filter is designed to optimize the detection of the occurrence of a prototype signal in the presence of noise. Assuming $n(t)$ is stationary with a noise power-spectral density $|N(w)|^2$, one can prove using the Schwarz inequality (Turin, 1960; Whalen, 1971) that the matched filter has a transfer function

$$H(w) = \frac{A^*(w)}{|N(w)|^2},\qquad(5.4)$$

where $A^*(w)$ is the complex conjugate of the Fourier transform of signal $a(t)$. If the noise is so-called white noise, its spectrum is flat, i.e., $|N(w)|^2 = n_w^2$, and the transfer function can be simplified to

$$H(w) = \frac{1}{n_w^2} A^*(w).\qquad(5.5)$$

For the model of Eq. (5.1), the filter output attains its maximum at the origin. If $b(t)$ is delayed by v, the corresponding output peak will shift by the same amount.

In image registration, the matched filtering technique can be used to determine the translational shift. Let $b(x, y)$ be a translated version of a reference image $a(x, y)$, then

$$b(x, y) = a(x - x_d, y - y_d) + n(x, y),\qquad(5.6)$$

where x_d and y_d denote the translational offsets in x and y directions, respectively. The Fourier transforms of $a(x, y)$ and $b(x, y)$ are

$$A(w_x, w_y) = F\{a(x, y)\} = \int_{-\infty}^{\infty}\int_{-\infty}^{\infty} a(x, y) \cdot e^{-j(w_x x + w_y y)} dx dy\qquad(5.7)$$

and

$$\begin{aligned} B(w_x, w_y) &= \int_{-\infty}^{\infty}\int_{-\infty}^{\infty} b(x, y) \cdot e^{-j(w_x x + w_y y)} dx dy \\ &= A(w_x, w_y) \cdot e^{-j(w_x x_d + w_y y_d)} + N(w_x, w_y), \end{aligned}\qquad(5.8)$$

where $F\{\cdot\}$ denotes the Fourier transform and $j^2 = -1$. For a known noise power spectrum, the transfer function of the matched filter is of the form (Vanderluht, 1969; Kumar and Pochapsky, 1986):

$$H(w_x, w_y) = \frac{A^*(w_x, w_y)}{|N(w_x, w_y)|^2}.\qquad(5.9)$$

For an unknown noise spectrum, it is reasonable to assume that $n(x, y)$ is a sample from a white noise process with power density N_0. The transfer function (5.9) then becomes

$$H(w_x, w_y) = \frac{1}{N_0} A^*(w_x, w_y).\qquad(5.10)$$

The filter output in the Fourier domain is

$$
M(w_x, w_y) = H(w_x, w_y) \cdot B(w_x, w_y)
$$
$$
= \frac{A^2(w_x, w_y)}{|N(w_x, w_y)|^2} \cdot e^{-j(w_x x_d + w_y y_d)} + \frac{A^*(w_x, w_y)}{N^*(w_x, w_y)}. \qquad (5.11)
$$

The inverse Fourier transform of $A^2(w_x, w_y)$, $F^{-1}\{A^2(w_x, w_y)\}$, is the image auto-correlation whose maximum output is at the origin. The inverse Fourier transform of $A^*(w_x, w_y)$ is the image $a(x, y)$ rotated by π. Assuming white noise, the maximum of $F^{-1}\{A^2(w_x, w_y)\}$ will dominate the output field $F^{-1}\{M(w_x, w_y)\}$. Note that the phase term $e^{-j(w_x x_d + w_y y_d)}$ will shift the output by (x_d, y_d) in the spatial domain. Thus, finding the image translation amounts to locating the maximum output away from the origin.

One limitation of the matched filter is that the output is primarily dependent on the energy of the input images rather than on image structures, such as edges, object shapes, image textures, etc. The output peak around (x_d, y_d) is usually broad (Chen, 1993; Chen *et al.*, 1994). This makes the detection of the maximum output, and hence the estimate of the image translation, unreliable in a noisy environment. Nevertheless, this problem can be properly solved by phase-only matched filtering techniques discussed below.

5.3 Importance of spectral phase

In the Fourier domain, the spectral magnitude and phase tend to play different roles. Oppenheim and Lim (1981) demonstrated that in many situations, the important features of an image are properly preserved in the spectral phase. Let $g(x, y)$ and $G(w_x, w_y)$ represent the Fourier transform pair of an image, i.e.,

$$
G(w_x, w_y) = F\{g(x, y)\} = |G(w_x, w_y)| \cdot e^{j\phi(w_x, w_y)}, \qquad (5.12)
$$

where $|G(w_x, w_y)|$ and $\phi(w_x, w_y)$ denote the spectral magnitude and phase of $g(x, y)$, respectively. The inverse Fourier transform of each term yields a new image (Kermisch, 1970; Kottzer *et al.*, 1992). An image reconstructed from the spectral magnitude is referred to as a magnitude-only image, given by

$$
g_m(x, y) = F^{-1}\{|G(w_x, w_y)|\}. \qquad (5.13)
$$

The reconstruction from the phase term, referred to as a phase-only image, is expressed by

$$
g_p(x, y) = F^{-1}\left\{M e^{j\phi(w_x, w_y)}\right\}, \qquad (5.14)
$$

Figure 5.1. A leaf image. The spatial dimension is 256 × 256. The intensity resolution is 8 bits. (Source: Chen *et al.*, 1994, © IEEE, reprinted with permission.)

where M is some scale constant for the purpose of image display. Figures 5.1
and 5.2 show a Fourier transform pair of a leaf image. Figures 5.3 and 5.4 are the
corresponding magnitude-only and phase-only images. These reconstructed images
demonstrate that the phase-only image retains some meaningful information of the
original image.

Another interesting demonstration of the importance of the spectral phase is
image synthesis from combining the spectral magnitude of one image and the phase
of another (Oppenheim and Lim, 1981). This is demonstrated on the cactus image
in Fig. 5.5. After the Fourier transformation, its spectral magnitude is extracted.
The product of the spectral magnitude of the cactus image and the spectral phase of
the leaf image forms the Fourier spectrum of a new image. The new image, which
is the inverse Fourier transform of the product, is displayed in Fig. 5.6. This image
most closely resembles the leaf image from which the phase information originates.
One explanation for this observation is that phase preserves the *location* of events,
and much of the *intelligibility* of the image may be related to the location of *events*
such as edges, lines, points, etc. A straightforward example is spatial translation
in an image, which has no effect on the spectral magnitude but modifies the
phase.

Figure 5.2. Spectral magnitude of the leaf image. The origin is shifted to the center of the image. (Source: Chen *et al.*, 1994, © IEEE, reprinted with permission.)

Figure 5.3. Magnitude-only reconstruction of the leaf image.

Figure 5.4. Phase-only reconstruction of the leaf image.

Figure 5.5. A cactus image.

Figure 5.6. A synthesized image from combining the phase of the leaf image and the magnitude of the cactus image.

5.4 Phase-only and symmetric phase-only matched filtering

Transfer function (5.10) can be reorganized as

$$H(w_x, w_y) = \frac{1}{N_0} A^*(w_x, w_y) = \frac{1}{N_0} \cdot \frac{A^*(w_x, w_y)}{|A(w_x, w_y)|} \cdot |A(w_x, w_y)|$$

or

$$H(w_x, w_y) = \frac{1}{N_0} \cdot \text{Phase}\{A^*(w_x, w_y)\} \cdot |A(w_x, w_y)|, \qquad (5.10a)$$

where Phase$\{A^*(w_x, w_y)\}$ denotes the phase term $\frac{A^*(w_x, w_y)}{|A(w_x, w_y)|}$. Assuming white noise, omitting the noise term does not change the information content of the output. If the magnitude term is also left out, the resulting transfer function $H(w_x, w_y)$ is known as the phase-only matched filter (Horver and Gianino, 1984), i.e.,

$$H(w_x, w_y) = \text{Phase}\{A^*(w_x, w_y)\} = e^{-j\phi_A(w_x, w_y)}. \qquad (5.15)$$

Since only the spectral phase is utilized, the filter output between identical images under the constraint of translation yields a much sharper peak than the classical matched filtering, which makes locating the maximum output more accurate and reliable. Defining the optical efficiency of the image matching as the ratio of the output energy at the maximum to the total input signal energy, Horver and Gianino reported that for two identical images to be registered, the optical efficiency of the phase-only matched filtering was 57.6 times higher than that of the classical matched filtering (Horver and Gianino, 1984). In image detection applications, the phase-only matched filtering also showed a much higher discrimination power. Even if noise contamination is significant, for example, the input SNR is close to 1, the discrimination ability still holds up well (Chen *et al.*, 1994).

A further improvement of the phase-only matched filtering can be achieved by extracting the phases of both the reference and sensed images. Consider two images related by translation, i.e., $b(x, y) = a(x - x_d, y - y_d)$. Their Fourier transforms are related by

$$B(w_x, w_y) = e^{-j(w_x x_d + w_y y_d)} \cdot A(w_x, w_y). \qquad (5.16)$$

The product of the phase terms of the two images is given by

$$M_{spm}(w_x, w_y) = \frac{A^*(w_x, w_y)}{|A(w_x, w_y)|} \cdot \frac{B(w_x, w_y)}{|B(w_x, w_y)|} = e^{-j(w_x x_d + w_y y_d)}. \qquad (5.17)$$

This expression is referred to as the symmetric phase-only matched filtering (spm) (Lam, 1985; De Castro *et al.*, 1989; Wu *et al.*, 1989). An even sharper output peak is expected in this case, since the inverse Fourier transform of the above function is a Dirac δ-function centered at (x_d, y_d). This technique is expected to be more robust with respect to noise. The symmetric phase-only matched filtering can be considered as a nonlinear two-step process. The first step is to extract the phase of the input image followed by the application of the phase-only matched filtering. If the two images are related by translation and their noise spectra are identical, Eq. (5.17) can be reorganized as a linear filtering operation, i.e.,

$$M_{spm}(w_x, w_y) = \frac{A^*(w_x, w_y)}{|A(w_x, w_y)|^2} \cdot B(w_x, w_y) \qquad (5.18)$$

with the transfer function

$$H(w_x, w_y) = \frac{A^*(w_x, w_y)}{|A(w_x, w_y)|^2}. \qquad (5.19)$$

Figure 5.7 shows the outputs of the different matched filters. Some further improvements of the symmetric phase-only filter can be achieved by preprocessing the spectral phases. For instance, in the technique known as binary phase-only filtering, the phases are binarized by passing the real or imaginary part of the phase through a hard limit (Horver and Gianino, 1985; Flannery *et al.*, 1988; Ersoy and Zeng, 1989; Wang *et al.*, 1993).

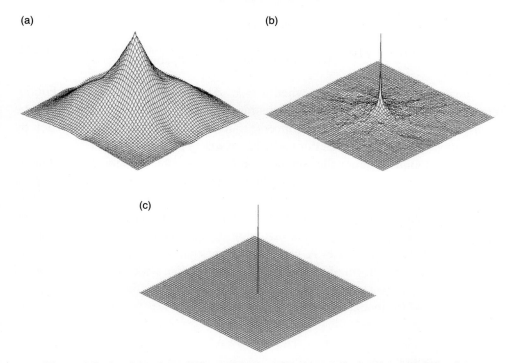

(a)   (b)

(c)

Figure 5.7. Outputs of the different matched filtering on the leaf image of Fig. 5.1. (a) Output due to matched filtering (the origin is shifted to the center of the output field); note the relative broad peak. (b), (c) Outputs due to the phase-only matched filter and the symmetric phase-only matched filtering, respectively. (Source: Chen *et al.*, 1994, © IEEE, reprinted with permission.)

5.5 Fourier-Mellin invariant transforms

The matched filters make use of the phase shift property of the Fourier transform. If the geometric change contains rotation and scaling in addition to translation, the phase shift property no longer holds and the filters will fail. To implement an image stabilization system for the observation of human retina, De Castro and Morandi (1987) extended the above technique to estimate both image translation and rotation. If $a(x, y)$ and $b(x, y)$ are related by translation (x_d, y_d) and rotation θ_r, i.e.,

$$b(x, y) = a(x \cos \theta_r + y \sin \theta_r - x_d, -x \sin \theta_r + y \cos \theta_r - y_d), \quad (5.20)$$

their corresponding transforms are related by

$$B(w_x, w_y) = e^{-j(w_x x_d + w_y y_d)} A(w_x \cos \theta_r + w_y \sin \theta_r, -w_x \sin \theta_r + w_y \cos \theta_r). \quad (5.21)$$

The ratio function

$$Q(w_x, w_y, \theta) = \frac{B(w_x, w_y)}{A(w_x \cos\theta + w_y \sin\theta, -w_x \sin\theta + w_y \cos\theta)} \quad (5.22)$$

is equal to $e^{-j(w_x x_d + w_y y_d)}$ only when $\theta = \theta_r$. The corresponding inverse Fourier transform is a Dirac δ-function. Registration is thus achieved by detecting the appearance of a sharp peak in the inverse transform of the above ratio function by continuously rotating the Fourier spectrum of $a(x, y)$.

This technique requires the computation for all possible rotation angles, that is, the search for θ_r is exhaustive. If scaling is also involved, the parametric space will expand to four dimensions, which will further exacerbate the already low efficiency. One alternative is to derive a transform that is invariant in certain geometric transformations, so the matching of the images can be carried out in a reduced parametric space. A well-known invariant is the translation-invariant Fourier-Mellin transform of a two-dimensional function, which represents rotation and scaling as translations along corresponding axes of the parametric space (Altmann and Reitbock, 1984; Sheng and Arsenault, 1986; Segman, 1992; Chen, 1993, Chen *et al.*, 1994).

Consider two images $a(x, y)$ and $b(x, y)$ related by translation, rotation, and scaling as follows:

$$b(x, y) = a(\sigma_s(x \cos\theta_r + y \sin\theta_r) - x_d, \sigma_s(-x \sin\theta_r + y \cos\theta_r) - y_d), \quad (5.23)$$

where σ_s is the isotropic scale factor, θ_r the rotation angle, and (x_d, y_d) are the translational offsets. The corresponding Fourier transforms are related by

$$B(w_x, w_y) = e^{-j\phi_b(w_x, w_y)}\sigma_s^{-2}|A(\sigma_s^{-1}(w_x \cos\theta_r + w_y \sin\theta_r),$$
$$\sigma_s^{-1}(-w_x \sin\theta_r + w_y \cos\theta_r))|, \quad (5.24)$$

where $\phi_b(w_x, w_y)$ is the spectral phase of $b(x, y)$. The phase term depends on the translation, scaling, and rotation, but the magnitude

$$|B(w_x, w_y)| = \sigma_s^{-2}|A(\sigma_s^{-1}(w_x \cos\theta_r + w_y \sin\theta_r), \sigma_s^{-1}(-w_x \sin\theta_r + w_y \cos\theta_r))|$$
$$(5.25)$$

is translation invariant. This function also shows that a rotation of the image rotates the spectral magnitude by the same angle, and that a scaling by σ_s magnifies the spectral magnitude by σ_s^{-1}. Another useful property is the invariance of the spectral origin with respect to rotation. Since the radial and angular directions are orthogonal, the rotation and scaling can be decoupled into the rectangular coordinate system

$$w_x = \rho \cos \theta, \quad w_y = \rho \sin \theta. \tag{5.26}$$

The spectral magnitudes of $a(x, y)$ and $b(x, y)$ in the new coordinate system are

$$a_p(\theta, \rho) = |A(\rho \cos \theta, \rho \sin \theta)|, \tag{5.27}$$

$$b_p(\theta, \rho) = |B(\rho \cos \theta, \rho \sin \theta)|. \tag{5.28}$$

Note that the spectral magnitude is a periodic function in the angular direction. If the original image is a real function, the magnitude of its Fourier transform is a twofold rotational symmetry. In other words, the spectral magnitude remains exactly the same if it is rotated by π.

$$a_p(\theta \pm n\pi, \rho) = a_p(\theta, \rho), \quad n = \pm 1, \pm 2, \pm 3, \ldots \tag{5.29}$$

Thus, instead of using the whole field, half of the magnitude field is sufficient to construct the filter transfer function. This is particularly helpful if computational efficiency is of concern. Using the trigonometric identities $\sin(\alpha + \beta) = \sin \alpha \cos \beta + \cos \alpha \sin \beta$ and $\cos(\alpha + \beta) = \cos \alpha \cos \beta - \sin \alpha \sin \beta$, one can prove (Chen, 1993) that

$$b_p(\theta, \rho) = \sigma_s^{-2} a_p(\theta - \theta_r, \rho/\sigma_s). \tag{5.30}$$

This coordinate transformation converts the image rotation to translation in the angular direction and the image scaling to coordinate scaling in the radial direction. The scale change also modifies the intensity of the spectral magnitude by a factor of σ_s^{-2}. It can be further simplified by a logarithmic conversion in the radial coordinate. Letting $\lambda = \log \rho$ and $\kappa_s = \log \sigma_s$, the spectral magnitudes of the images in the polar-log coordinate system will have the forms

$$a_{pl}(\theta, \lambda) = a_p(\theta, \rho) \tag{5.31}$$

and

$$b_{pl}(\theta, \lambda) = b_p(\theta, \rho) = \sigma_s^{-2} a_{pl}(\theta - \theta_r, \lambda - \kappa_s). \tag{5.32}$$

Both rotation and scaling are reduced to translation, as proposed by Pratt (1978) and Schalkoff (1989). This kind of polar-log mapping of the spectral magnitude is in fact a physical realization of the Fourier-Mellin transform of a two-dimensional function. Figure 5.8 displays the Fourier-Mellin transform of the leaf image. The transformation decouples the image rotation, scaling, and translation, and is therefore computationally efficient in image registration applications.

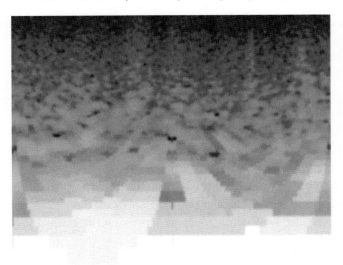

Figure 5.8. Fourier-Mellin transform of the leaf image of Fig. 5.1. The angular direction is mapped to the abscissa and the radial direction to the ordinate.

Biological visual systems appear to have some similarities with the polar-log mapping (Schwartz, 1980; Cavanagh, 1981; Schwartz, 1981). Schwartz suggested that a polar-log mapping of the visual field could provide an approximation to the striate cortex topography (Schwartz, 1981) and Cavanagh hypothesized that a polar-log frequency representation might be directly computed by the striate cortex (Cavanagh, 1981). Despite the controversy over the existence of the frequency representation in the retina-cortex system, the polar-log mapping in the visual system has been commonly accepted.

5.6 Fourier-Mellin matched filtering

As explained in the previous section, the translation-invariant Fourier-Mellin transform of an image represents the image rotation and scaling as translational shifts in some equivalent coordinate system. In an image registration application, one can first make use of this property to estimate the image rotation and scaling. After correcting for the rotation and scaling, the image translation can be determined using the matched filtering technique (Chen, 1993; Chen *et al.*, 1994; Reddy and Chatterji, 1996).

For a pair of images $a(x, y)$ and $b(x, y)$, the registration algorithm based on the Fourier-Mellin matched filtering is summarized as follows:

(1) Compute the Fourier transforms $A(w_x, w_y)$ and $B(w_x, w_y)$ of the images.
(2) For rotation and scaling estimation, extract the spectral magnitudes $|A(w_x, w_y)|$ and $|B(w_x, w_y)|$.
(3) Map the magnitudes into polar-log coordinates $a_{pl}(\theta, \lambda)$ and $b_{pl}(\theta, \lambda)$.
(4) Apply a one-dimensional window function in the radial direction on $a_{pl}(\theta, \lambda)$ and $b_{pl}(\theta, \lambda)$. Then, compute the respective Fourier transforms $A_{pl}(w_\theta, w_\lambda)$ and $B_{pl}(w_\theta, w_\lambda)$.
(5) Construct the phase-only filter transfer function, $H(w_\theta, w_\lambda) = \text{Phase}\{A_{pl}^*(w_\theta, w_\lambda)\}$.
(6) Compute the product of $H(w_\theta, w_\lambda)$ and $B_{pl}(w_\theta, w_\lambda)$ for phase-only matched filtering, or the product of $H(w_\theta, w_\lambda)$ and $\text{Phase}\{B_{pl}(w_\theta, w_\lambda)\}$ for symmetric phase-only matched filtering.
(7) Compute the inverse Fourier transform of the product.
(8) Detect the location of the maximum in the output field, which corresponds to the image rotation and scaling (θ_r, σ_s).
(9) Rotate and scale $b(x, y)$ by (θ_r, σ_s) to get a corrected image $r_b(x, y)$ for rotation and scale.
(10) For translation estimation, calculate the phase-only filter transfer function, $H(w_x, w_y) = \text{Phase}\{A^*(w_x, w_y)\}$.
(11) Compute the Fourier transform $R_b(w_x, w_y)$ of $r_b(x, y)$.
(12) Compute the product of $H(w_x, w_y)$ and $R_b(w_x, w_y)$ for phase-only matched filtering, or of $H(w_x, w_y)$ and $\text{Phase}\{R_b(w_x, w_y)\}$ for symmetric phase-only matched filtering.
(13) Compute the inverse Fourier transform(s) of the above product(s).
(14) Detect the location of the maximum in the output field, which corresponds to the image translation (x_d, y_d).
(15) Shift image $r_b(x, y)$ by (x_d, y_d).

The result is a registered image to $a(x, y)$.

5.7 Experimental results

The algorithm requires multiple forward and inverse Fourier transforms. The discrete implementation must be done with care to avoid the introduction of artifacts due to sampling and truncation aliasing. Severe artifacts could overshadow the signal spectrum and result in a spurious detection of the image shift (Chen, 1993; Chen *et al.*, 1994). Eliminating the aliasing artifacts is thus crucial to the success of the algorithm. Windowing the image before each Fourier transform can minimize the aliasing artifacts and is thus recommended. For example, a two-dimensional window function (Pratt, 1978; Chen *et al.*, 1999) could be applied to the images before the Fourier transformation in Step 1 and Step 11 of the algorithm. The

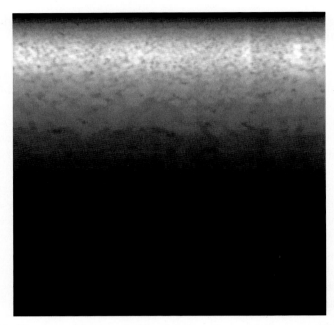

Figure 5.9. The same image as in Fig. 5.8, but after applying a one-dimensional Hanning window along the radial coordinate. The window was applied before the logarithmic conversion.

Figure 5.10. The leaf image was rotated by −100° and scaled by 90%. White noise was also added. The power SNR is 10. (Source: Chen *et al.*, 1994, © IEEE, reprinted with permission.)

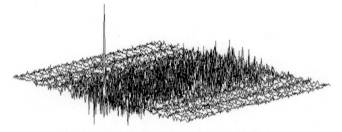

Figure 5.11. The output of symmetric phase-only matched filtering on the Fourier-Mellin transforms of the images in Fig. 5.1 and Fig. 5.10. The output dimension is 256 × 256. The peak is located at (124, 243), corresponding to a rotation angle of −99.84° and a scale change of 89.7%. Defining the peak as the output signal and the mean output energy excluding the peak area as the output noise, the output power SNR = 361.4. (Source: Chen *et al.*, 1994, © IEEE, reprinted with permission.)

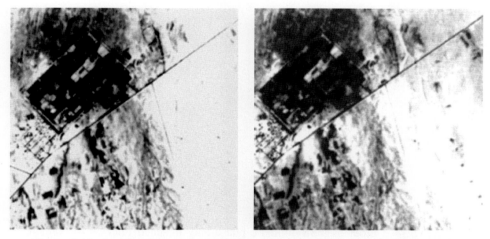

Figure 5.12. A pair of remote sensing images acquired by the SPOT-1 satellite on February 24, 1986 and February 25, 1986, respectively. (Source: Chen *et al.*, 1994, © IEEE, reprinted with permission.)

Fourier-Mellin transform of an image is a continuous function in the angular coordinate, but the magnitude in the radial coordinate can decrease tenfold or more from the origin to the periphery, as shown in Fig. 5.8. This kind of discontinuity between the two ends of the radial coordinate can introduce severe artifacts in the Fourier transform. It is imperative to apply a one-dimensional window function in the radial direction prior to the computation of the Fourier transform in Step 4. Figure 5.9 shows the Fourier-Mellin transform after a one-dimensional Hanning windowing in the radial direction (Chen *et al.*, 1994). The window function is applied before the logarithmic scale conversion. Figure 5.10 shows a rotated and

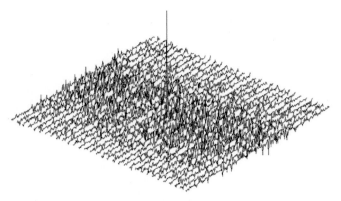

Figure 5.13. Output of the symmetric phase-only matched filter on the Fourier-Mellin transforms of the remote sensing images in Fig. 5.11. The peak corresponds to a rotation of 7° and a scaling rate of 100%. (Source: Chen *et al.*, 1994, © IEEE, reprinted with permission.)

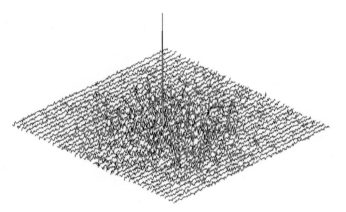

Figure 5.14. Output of the symmetric phase-only matched filter on the remote sensing images after rotation correction. The peak shifts away from the origin by (12, −1) pixels. The corresponding ground displacement is 240 m horizontally.

scaled version of the leaf image with additive white noise. Figure 5.11 is the output of the symmetric phase-only matched filter on the Fourier-Mellin transforms of the images in Figs. 5.1 and 5.10. The peak location corresponds to the image rotation and scaling. Figure 5.12 displays a pair of remote sensing images. Figure 5.13 demonstrates the detection of the rotation and scaling changes and Fig. 5.14 the detection of the image translation. In addition to image registration, this technique can also be used in other image analysis tasks, such as kinoform computation (Gallagher and Liu, 1973), image identification (Chen, 1993; Chen *et al.*, 1994), optical object detection (Horver and Gianino, 1984), and motion tracking (Noyer *et al.*, 2004; Keller and Averbuch, 2007).

References

Altmann, J. and Reitbock, H. J. O. (1984). A fast correlation method for scale-, and translation-invariant pattern recognition. *IEEE Transactions on Pattern Analysis and Machine Intelligence*, **6**, 46–57.

Cavanagh, P. (1981). Size and position invariance in the visual system. *Perception*, **10**, 491–499.

Chen, Q. S. (1993). Image Registration and Its Applications in Medical Imaging. Ph.D. thesis, Vrije Universiteit Brussel, Brussels, Belgium.

Chen, Q. S., Crownover, R., and Weinhous, M. S. (1999). Subunity coordinate translation with Fourier transform to achieve efficient and quality three-dimensional medical image interpolation. *Medical Physics*, **26**, 1776–1782.

Chen, Q. S., Defrise, M., and Deconinck, F. (1994). Symmetric phase-only matched filtering of Fourier-Mellin transforms for image registration and recognition. *IEEE Transactions on Pattern Analysis and Machine Intelligence*, **16**, 1156–1168.

De Castro, E. and Morandi, C. (1987). Registration of translated and rotated images using finite Fourier transforms. *IEEE Transactions on Pattern Analysis and Machine Intelligence*, **9**, 700–703.

De Castro, E., Cristini, G., Martelli, A., Morandi, C., and Vascotto, M. (1989). Compensation of random eye motion in television ophthalmoscopy: Preliminary results. *IEEE Transactions on Medical Imaging*, **6**, 74–81.

Ersoy, O. K. and Zeng, M. (1989). Nonlinear matched filtering. *Journal of Optical Society of America A,* **6**, 636–648.

Flannery, D. L., Loomis, J. S., and Milkovich, M. E. (1988). Transform-ratio ternary phase-amplitude filter formulation for improved correlation discrimination. *Applied Optics,* **27**, 4079–4083.

Gallagher, N. C. and Liu, B. (1973). Method for computing kinoforms that reduces image reconstruction error. *Applied Optics*, **12**, 2328–2335.

Horver, J. H. and Gianino, P. D. (1984). Phase-only matched filtering. *Applied Optics*, **23**, 812–816.

Horver, J. H. and Gianino, P. D. (1985). Pattern recognition with binary phase-only filters. *Applied Optics*, **24**, 609–611.

Keller, Y. and Averbuch, A. (2007). A project-based extension to phase correlation image alignment. *Signal Processing*, **87**, 124–133.

Kermisch, D. (1970). Image reconstruction from phase information only. *Journal of Optical Society of America A*, **60**, 15–17.

Kottzer, T., Rosen, J., and Shamir, J. (1992). Phase extraction pattern recognition. *Applied Optics*, **31**, 1126–1137.

Kumar, B. V. K. and Pochapsky, E. (1986). Signal-to-noise ratio considerations in modified matched spatial filtering. *Journal of Optical Society of America A*, **3**, 777–786.

Lam, K. P. (1985). Contour map registration using Fourier descriptions of gradient codes. *IEEE Transactions on Pattern Analysis and Machine Intelligence*, **7**, 332–338.

Noyer, J.–C., Lanvin, P., and Benjelloun, M. (2004). Non-linear matched filtering for object detection and tracking. *Pattern Recognition Letters*, **25**, 655–668.

Oppenheim, A. V. and Lim, J. S. (1981). The importance of phase in signals. *Proceedings of the IEEE*, **69**, 529–541.

Pratt, W. K. (1978). *Digital Image Processing*. New York: John Wiley & Sons, pp. 526–566.

Reddy, B. S. and Chatterji, B. N. (1996). An FFT-based technique for translation, rotation and scale-invariant image registration. *IEEE Transactions on Image Processing*, **5**, 1266–1271.

Schalkoff, R. J. (1989). *Digital Image Processing and Computer Vision*. New York: John Wiley & Sons, pp. 279–286.

Schwartz, E. L. (1980). Computational anatomy and functional structure of striate cortex: A spatial mapping approach to perceptual coding. *Vision Research,* **20**, 645–669.

Schwartz, E. L. (1981). Cortical anatomy and spatial frequency analysis. *Perception*, **10**, 445–468.

Segman, J. (1992). Fourier cross correlation and invariant transformations for an optimal recognition of functions deformed by affine groups. *Journal of Optical Society of America A*, **9**, 895–902.

Sheng, Z. Y. and Arsenault, H. H. (1986). Experiments on pattern recognition using invariant Fourier-Mellin descriptors. *Journal of Optical Society of America A*, **3**, 771–776.

Turin, G. L. (1960). An introduction to matched filters. *IRE Transactions on Information Theory*, **6**, 311–29.

Vanderluht, A. (1969). Signal detection by complex spatial filtering. *IEEE Transactions on Information Theory,* **10**, 130–145.

Wang, Z. Q., Gillespie, W. A., Cartwright, C. M., and Soutar, C. (1993). Optical pattern recognition using a synthetic discriminant amplitude-compensated matched filter. *Applied Optics*, **32**, 184–189.

Whalen, A. D. (1971). *Detection of Signal in Noise*. New York: Academic Press, pp. 167–178.

Wu, Q. X., Bones, P. J., and Bates, R. H. T. (1989). Translation motion compensation for coronary angiogram sequences. *IEEE Transactions on Medical Imaging*, **8**, 276–282.

6

Image registration using mutual information

ARLENE A. COLE-RHODES AND PRAMOD K. VARSHNEY

Abstract

This chapter provides an overview of the use of *mutual information* (MI) as a similarity measure for the registration of multisensor remote sensing images. MI has been known for some time to be effective for the registration of monomodal, as well as multimodal images in medical applications. However, its use in remote sensing applications has only been explored more recently. Like correlation, MI-based registration is an area-based method. It does not require any preprocessing, which allows the registration to be fully automated. The MI approach is based on principles of *information theory*. Specifically, it provides a measure of the amount of information that one variable contains about the other. In registration, we are concerned with maximizing the dependency of a pair of images. In this context, we discuss the computation of mutual information and various key issues concerning its evaluation and implementation. These issues include the estimation of the probability density function and computation of the joint histogram, normalization of MI, and use of different types of interpolation, search and optimization techniques for finding the parameters of the registration transformation (including multiresolution approaches).

6.1 Introduction

In this chapter, we discuss the application of *mutual information* (MI) as a similarity metric for cross-registering images produced by different imaging sensors. These images may be from different sources taken at different times and may be produced at different spectral frequencies and/or at different spatial resolutions. The use of mutual information as a similarity measure for registration was developed independently in 1995 by Viola and Wells (1995) and Collignon *et al.* (1995a). It is based on principles of information theory (Cover and Thomas, 1991), and it has been widely applied to the registration of medical images (Maes *et al.*, 2003; Pluim *et al.*, 2003). More recently, mutual information has been applied to the registration of remotely sensed images (Le Moigne *et al.*, 2002; Cole-Rhodes *et al.*, 2003; Chen *et al.*, 2003a; Chen *et al.*, 2003b; Chen and Varshney, 2004).

Like correlation, MI is an area-based method. It uses the raw intensities present in the images without assuming any specific relationship between the given images (Brown, 1992; Zitová and Flusser, 2003). No preprocessing of the image is required, which allows the registration process to be fully automated. MI has been used extensively in the registration of 3D volumetric medical images, and it is currently the leading technique for the registration of both monomodal and multimodal images. It has been found to be robust, accurate, and especially effective in multimodal registration where many other, more common similarity measures have failed. In medical applications, it may be necessary to register anatomical images, such as those produced by Magnetic Resonance Imaging (MRI) or Computed Tomography (CT) scan, to a complementary functional image of a patient, such as from a Positron Emission Tomography (PET) or functional MRI (fMRI). For remote sensing applications, complementary data due to Synthetic-Aperture Radar (SAR), or multispectral/hyperspectral data acquired by sensors placed on airborne/spaceborne platforms, must be registered. In general, the correlation similarity metric provides reliable registration when the relationship between the intensities of the two images to be matched is linear, but mutual information is theoretically more robust to complex variations between the intensities of the image pairs, such as those that can occur between pairs of multimodal images (Collignon *et al.*, 1995b). We start in Section 6.2 with a definition of mutual information from information theory, and show how it is adapted for the problem of image registration.

Some implementation issues which influence the accuracy of the registration results include the estimation of the probability density functions (PDFs), interpolation methods for the PDFs and/or the images, and optimization methods for the metric. Also, the use of a multiresolution approach to speed up the computation needs to be considered. The effect of the choices made with regards to the above issues on the performance (as far as speed and accuracy/precision) of the mutual information-based registration scheme has been studied by Zhu and Cochoff (2002) and will be further discussed in this chapter. Specifically, we discuss implementation details which affect the accuracy of the estimated probability density functions, such as the approach to histogram binning and the total number of bins used in the histogram. The smoothness (or ruggedness) of the MI surface affects our success in determining the correct registration parameters. In particular, the presence of concave artifacts may introduce local maxima which could trap an optimization scheme, or cause the correct registration parameters to appear as a local, rather than a global optimum. Such artifacts may be caused by the interpolation scheme used to generate the MI surface. Many researchers have proposed methods for reducing these artifacts (Pluim *et al.*, 2000; Chen and Varshney, 2003). In Pluim *et al.* (2000), the authors suggest that partial volume interpolation can lead to dispersion of the joint histogram producing concave artifacts, which may prohibit subpixel accuracy. A discussion on these implementation details is provided in Sections 6.3 to 6.5.

The global optimum of the objective function can be found by performing an exhaustive search over the space of allowable transformations, but this method is computationally expensive and very inefficient. Search strategies, which have been proposed for optimizing MI include Powell's method (Powell, 1964), the simplex method (Nelder and Mead, 1965), and gradient-based optimization (see Chapter 12 of this book), with some heuristics to determine the global optimum from local optima. Others attempt to speed up the search procedure by the use of a multiresolution strategy, which is based on a pyramid structure. Researchers have also proposed sampling approaches to estimating the relevant PDFs for MI. All of the above have been successfully applied for automating the registration process using MI. Optimization algorithms are discussed in Section 6.6, followed by some concluding remarks in Section 6.7.

6.2 MI as a similarity measure for registration

Mutual information is based on principles of information theory and has been recently proposed as a similarity measure for image registration applications. It is more robust than other similarity measures, such as correlation, as it does not assume any particular relationship between the intensity values of the two images to be registered.

For two variables A and B, the mutual information $I(A, B)$ can be defined as

$$
\begin{aligned}
I(A, B) &= H(B) - H(B|A) \\
&= H(A) + H(B) - H(A, B) \\
&= H(A) - H(A|B),
\end{aligned}
\tag{6.1}
$$

where $H(A)$ and $H(B)$ are the entropies of A and B, respectively, $H(A, B)$ denotes the *joint entropy* of A and B, and $H(B|A) = H(A, B) - H(A)$ is the *conditional entropy* of B given A. In the context of image registration, we consider A and B to be random variables representing the gray values (intensities) in the two images to be registered (i.e., the *reference and sensed* images), whose gray level PDFs are given by $p_A(a)$ and $p_B(b)$, respectively.

MI can be equivalently defined in terms of the Shannon entropy, used in the definition of MI in Eq. (6.1) and given by

$$
H(A) = -E_A[\log p_A(a)],
$$

where $E_A[.]$ represents the expected value with respect to random variable A. More specifically, Shannon's entropy is defined as

$$H(A) = \sum_a p_A(a) \cdot \log \frac{1}{p_A(a)} = - \sum_a p_A(a) \cdot \log p_A(a). \qquad (6.2)$$

It is a measure of the uncertainty of a certain event. If we consider that $\log(1/p_A(a))$ specifies the amount of information gained from an event $a \in A$ with probability $p_A(a)$ (i.e., the more rare the event, the more information that is gained), then entropy can be defined by weighting this information by the probability of occurrence and summing it over all events. Thus, entropy can be interpreted as the average amount of information to be gained from the specified set of events. When all events have the same probability of occurrence, the entropy reaches its maximum, as we are uncertain of which particular event will occur. An event with a higher chance of occurrence provides less information and thus lower uncertainty, as we have already expected it to occur. On the other hand, the uncertainty (or entropy) of a rare event will be above average. Thus, given a random variable A with PDF $p_A(a)$, Shannon's entropy provides a measure of the dispersion of the probability density function. In other words, a sharp-peaked PDF would have a relatively low entropy, while that of a dispersed PDF would be relatively high.

Thus, the mutual information $I(A, B)$ between two random variables A and B can be defined as

$$I(A, B) = \sum_a \sum_b p_{A,B}(a, b) \log \frac{p_{A,B}(a, b)}{p_A(a) p_B(b)}. \qquad (6.3)$$

It measures the Kullback-Leibler distance (Cover and Thomas, 1991) between the joint PDF $p_{A,B}(a, b)$ of two random variables A and B and the product of their marginal PDFs, $p_A(a) \cdot p_B(b)$. Thus, it measures the distance of the joint PDF of the random variables A and B from that associated with complete independence of A and B, implying that mutual information is zero if and only if the two random variables are independent. Thus we state that MI is a measure of the statistical dependence between two datasets indicating the amount of information that random variable A contains about random variable B, and vice versa.

Mutual information (MI) was proposed as a registration measure for image registration concurrently by Viola and Wells (1995), Collignon *et al.* (1995a), and Studholme *et al.* (1995). It has been found to be very successful and it is now the most widely studied entropic registration measure. MI provides a measure of the strength of dependence between the intensities of two images, and the registration criterion is based on the premise that mutual information of the image intensity pairs is maximized when the two images are geometrically aligned. The maximization of mutual information (MMI) criterion is applied based on the assumption that there is an underlying statistical relationship between image pairs, each of which is a function of the imaging device used on an identical scene. General multisensor images have intensity maps that are unique to the imaging devices used to acquire them. Other information theoretic measures, such as *Jumarie entropy* (Rodriguez-Carranza and Loew, 1998) and *Jensen-Rényi divergence* (JRD) (Hamza *et al.*, 2001; He *et al.*, 2003), have also been used as similarity metrics for registration. The latter JRD measure is a more general form of MI; under certain conditions it reduces to MI, but in general, it has been shown to be more robust for image registration (see Hamza *et al.* (2001) and Knops *et al.* (2003) for further details). At the center of the calculation of mutual information and other similar information-theoretic similarity metrics is the determination of the joint histogram of a pair of images. Since this method is widely used for estimating the required probability density functions, it is described next.

6.3 Joint histogram and PDF estimation

Mutual information has been defined as a function of the PDFs of the gray values (intensities) in the two images, which are to be registered. These probability density functions can be computed by determining the joint histogram of the two images. The accuracy of these estimates depends on the availability of enough pixels to occupy each histogram bin, so that the statistical description obtained from the samples is accurate. Factors which can affect this accuracy are the number of bins used in the histogram and the method of binning, for example, simple binning (Maes *et al.*, 1997; Maes *et al.*, 2003) or binning based on the use of Parzen windows (Thévenaz *et al.*, 1998; Thévenaz and Unser, 2000).

Mutual information of two images A and B can be computed by estimating the corresponding probability density functions by normalizing each of the histograms, $h_A(a)$ and $h_B(b)$, respectively, as well as their joint histogram $h_{A,B}(a, b)$. Thus, it is expected that the dispersion of the joint histogram will be minimized by a registration process based on maximization of MI. When a simple binning procedure is used, MI is defined as (Collignon *et al.*, 1995a):

$$I(A, B) = \frac{1}{M} \sum_a \sum_b h_{A,B}(a, b) \log \frac{M h_{A,B}(a, b)}{h_A(a) h_B(b)}, \tag{6.4}$$

where M is the sum of all the entries in the histogram, which is the total number of pixels in the overlapping area of the two images.

Alternatively, the probability density functions can be estimated using Parzen windows as follows. Let $W(.)$ be a function with a unit integral, and let $\{a_i\}$ be a subset of N samples from the set A with probability density function $p_A(a)$, then the Parzen estimate of the PDF $p(.)$ is given by

$$p_N(a) = \frac{1}{N} \sum_{i=1}^{N} W(a - a_i). \qquad (6.5)$$

Parzen windows were used by Viola and Wells (1995) to estimate the image intensity PDFs required in the computation of MI. Using a Gaussian window function, the Parzen estimate of the required joint and marginal probability density functions is computed on a small number of samples drawn from the overlapping part of the images. A second set of samples is then used to estimate the entropies defined in the mutual information expression of Eq. (6.3). This stochastic sampling increases the computational speed of the algorithm, but may reduce the registration robustness due to the use of a limited number of samples, which may introduce local maxima (Unser *et al.*, 1993). Parzen windows were also used to estimate joint probability functions by Thévenaz and Unser (2000), although these were based on B-splines. (Cubic spline interpolation was also applied to the image transformation in that work.)

In practice, simple binning that involves uniform or equidistant binning is often employed. It was observed, though, by Knops *et al.* (2003) that uniform binning disregards features in remote sensing images and anatomical structures in medical imagery. A coherent structure may end up in two or more regions, which is not desirable. In contrast, it was observed that nonuniform binning, that is, one with a variable bin size, can achieve natural clustering and also less dispersion in the joint histogram, yielding thereby better performance of image registration algorithms. Successful approaches like *K*-means clustering (Knops *et al.*, 2003) and histogram preservation binning (Camp and Robb, 1999) have been used to achieve this natural clustering via variable bin size clustering. These methods have resulted in significant reduction of noise and spurious local maxima without causing significant displacement or smoothing of global maxima.

6.4 Normalized MI

The MI-based registration criterion assumes that the required probability densities can be reliably estimated. Mutual information is determined from a joint histogram, which is computed using the overlapping areas of the images only. The computational load thus varies with the size of the overlap between the two images, and hence MI is not normalized with respect to this overlap size. The sensitivity of the mutual information measure to the size of the overlap of the two images is a limitation of this metric. It can be observed, based on its definition in terms of the joint histogram, that the size of the image overlap (which can vary) is directly related to the sample size, M, used to compute the histogram. If the sample size is not large enough, the reliability of the estimate using the PDF may be compromised due to the use of too few samples. But less overlap, which implies fewer samples, could yield a relatively high MI value, even as the statistical power of the joint histogram becomes unreliable. Indeed, as the images are moved away from the optimal registration point, MI could still increase if the increase in marginal entropies of the images exceeds the decrease in the joint entropy between them.

Different solutions have been proposed to overcome this problem, such as normalizing MI using a weight based on the size of the region of overlap (Rodriguez-Carranza and Loew, 1998), and modifying the MI measure in a manner that makes it invariant to the overlap size (Studholme *et al.*, 1999). The *normalized mutual information* measure proposed by the latter is

$$\overline{H}(A, B) = \frac{H(A) + H(B)}{H(A, B)}. \tag{6.6}$$

The *entropy correlation coefficient* defined by

$$\overline{E}(A, B) = 2\frac{I(A, B)}{H(A) + H(B)} \tag{6.7}$$

is an alternative measure, which was proposed by Astola and Virtanen (1982) and experimented with by Maes *et al.* (1997). This measure is considered a normalized form of MI in the range [0, 1], where 0 corresponds to full independence and 1 corresponds to complete dependence between the variables A and B. The above two measures are related by $\overline{E}(A, B) = 2(1 - 1/\overline{H}(A, B))$. In Studholme *et al.* (1999), it is stated that normalized MI, $\overline{H}(A, B)$, outperforms its non-normalized version, $H(A, B)$, when the overlap region between the two images is small.

6.5 Interpolation methods and MI surface artifacts

For remote sensing images, the domain of transformations over which the similarity metric is to be optimized is two-dimensional (2D), and is referred to as the parameter space. This may range from a rigid body transformation with 3 parameters (or 4, if isometric scaling is included) to an affine transformation with 6 parameters (which includes stretching and shearing) to curved or elastic transformations (which locally deforms one image into another). The transformation to be estimated is continuous in nature, and we would like to produce registration results which are of *subpixel* accuracy. Depending on various factors, the MI function may appear as a rugged surface with many local maxima, which makes the optimization problem difficult. When a geometric transformation is applied to one image to spatially align it with another, interpolation is applied to estimate the gray values of the transformed image at nongrid points. Thus during registration, image interpolation may be used to oversample the transformed image and estimate gray values at all locations that lie on grid points of the fixed image, when the grids of the two images do not line up exactly (Pluim *et al.*, 2000). The most common method of interpolation is linear interpolation, but other methods, for example, nearest neighbor (Cole-Rhodes *et al.*, 2002), trilinear (Maes *et al.*, 2003), and cubic spline (Thévenaz *et al.*, 1998) interpolation, may also be used. The effect of interpolation and the number of histogram bins on the MI similarity metric was investigated by Pluim *et al.* (2000).

A smooth MI function is generally produced if the two images to be registered have different spatial resolutions, for example, the images were acquired by different sensors. For multitemporal registration, on the other hand, the images often have the same spatial resolution, so interpolation-induced artifacts are likely to be present. Depending on the type of interpolation used, a pattern of artifacts can be introduced in the mutual information function, which can affect the optimization process and hence the accuracy of the final result. Indeed, interpolation-induced artifacts have been found to occur in MI-based registration when both images have equal sample spacing in at least one dimension.

One way of reducing these artifacts is by resampling one of the two images, such that the ratio of the two sample spacings along a certain dimension is irrational, as was suggested by Tsao (2003). Resampling of the image was shown to improve the smoothness of the MI surface, since it establishes a difference in the pixel sizes, that is, grid alignment cannot occur. However, the registration accuracy might decrease due to extra rounding-off operations during the resampling procedure. A resampling approach dependent on image size was proposed by Pluim *et al.* (2000), but even with the resulting smoothing of the MI surface, it was inconclusive whether the improved registration accuracy was due to this approach. It was shown by Cole-Rhodes *et al.* (2003) that the use of cubic spline interpolation to geometrically transform the sensed image can produce a smooth MI surface, even when the histogram is generated using simple binning.

An interpolation method for the joint histogram, which was proposed by Collignon *et al.* (1995a), is the *partial volume interpolation* (PVI). This method is not used to create an interpolated image, but was designed to update the joint histogram of the two images. This method might also produce interpolation artifacts as noted by Pluim *et al.* (2000). When grid points are not aligned with the transformed image, the method updates multiple entries of the joint histogram in the neighborhood of a transformed point using assigned weights; otherwise, a single histogram entry is updated. PVI introduces artifacts when the images have identical pixel grids, or grids that are multiples of each other. The partial volume weights become zero if all the grids align and nonzero as the grids are misaligned. Thus, local maxima of the objective function occur at such grid aligning positions. The impact of artifacts caused by trilinear interpolation and partial volume interpolation on MI-based registration accuracy was investigated by Pluim *et al.* (2000). Partial intensity interpolation for histogram updating was applied by Maes *et al.* (1997). This method computes a weighted average of the neighboring gray values and then updates multiple entries of the joint histogram.

The effect of the different types of artifacts introduced by the different methods of interpolation were characterized and investigated by Tsao (2003). Various approaches have been suggested in order to mitigate or reduce the problems caused by artifacts. In Chen and Varshney (2003), a *generalized partial volume joint histogram estimation* (GPVE) method was proposed. This method is a generalization of PVI in that it uses a kernel of higher order, rather than the linear kernel used for PVI. This has been shown to reduce the interpolation artifacts associated with PVI (see Fig. 6.1). To speed up the computation, the GPVE method can be applied selectively to alignment parameters, such as translation or rotation, that are likely to result in artifacts (which can compromise the subpixel accuracy of the registration). Some comparison results are available in Chen *et al.* (2003c).

Since the registration algorithms must perform an optimization task, it is desirable to have a smooth and differentiable MI function. In addressing the differentiability of the MI function and the computation of the underlying joint histogram $h_{A,B}(a, b)$, which is used to estimate the joint probability density function, Parzen windowing is applied for the interpolation of the transformed image and/or the joint histogram. Note that the partial volume approach (both PVI and GPVE) for building a joint histogram avoids the determination of an explicit transformed image by filling in entries of the joint histogram based on a weighted sum of the intensities of the surrounding pixels.

6.6 Search strategy and optimization of mutual information

6.6.1 Brief overview

MI is a multivariable function, which needs to be optimized over a set of transformation parameters that defines the search space for the optimization scheme. A number of search strategies have been proposed for finding the maximum of the multivariable MI function. For 2D images, the search space may be rigid (3 or 4 parameters if scale is included), affine (6 parameters), or elastic (using, for example, polynomial interpolation, interpolations based on thin-plate splines or B-splines, etc., through a specified set of control points). Different optimization schemes have been used by different researchers. These include Powell's method, the simplex method, a steepest ascent scheme using the gradient (which may be calculated or approximated), and the Levenberg-Marquardt algorithm (Levenberg, 1944; Marquardt, 1963), which uses a second-order gradient.

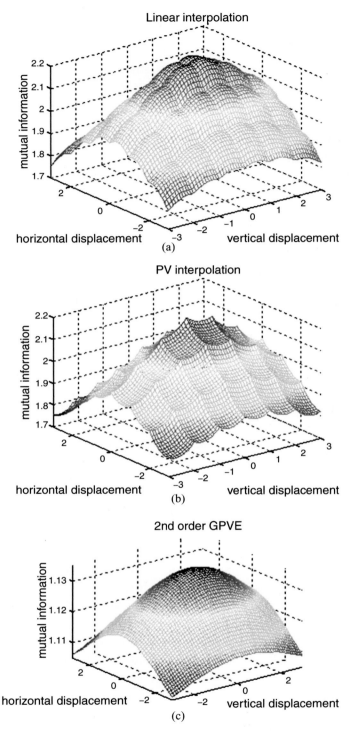

Figure 6.1. (a–c) Typical artifacts resulting from linear interpolation, partial volume interpolation and generalized partial volume estimation. See color plates section.

Powell's multidimensional search method of optimizing MI was used by Maes *et al.* (1997). For an *n*-dimensional search space, Powell's method performs a one-dimensional optimization along each of the *n* dimensions in turn. Namely, the objective function is optimized, at each iteration, along each of the *n* standard base vectors, generating a sequence of *n* points. The orientation of the vector from the starting point to the end point of the sequence is considered the average direction of movement during the iteration in question. Brent's method was used to perform the required one-dimensional optimizations. It does not require the calculation of function derivatives but is sensitive to local optima in the function. The simplex method, too, does not require computing derivatives; it optimizes over all dimensions simultaneously but has slow convergence. Steepest ascent (or hill climbing) algorithms have also been used with good success.

In Thévenaz and Unser (2000) the Levenberg-Marquardt technique was used to optimize MI, applying it to medical imagery. The required derivatives are explicitly calculated based on a partial volume interpolation of the MI criterion, and the implementation is carried out in a multiresolution framework. An optimizer was designed specifically for the MI criterion, which requires the function to be differentiable, so that the gradient and the Hessian matrix can be computed. Differentiability is achieved by the use of Parzen windows. On the other hand, the technique presented in the following section for optimizing MI, which is a multi-resolution scheme based on wavelet decomposition, does not require explicit derivation of the required gradient vector. This is a stochastic gradient technique (Cole-Rhodes *et al.*, 2003) that uses, instead, an approximation to the gradient. We describe below how this technique was applied to image registration.

6.6.2 Stochastic gradient optimization for image registration

The optimization technique described here is the *simultaneous perturbation stochastic approximation* (SPSA) algorithm, which was first introduced by Spall (1992). The technique has recently attracted considerable attention for solving challenging optimization problems, for which it is difficult to obtain directly the gradient of the objective function with respect to the parameters being optimized. SPSA is based on an easily implemented and highly efficient gradient approximation that relies only on measurements of the objective function. It does not rely on explicit knowledge of the gradient of the objective function, or on measurements of this gradient.

Let L be the objective function to be optimized (the MI similarity measure in this case). We consider a parameter search space of two-dimensional rigid transformations, consisting of rotation, translation in the x and y directions, and isometric scaling. These four parameters at iteration k are represented by the vector

$p_k = [t_x\ t_y\ \theta\ s]^T$. At each iteration, the gradient approximation is based only on two function measurements (regardless of the dimension of the parameter space). An additional measurement is made at each newly computed point, in order to decide whether to accept this new update or discard it and generate an alternative random point. Namely, at iteration k, the parameters are updated according to the steepest ascent rule

$$\mathbf{p}_{k+1} = \mathbf{p}_k + a_k \mathbf{g}_k,$$

where the gradient vector $\mathbf{g}_k = (g_k^{(1)} g_k^{(2)} \ldots g_k^{(m)})^T$ for the m-dimensional parameter space is determined by

$$g_k^{(i)} = \left[L(\mathbf{p}_k + c_k \mathbf{g}_k) - L(\mathbf{p}_k - c_k \mathbf{g}_k)\right] / \left(2c_k \Delta_k^{(i)}\right), \quad i = 1, 2, \ldots m.$$

Each element $\Delta_k^{(i)}$ of the perturbation vector $\mathbf{\Delta}_k$ takes on a value of $+1$ or -1, as generated by a Bernoulli distribution, and a_k and c_k are positive sequences of the form

$$a_{k+1} = a/(k + A + 1)^\alpha,$$
$$c_k = c/(k + 1)^\gamma,$$

with $0 < \gamma < \alpha < 1$, and some constants a, c, and A. (Note that a_k and c_k decrease to zero, as $k \to \infty$.) Further details can be found in Cole-Rhodes *et al.* (2003), in which three parameters were updated at each iteration (i.e., $m = 3$).

The SPSA technique is very powerful. Due to the stochastic nature of the gradient approximation, the method can overcome local maxima of the objective function to reach the global maximum. All the components of $\mathbf{p}_k$ are randomly perturbed to obtain two measurements of $L(.)$. Each component of the gradient vector is then formed using the ratio of the difference in these measurements and the individual components in the perturbation vector. The algorithm starts from an initial guess $\mathbf{p}_0$, and computes the next estimate of the optimal parameters at each iteration using the above described gradient approximation. In Spall (1992), sufficient conditions for the convergence of the iterative SPSA process in the stochastic almost sure sense were presented, using a differential equation approach that is well known in general stochastic approximation theory. Convergence was established by requiring $L(.)$ to be sufficiently smooth (i.e., three times continuously differentiable) near the optimum, and imposing the following conditions on the gain sequences a_k and c_k to ensure that they approach zero at rates that are neither too fast nor too slow, i.e.,

$$a_k, c_k > 0\ \forall k; \quad a_k \to 0, c_k \to 0 \quad \text{as} \quad k \to \infty;$$
$$\sum_k a_k = \infty, \sum_k (a_k/c_k)^2 < \infty.$$

(a) 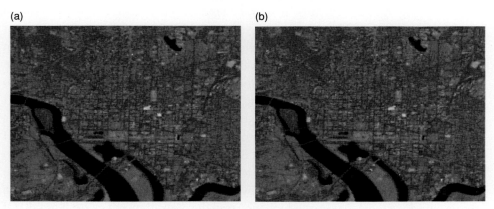 (b)

Figure 6.2. A pair of Landsat test images over Washington, DC: (a) Band 4 of a 1997 Landsat-5 image, and (b) band 4 of a 1999 Landsat-7 image.

(a) 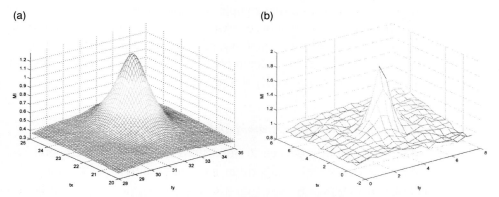 (b)

Figure 6.3. Mutual information surfaces obtained due to B-spline interpolation: (a) Subpixel MI surface for image pair of Fig. 6.2 (original images), and (b) MI surface for coarse, level 4 images (derived from image pair of Fig. 6.2). See in color plates section. (Source: Cole-Rhodes *et al.*, 2003, © IEEE, reprinted with permission.)

The components of the perturbation vector $\mathbf{\Delta}_k$ are required to be independent and distributed symmetrically about 0 with finite inverse moments $E[1/|\Delta_k^{(i)}|]$ for all k, i. These conditions on $\mathbf{\Delta}_k$ make the gradient approximation $\mathbf{g}_k(.)$ an almost unbiased estimator of the true gradient $\mathbf{g}(.)$, i.e., $E[\mathbf{g}_k(\mathbf{p}_k)] = \mathbf{g}(\mathbf{p}_k) + O(c_k^2)$. For small c_k, these misdirections act like random errors which average and cancel out over a number of iterations.

Figure 6.2 shows a pair of Landsat images over the Washington, DC, area, which were used as test images for our experiment. Figures 6.2(a)–(b) show images extracted from band 4 of a 1997 Landsat-5 image, and band 4 of a 1999 Landsat-7 image, respectively. Figure 6.3(a) shows the MI surface obtained for the original images of Fig. 6.2. Specifically, it shows a smooth MI surface that was produced by using a B-spline interpolation (Unser *et al.*, 1993) in the transformation of the reference image of Fig. 6.2(a). An important consideration for the application of the optimization scheme is that the further away the initial guess is from the global maximum, the more local maxima the algorithm may need to overcome to reach the global maximum, and thus the more likely it is to fail. Note that the MI surface is less smooth for coarser images, which correspond to the deeper levels of the wavelet decomposition. For example, Fig. 6.3(b) illustrates the rugged MI surface obtained for the level 4 wavelet images due to Simoncelli's decomposition (Simoncelli *et al.*, 1992) of the Landsat image pair of Fig. 6.2. Note the ripple appearance on the MI surface as we move away from the global maximum. These ripples produce local maxima, which may trap the algorithm and cause it to fail. Of course, failure at any of the coarser levels can be catastrophic to the optimization algorithm. Thus, to ensure significant smoothing of the MI surface at the coarsest decomposition level we may use, for example, a histogram with 64 bins rather than 256 bins. It is important to note that in searching for the maximum MI value (in the case of a more general affine transformation), rotation and scale could also vary, so although a definite peak in the MI value probably exists in this three-dimensional parameter space, it is not easy to verify this graphically.

6.6.3 Multiresolution pyramid and wavelet-based schemes

A multiresolution pyramid approach is often used to reduce the computational cost of an MI-based registration algorithm and the sensitivity of the algorithm to local maxima. It is typically based on a coarse-to-fine hierarchical structure, where images at the coarsest resolution level are matched first, using relatively large features. The match is subsequently refined as one climbs up the pyramid, using higher-resolution images and finer details at each stage. In other words, the search for an optimal match is performed initially on the lowest-resolution level of the pyramid, and the optimization result at each level is passed up to the next resolution level of the pyramid. Multiresolution approaches provide a more reliable local model of the similarity measure, which can be helpful in focusing on the optimum (at different resolutions) in cases where the objective function has multiple local maxima corresponding to incorrect spatial alignments. Figure 6.4 depicts a typical block diagram of a pyramid-based, wavelet registration scheme.

Pyramid-like schemes can be obtained by subsampling or filtering the images using, for example, simple averaging, a Gaussian pyramid, a spline-based pyramid (Thévenaz and Unser, 2000), or a wavelet decomposition. The multiresolution nature of the wavelet decomposition and the ability of the wavelet basis to capture spatial and spectral information in a hierarchical setting makes it an attractive choice

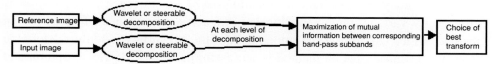

Figure 6.4. Wavelet-based mutual information image registration scheme. (Source: Cole-Rhodes *et al.*, 2003, © IEEE, reprinted with permission.)

for use in a pyramid-like structure. Wavelets have been used in multisensor image registration to bring the multisource data to the same resolution while preserving the important features of the data. The use of this iterative filtering can be crucial in dealing with the differing intensities which occur in multisensor image pairs. Many different types of wavelets have been applied in a multiresolution approach ranging from Daubechies (Chen and Varshney, 2000; Le Moigne and Zavorin, 2000; Le Moigne *et al.*, 2002; Zavorin *et al.*, 2002) to Simoncelli (Johnson *et al.*, 2001; Le Moigne *et al.*, 2001; Cole-Rhodes *et al.*, 2003). In multidimensional wavelet analysis, local feature extraction is done using a wavelet basis projection. Such projections allow the use of different sets of features for matching.

6.7 Conclusions

Mutual information is a similarity measure which is based on information theory, and it depends only on the intensity values within an image. It has been shown to be a powerful and robust similarity measure for a wide variety of registration applications, and it tends to outperform other intensity-based similarity measures, such as correlation and mean square difference for multimodal registration (Collignon *et al.*, 1995a; Zavorin *et al.*, 2002; Zhu and Cochoff, 2002). MI has been widely used for the registration of medical images, but its application to the registration of remote sensing images is more recent (see, for example, Chen and Varshney, 2000; Johnson *et al.*, 2001; Cole-Rhodes *et al.*, 2003). Johnson *et al.* (2001) presented some of the first results comparing correlation to MI. They showed that MI produced consistently sharper peaks at the correct registration than correlation, which allows for better registration accuracy. Also MI was shown to be more robust for low signal to (Gaussian) noise ratios. In addition, MI outperforms other measures even in cases of contrast reversal where the gray levels of corresponding pixels (in the images to be registered) differ significantly. Such contrast reversal can occur in multispectral, remotely sensed images due to differences in material reflectance. MI draws on the assumption that regions of similar gray values in one image will correspond to regions of similar gray level values in the other image, even if those values are different.

The accuracy of a mutual information-based image registration scheme is affected by various implementation issues, such as the choice of interpolation method, which can impact the smoothness of the MI surface, and thus determine whether the chosen search method is successful in finding the global optimum at subpixel precision. The various approaches can be evaluated based on the final registration results, and the quality of any improvements can be assessed visually using 2D or 3D plots of the MI function for the case of a low-dimensional search space. The final registration result is often assessed by visual inspection using a mosaic. Thus, performance evaluation of the final registration parameters is an important task. One useful performance measure is the mean squared error, for cases where ground truth is available. In Kumar *et al.* (2006), a metric is proposed for evaluating the performance of competing components of registration based on the quality of the MI landscape/surface. This metric, called the fitness distance correlation, evaluates the correlation between the distance of each solution (transformation parameter set) from the true solution and the fitness value (MI) corresponding to that solution. Development of additional appropriate performance metrics is an open research issue. Also of interest are methods which will further decrease the computational cost of the registration process, in addition to the speedup gained by a multiresolution, pyramid-like scheme.

References

Astola, J. and Virtanen, I. (1982). Entropy correlation coefficient, a measure of statistical dependence for categorized data. In *Proceedings of the University of Vaasa, Discussion Papers*, Finland, No. 44.

Brown, L. (1992), A survey of image registration techniques, *ACM Computing Surveys*, **24**(4), 325–376.

Camp, J. and Robb, R. (1999). A novel binning method for improved accuracy and speed of volume image coregistration using normalized mutual information. In *Proceedings of the SPIE Conference on Medical Imaging: Image Processing*, San Diego, CA, Vol. 3661, pp. 24–31.

Chen, H. M. and Varshney, P. K. (2000). A pyramid approach for multimodality image registration based on mutual information. In *Proceedings of the Third International Conference on Information Fusion*, Paris, France, Vol. 1, pp. MOD3/9–MOD315.

Chen, H. M. and Varshney, P. K. (2003). Mutual information based CT-MR brain image registration using generalized partial volume joint histogram estimation. *IEEE Transactions on Medical Imaging*, **22**(9), 1111–1119.

Chen, H. M. and Varshney, P. K. (2004). MI based registration of multi-sensor and multi-temporal images. In P. K. Varshney and M. K. Arora, eds., *Advanced Image Processing Techniques for Remotely Sensed Hyperspectral Data*. New York: Springer-Verlag, pp. 181–198.

Chen, H. M., Varshney, P. K., and Arora, M. K. (2003a). Performance of mutual information similarity registration of multitemporal remote sensing images. *IEEE Transactions on Geoscience and Remote Sensing*, **41**(11), 2445–2454.

Chen, H., Varshney, P. K., and Arora, M. K. (2003b). Mutual information based image registration for remote sensing data. *International Journal of Remote Sensing*, **24**(18), 3701–3706.

Chen, H. M., Varshney, P. K., and Arora, M. K. (2003c). A study of joint histogram estimation methods to register multi-sensor remote sensing images using mutual information. In *Proceedings of the IEEE International Geoscience and Remote Sensing Symposium*, Toulouse, France, Vol. 6, pp. 4035–4037.

Cole-Rhodes, A., Johnson, K., and Le Moigne, J. (2002). Multiresolution registration of remote sensing images using stochastic gradient. In *Proceedings of the SPIE Conference on Wavelet and Independent Component Analysis Applications IX*, Orlando, FL, Vol. 4738, pp. 44–55.

Cole-Rhodes, A., Johnson, K., Le Moigne, J., and Zavorin, I. (2003). Multiresolution registration of remote sensing imagery by optimization of mutual information using a stochastic gradient. *IEEE Transactions on Image Processing*, **12**(12), 1495–1511.

Collignon, A., Maes, F., Delaere, D., Vandermeulen, D., Suetens, P., and Marchal, G. (1995a). Automated multimodality image registration based on information theory. In *Proceedings of the 14th International Conference on Information Processing in Medical Imaging*, Ile de Berder, France; Y. Bizais, C. Barillot, and R. Di Paula, eds. Dordrecht, The Netherlands: Kluwer Academic, pp. 263–274.

Collignon, A., Vandermeulen, D., Suetens P., and Marchal, G. (1995b). 3D multimodality medical image registration using feature space clustering. In *Proceedings of the First International Conference on Computer Vision, Virtual Reality, and Robotics in Medicine*, Nice, France; *Lecture Notes in Computer Science*, N. Ayache, ed. Berlin: Springer, Vol. 905, pp. 195–204.

Cover, T. M. and Thomas, J. A. (1991). *Elements of Information Theory*. New York: Wiley.

Hamza, A. B., He, Y., and Krim, H. (2001). An information divergence measure for ISAR image registration. In *Proceedings of the IEEE Workshop on Statistical Signal Processing*, Singapore, pp. 130–133.

He, Y., Hamza, A. B., and Krim, H. (2003). A generalized divergence measure for robust image registration. *IEEE Transactions on Signal Processing*, **51**, 1211–1220.

Johnson, K., Cole-Rhodes, A., Zavorin, I., and Le Moigne, J. (2001). Mutual information as a similarity measure for remote sensing image registration. In *Proceedings of the SPIE Conference on Geo-Spatial Image and Data Exploitation II*, Orlando, FL, Vol. 4383, pp. 51–61.

Knops, Z. F., Maintz, J. B. A., Viergever, M. A., and Pluim, J. P. W. (2003). Normalized mutual information based registration using *K*-means clustering based histogram binning. In *Proceedings of the SPIE Conference on Medical Imaging*, San Diego, CA, Vol. 5032, pp. 1072–1080.

Le Moigne, J. and Zavorin, I. (2000). Use of wavelets for image registration. In *Proceedings of the SPIE Conference on Wavelet Applications VII*, Orlando, FL, Vol. 4056, pp. 99–108.

Le Moigne, J., Campbell, W. J., and Cromp, R. F. (2002). An automated parallel image registration technique of multiple source remote sensing data. *IEEE Transactions on Geoscience and Remote Sensing*, **40**(8), 1849–1864.

Le Moigne, J., Cole-Rhodes, A., Eastman, R., Johnson, K., Morisette, J., Netanyahu, N. S., Stone, H., and Zavorin, I. (2001). Multi-sensor registration of remotely sensed imagery. In *Proceedings of the Eighth International Symposium on Remote Sensing*, Vol. 4541, Toulouse, France.

Levenberg, K. (1944). A method for the solution of certain non-linear problems in least squares. *The Journal of Applied Mathematics*, **2**, 164–168.

Maes, F., Collignon, A., Vandermeulen, D., Marchal, G., and Suetens, P. (1997). Multimodality image registration by maximization of mutual information. *IEEE Transactions on Medical Imaging*, **16**(2), 187–198.

Maes, F., Vandermeulen, D., and Suetens, P. (2003). Medical image registration using mutual information. *Proceedings of the IEEE*, **91**(10), 1699–1722.

Marquardt, D. (1963). An algorithm for least-squares estimation of nonlinear parameters. *SIAM Journal on Applied Mathematics*, **11**, 431–441.

Nelder, J. A. and Mead, R. (1965). A simplex method for function minimization. *The Computer Journal*, **7**(4), 308–313.

Pluim, J. P., Maintz, J. B., and Viergever, M. A. (2000). Interpolation artefacts in mutual information-based registration. *Computer Vision and Image Understanding*, **77**(11), 211–232.

Pluim, J. P., Maintz, J. B., and Viergever, M. A. (2003). Mutual information-based registration of medical images: A survey. *IEEE Transactions on Medical Imaging*, **22**(8), 986–1004.

Powell, M. J. D. (1964). An efficient method for finding the minimum of a function of several variables without calculating derivatives. *The Computer Journal*, **7**(2), 155–162.

Rodriguez-Carranza, C. E. and Loew, M. H. (1998). A weighted and deterministic entropy measure for image registration using mutual information. In *Proceedings of the SPIE Conference on Image Processing*, San Diego, CA, Vol. 3338, pp. 155–166.

Simoncelli, E., Freeman, W., Adelson, E., and Heeger, D. (1992). Shiftable multiscale transforms. *IEEE Transactions on Information Theory*, **38**(2), 587–607.

Spall, J. C. (1992). Multivariate stochastic approximation using a simultaneous perturbation gradient approximation. *IEEE Transactions on Automatic Control*, **37**(3), 332–341.

Studholme, C., Hill, D. L., and Hawkes, D. J. (1995). Automated 3D registration of truncated MR and CT images of the head. In *Proceedings of the Sixth British Machine Vision Conference*, Birmingham, UK, pp. 27–36.

Studholme, C., Hill, D., and Hawkes, D. (1999). An overlap invariant entropy measure of 3D medical image alignment. *Pattern Recognition*, **32**(1), 71–86.

Thévenaz, P., and Unser, M. (2000). Optimization of mutual information for multiresolution image registration. *IEEE Transactions on Image Processing*, **9**(12), 2083–2099.

Thévenaz, P., Ruttimann, U. E., and Unser, M. (1998). A pyramid approach to subpixel registration based on intensity. *IEEE Transactions on Image Processing*, **7**(1), 27–41.

Tsao, J. (2003). Interpolation artifacts in multimodality image registration based on maximization of mutual information. *IEEE Transactions on Medical Imaging*, **22**(7), 845–846.

Unser, M., Aldroubi, A., and Eden, M. (1993). B-spline signal processing: Part 1. Theory. *IEEE Transactions on Signal Processing*, **41**(2), 821–833.

Viola, P. and Wells, W. M., III (1995). Alignment by maximization of mutual information. In *Proceedings of the Fifth IEEE International Conference on Computer Vision*, Cambridge, MA, pp. 16–23.

Zavorin, I., Stone, H., and Le Moigne, J. (2002). Iterative pyramid-based approach to subpixel registration of multisensor satellite imagery. In *Proceedings of the SPIE Conference on Earth Observing Systems VII*, Seattle, WA, Vol. 4814, pp. 435–446.

Zhu, Y. and Cochoff, S. (2002). Influence of implementation parameters on registration of MR and SPECT brain images by maximization of mutual information. *Journal of Nuclear Medicine*, **43**(2), 160–166.

Zitová, B., and Flusser, J. (2003). Image registration methods: A survey. *Image and Vision Computing*, **21**, 977–1000.

Part III

Feature Matching and Strategies for Image Registration

7

Registration of multiview images

A. ARDESHIR GOSHTASBY

Abstract

Multiview images of a 3D scene contain sharp local geometric differences. To register such images, a transformation function is needed that can accurately model local geometric differences between the images. A weighted linear transformation function for the registration of multiview images is proposed. Properties of this transformation function are explored and its accuracy in image registration is compared with accuracies of other transformation functions.

7.1 Introduction

Due to the acquisition of satellite images at a high altitude and relatively low resolution, overlapping images have very little to no local geometric differences, although global geometric differences may exist between them. Surface spline (Goshtasby, 1988) and multiquadric (Zagorchev and Goshtasby, 2006) transformation functions have been found to effectively model global geometric differences between overlapping satellite images.

High-resolution multiview images of a scene captured by low-flying aircrafts contain considerable local geometric differences. Local neighborhoods in the scene may appear differently in multiview images due to variation in local scene relief and difference in imaging view angle. Global transformation functions that successfully registered satellite images might not be able to satisfactorily register multiview aerial images.

A new transformation function for the registration of multiview aerial images is proposed. The transformation function is defined by a weighted sum of linear functions, each containing information about the geometric difference between corresponding local neighborhoods in the images. This transformation function models a scene by a composite of linear patches. The weighted linear formulation smoothly combines the local linear functions to create a smooth and continuous mapping from one image to another. Monotonically decreasing and asymmetric weight functions are used to blend the linear functions. The monotonically decreasing nature of the weight functions ensures that when the geometry of one image is transformed to locally resemble that of another image, this local transformation does change the geometry of distant neighborhoods. The asymmetric nature of the weight functions stretches them towards gaps between the control points, adapting the transformation to the arrangement of the control points.

A piecewise-linear transformation function (Goshtasby, 1986) may be used to register images with local geometric differences. In a piecewise-linear transformation, a linear mapping is considered at each triangular region obtained by three control points, while in the weighted linear transformation, a linear mapping is considered at each control point. Whereas piecewise-linear is continuous only, weighted-linear is both continuous and smooth. A piecewise-cubic transformation (Goshtasby, 1987) is also continuous and smooth, but similarly to piecewise-linear, it can only map to each other image regions that are contained in the convex hulls of the control points. The weighted-linear transformation maps entire images to one another.

We assume that a set of corresponding control points in the images is available. Control points are unique landmarks in images (Goshtasby, 2005). From the spacing difference between corresponding control points, local geometric differences between the images are linearly estimated. The linear functions are then blended using rational Gaussian weights to create a continuous and smooth transformation over the image domain. The issues of control point selection from the images and the determination of correspondence between the point sets are addressed elsewhere; see, for example, Umeyama (1993), Olson (1997), Mount *et al.* (1999), and Chui and Rangarajan (2003). The process first involves locating a number of control points in one image and searching for the corresponding points in the second image via template matching (Kanade and Okutomi 1994; O'Neill and Denos 1995). Alternatively, a number of control points can be found in each image independently and correspondence between them can then be established via clustering (Stockman *et al.*, 1982), coherence (Orrite and Herrero, 2004), and invariance (Mundy and Zisserman, 1992). Details of these methods can be found in Goshtasby (2005).

Given a pair of images to be registered, we will refer to one of the images as the *reference* image and to the other image as the *sensed* image. The reference image will be kept intact and the sensed image will be geometrically transformed to spatially align with the reference image.

7.2 Problem description

Given N corresponding control points in the images,

$$\{(x_i, y_i), (X_i, Y_i) : i = 1, \ldots, N\}, \tag{7.1}$$

where (x_i, y_i) denotes the coordinates of the ith control point in the reference image and (X_i, Y_i) denotes the coordinates of the corresponding control point in the sensed image, we would like to determine functions $f_x(x, y)$ and $f_y(x, y)$ that satisfy

$$X_i \approx f_x(x_i, y_i), \tag{7.2}$$

$$Y_i \approx f_y(x_i, y_i), \qquad i = 1, \ldots, N, \tag{7.3}$$

where functions f_x and f_y represent the components of the transformation.

If we rearrange the coordinates of the corresponding control points in the images into the following two sets of 3D points:

$$\{(x_i, y_i, X_i) : i = 1, \ldots, N\}, \tag{7.4}$$

$$\{(x_i, y_i, Y_i) : i = 1, \ldots, N\}, \tag{7.5}$$

we can view Eqs. (7.2)–(7.3) as single-valued surfaces that approximate the 3D point sets given by (7.4)–(7.5). The problem of finding the transformation function for image registration can be considered, therefore, as the problem of finding two single-valued surfaces that approximate two 3D point sets. A large number of functions satisfy Eqs. (7.2)–(7.3). We are interested in those functions that can accurately model local geometric differences between multiview images. In the following sections, we first review surface approximation methods that are suitable for the registration of images with local geometric differences. We then describe a new weighted linear transformation, and discuss how to determine its approximation accuracy. The transformation's performance is compared with performances of other competitive transformations, and examples of image registration using the proposed weighted linear transformation are provided. Finally, concluding remarks are made.

7.3 Surface approximation methods

Surface approximation to irregularly spaced points has been studied extensively in *approximation theory*. For specific details, the reader is referred to the excellent reviews by Schumaker (1976), Barnhill (1977), Sabin (1980), Dyn (1987), Franke (1987), Alfeld (1989), Grosse (1990), and LeMéhauté *et al.* (1996).

Shepard's weighted mean method (Shepard, 1968) is perhaps the simplest and most intuitive approximation method known. The method estimates the data value at a point in the approximation domain from a weighted sum of data values at the control points. The weights are set proportional to the inverse distances of the control points to the approximation point and are normalized to have a sum of 1 everywhere in the approximation domain. That is,

$$f(x, y) = \sum_{i=1}^{N} f_i W_i(x, y), \tag{7.6}$$

where

$$W_i(x, y) = \frac{G_i(x, y)}{\sum_{j=1}^{N} G_j(x, y)}, \tag{7.7}$$

and

$$G_i(x, y) = \left\{ (x - x_i)^2 + (y - y_i)^2 \right\}^{-\frac{1}{2}}. \tag{7.8}$$

The coefficient f_i in Eq. (7.6) is either X_i or Y_i. Gordon and Wixom (1978) showed that $f(x, y)$ is continuous over the approximation domain and that it reduces to f_i at (x_i, y_i) for $i = 1, \ldots, N$. Therefore, $f(x, y)$ actually interpolates the data. Moreover, function $f(x, y)$ never exceeds the largest data value or becomes smaller than the smallest data value. This is actually one of the disadvantages of Shepard's method, which makes the approximating surface extend horizontally away from the control points.

$G_i(x, y)$ can actually be any monotonically decreasing radial function. If $G_i(x, y)$ is not an inverse distance metric, then $f(x, y)$ will not interpolate the data but rather approximate it. We will study the behavior of $f(x, y)$ for a Gaussian $G_i(x, y)$.

Since the sum of the weights in Eq. (7.7) is equal to 1 everywhere in the approximation domain, the weight functions stretch toward the gaps and adjust to the spacing between the control points. The approximation domain can be considered the space of the reference image, and the data value at a control point can be considered the X or Y component of the corresponding control point in the sensed image. The width of $G_i(x, y)$ determines the local/global effect of a data value on the approximation. The narrower the function $G_i(x, y)$ is, the closer the approximation $f(x_i, y_i)$ will be to the data value at (x_i, y_i). As the width of G_i decreases, the approximation produces a flat spot at (x_i, y_i). This is demonstrated in the 1D example in Fig. 7.1, using the 1D Gaussian

$$G_i(x) = \exp \left\{ -\frac{(x - x_i)^2}{2\sigma_i^2} \right\}. \tag{7.9}$$

The horizontal axis in Fig. 7.1 gives the locations of the control points in the reference image and the vertical axis gives the data values at the control

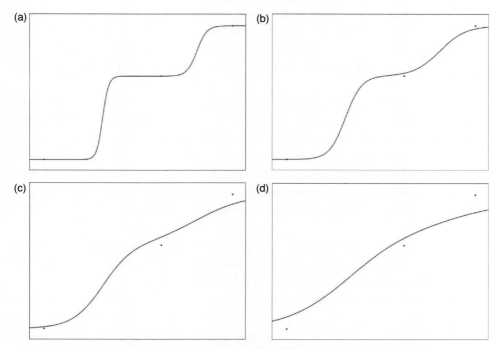

Figure 7.1. (a)–(d) Weighted mean approximation as a function of increasing width of the weight functions.

points, which in 1D are the corresponding coordinates of the control points in the sensed image. We observe that as the widths (standard deviations) of the Gaussians (of Eq. (7.9)) decrease, the function obtained produces horizontal flat spots at the control points, and a growing error away from the control points. As the widths of the Gaussians increase, the flat spots disappear. However, the function deviates from the data, and the error grows at the control points and their neighborhoods.

In order to remove the flat spots from the weighted mean approximation, Franke and Nielson (1980) replaced f_i by a quadratic function that interpolates the data at (x_i, y_i) as follows:

$$f(x, y) = \sum_{i=1}^{N} Q_i(x, y) W_i(x, y). \tag{7.10}$$

This revised Shepard formulation is known as the *modified Shepard interpolation*.

The coefficient f_i in Eq. (7.6) can be considered a function of degree 0. In 1D it is a line parallel to the x-axis, and the function defined by this equation becomes a weighted sum of horizontal lines in the xy plane. As such, horizontal spots will appear in the approximation as narrower weight functions are used. If instead of horizontal lines, quadratic functions are used, these functions will determine the local shape of the approximation, such that for very narrow weight functions the approximation will tend to coincide locally with the shape of the quadratic functions.

Although the weighted mean transformation and its modified version use the same width radial functions G_i, irrespective of the spacing between the control points, Renka (1988) adapted the widths of the radial functions (and, thus, the weight functions themselves) to the spacing between the control points. This is done by using a narrower radial function in areas where the density of the control points is relatively high. This mechanism adapts the approximation to the density of the control points locally, thereby improving the approximation accuracy Renka (1988).

Because of their locally sensitive nature both modified Shepard and its adaptive version are suitable for representing the transformation components required for registering images with local geometric differences.

The local shape of an approximating surface can be estimated from the gradients in data at the control points. To ensure that an approximating surface produces desired local shapes, we revise Shepard's formula to use data gradients as well as data values at the control points. Data gradients will reflect local geometric differences between the images and will make it possible to register images with local geometric differences.

7.4 Weighted linear transformation

Given a set of corresponding control points in the images, we first triangulate the control points in the reference image. If n triangles are obtained, knowing the correspondence between the control points in the images, we will obtain n triangles in the sensed image. Let us denote the vertices of the jth triangle in the reference and sensed images by

$$\{(x_{jk}, y_{jk}) : k = 1, 2, 3\} \tag{7.11}$$

and

$$\{(X_{jk}, Y_{jk}) : k = 1, 2, 3\}, \tag{7.12}$$

respectively. Note that these vertices simply rename the N corresponding control points given in (7.1). Let us also denote the coordinates of the centroid of the jth triangle in the reference image by (C_{x_j}, C_{y_j}) and that in the sensed image by

(C_{X_j}, C_{Y_j}). We then define a component of the transformation by a weighted sum of linear functions over triangular regions in the reference image. That is,

$$X(x, y) = \sum_{j=1}^{n} W_j(x, y) X_j(x, y), \tag{7.13}$$

$$Y(x, y) = \sum_{j=1}^{n} W_j(x, y) Y_j(x, y), \tag{7.14}$$

where

$$W_j(x, y) = \frac{G_j(x, y)}{\sum_{k=1}^{n} G_k(x, y)}, \qquad j = 1, \ldots, n, \tag{7.15}$$

are barycentric rational weights and

$$G_j(x, y) = \exp\left\{-\frac{(x - C_{x_j})^2 + (y - C_{y_j})^2}{2\sigma_j^2}\right\} \tag{7.16}$$

is a Gaussian of height 1 and standard deviation σ_j centered at (C_{x_j}, C_{y_j}) in the reference image. (The functions $X_j(x, y)$ and $Y_j(x, y)$ are defined below.) The standard deviation of a Gaussian is set proportional to the size of the triangle it belongs to. That is, we let

$$\sigma_j = s r_j, \tag{7.17}$$

where s is the proportionality term and r_j is the radius of the circle inscribing the jth triangle in the reference image. The parameter s is a global parameter that can be varied to change the smoothness of the transformation. By decreasing it, the transformation will follow the individual triangles more closely, reproducing sharp edges and corners. By increasing it, the transformation will smooth out noisy geometric differences between the images. The parameter s is assigned typically a value between 2 and 3. A very large value of s will excessively smooth the approximation, and thus increase the error, while a very small value of s will result in an approximation that is close to piecewise-linear.

Since the sum of the weight functions everywhere in the approximation domain should be equal to 1, and since the weights are monotonically decreasing, the weight functions automatically adjust to the size and shape of the triangles, and consequently to the spacing between the control points. In areas where the density of the control points is high, the triangles will be smaller and the weight functions will be narrower, while in areas where only sparse control points are available, the weight functions will stretch toward the gaps.

The linear functions $X_j(x, y)$ and $Y_j(x, y)$ in Eqs. (7.13)–(7.14) are defined by

$$X_j(x, y) = A_j x + B_j y + C_j, \tag{7.18}$$

$$Y_j(x, y) = D_j x + E_j y + F_j, \tag{7.19}$$

so that the jth triangular region in the reference image is mapped to the corresponding triangular region in the sensed image. Thus, A_j, B_j, and C_j are determined by fitting the plane

$$X = A_j x + B_j y + C_j \tag{7.20}$$

to the points $\{(x_{jk}, y_{jk}, X_{jk}) : k = 1, 2, 3\}$. Similarly, D_j, E_j, and F_j are determined by fitting the plane

$$Y = D_j x + E_j y + F_j \tag{7.21}$$

to the points $\{(x_{jk}, y_{jk}, Y_{jk}) : k = 1, 2, 3\}$.

The transformation function defined by Eqs. (7.13)–(7.14) represents a weighted sum of planes. Compared to the weighted sum of points, the former contains information about geometric gradients of the sensed image with respect to the reference image. This information is not used in transformation functions that are formulated in terms of control points only. Without such information, the transformation obtained cannot accurately represent local geometric differences between images captured from different views of a 3D scene.

The use of asymmetric weight functions that adapt to the varying spacing between the control points and the use of linear functions that contain information about the geometric gradients of the sensed image with respect to the reference image make it possible to create a transformation function that adapts to the organization of the control points, as well as to the geometric gradient of the sensed image with respect to the reference image.

A 1D example of this basic premise is illustrated in Fig. 7.2. The specified points and corresponding gradients are shown by the lines in Fig. 7.2(a). The objective is to find a curve that approximates the points and whose gradients agree with the specified ones at the points. The process of obtaining such a curve is demonstrated in Fig. 7.2(b). The dotted curves show the weight functions. The solid curves in Fig. 7.2(b) show the weights when multiplied by the linear functions representing the geometric gradients at the control points. The sum of the solid curves in Fig. 7.2(b) produces the approximating curve shown in Fig. 7.2(a). This curve represents a 1D transformation that produces the desired geometric gradients at the control points.

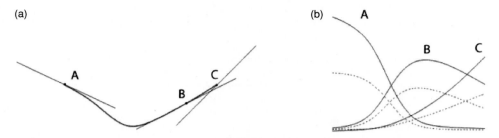

Figure 7.2. (a) Curve approximating three points and their corresponding gradients, and (b) computation of approximation curve as a weighted sum of the lines; dotted curves show the rational weight functions, and solid curves show the lines multiplied by their corresponding weight functions, such that their sum gives the approximating curve with the desired local gradients at the control points.

When registering 2D images, planes defined by Eqs. (7.20)–(7.21) are used. Note that each component of a transformation uses a linear function that represents the gradient of that component in the x or y direction in the local area defined by a triangle. By using the gradient at a control point, it becomes possible to properly warp the sensed image to locally take the geometry of the reference image.

Transformation functions such as surface splines (Harder and Desmarais, 1972; Goshtasby, 1988) and multiquadrics (Hardy, 1990; Zagorchev and Goshtasby, 2006) that use monotonically increasing and radially symmetric basis functions work well when the images have negligible local geometric differences and the spacing between the control points is rather uniform. If, however, spacing between the control points varies across the image domain, monotonically decreasing rational weight functions should be employed, as they adapt well to the spacing between the control points, and can model local geometric differences better than monotonically increasing radial functions.

To demonstrate the difference between transformation components that are formulated in terms of data values at the control points and transformation components that are formulated in terms of both data values and data gradients at the control points, consider the 1D example shown in Fig. 7.3. Defining a transformation component by a weighted sum of data values amounts to finding the weighted sum of lines of slope 0 (planes of slope 0 in 2D) as illustrated in Fig. 7.3(a). If a transformation component is defined by a weighted sum of horizontal lines and planes, it becomes very difficult to produce the desired local shapes in the transformation component. Therefore, sharp changes in data at adjacent control points, which represent large local geometric differences between the images, can result in fluctuations and overshoots in the transformation. If a transformation component is defined by a weighted sum of lines (planes in 2D) with specified slopes, it

Table 7.1 *Coordinates of nine uniformly spaced control points in the reference image with the X coordinates of the corresponding control points in the sensed image*

i	1	2	3	4	5	6	7	8	9
x_i	0	1	2	0	1	2	0	1	2
y_i	0	0	0	1	1	1	2	2	2
X_i	0	1	2	0	1	1	2	0	1	2

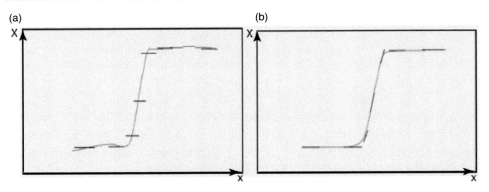

Figure 7.3. (a) Curve defined by a weighted sum of points or, equivalently, by a weighted sum of lines of slope 0 passing through the points, and (b) curve obtained by a weighted sum of lines of desired slopes passing through the same points.

becomes much easier to reproduce desired local shapes in the component of the transformation, as illustrated in Fig. 7.3(b).

A 2D example which demonstrates the difference between transformation functions that do not use geometric gradients at the control points, and the weighted linear transformation that does use geometric gradients at the control points is given in Fig. 7.4. Consider the nine uniformly spaced control points in the reference image with the X component of the corresponding control points in the sensed image given in Table 7.1. Since X changes linearly with x, it is anticipated that the approximating function represents a plane that passes through or very close to the 3D points. Approximation results without and with the use of geometric gradients are depicted in Fig. 7.4. For Gaussians with low standard deviations, the surface passes close to the data but produces flat horizontal spots. This results in large errors away from the control points. For Gaussians with high standard deviations, the staircase effect disappears, but the surface moves away from the points. This increases the error at and near the control points, which is translated to registration error accordingly.

Using the weighted linear method, the approximating surface closely follows the data for a wide range of standard deviations of Gaussians. Also noticeable is

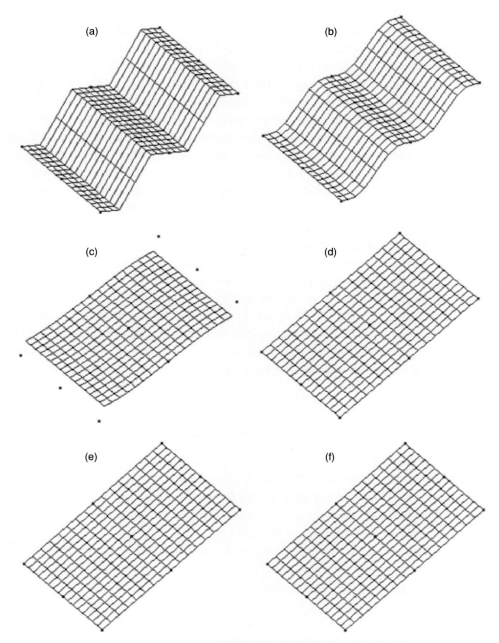

Figure 7.4. (a)–(c) Approximation of data in Table 7.1 using the weighted mean method as widths of the weight functions increase; (d)–(f) approximation using the weighted linear method and the same data and weights.

the change in spacing between the grid lines as the standard deviations vary in the weighted mean method, but not in the weighted linear method. For smaller standard deviations, although the surface obtained by the weighted mean approximation passes close to the points, the spacing between the grid lines becomes more nonuniform. Nonuniform spacing of the surface points under uniform spacing of points in the approximation domain implies change in the density of the surface points, which results in large registration errors away from the control points. Using, on the other hand, the weighted linear approximation, grid spacing remains uniform over a wide range of standard deviations, maintaining a rather uniform density of points over the approximation domain. This keeps the registration error down.

In applying the weighted linear transformation, the components defined by Eqs. (7.13)–(7.14) are obtained immediately after substituting the coordinates of corresponding control points and their gradients into the equations of the transformation without solving a system of equations. This is very desirable when a very large set of control points is given. A weighted linear transformation maps corresponding control points to each other approximately. As the widths of the weight functions decrease, the approximation locally converges to local linear functions that map triangles in the sensed image to the corresponding triangles in the reference image. As the widths of the weight functions increase, the transformation becomes smoother but moves farther from the approximating points. An example showing a transformation component defined by five control points (i.e., four triangles) is depicted in Fig. 7.5.

7.5 Properties of the weighted linear transformation

The transformation defined by Eqs. (7.13)–(7.14) uses monotonically decreasing weight functions. Thus, the effect of a local deformation or error will remain local. In contrast, transformation functions that use monotonically increasing radial functions, such as surface spline and multiquadric, can spread a local error or deformation over the entire image domain.

The widths of the weight functions in the weighted linear transformation are adjusted to the sizes of the triangles. In an area of higher density of the control points, since the triangles are smaller, the weight functions will be narrower. On the other hand, in areas where the control points are more sparse, the triangles are larger and thus the weight functions will be wider. This results in filling the gaps (see Fig. 7.6 for an illustration). For surface spline approximation, the same radial function is used irrespective of the spacing between the control points. Using the same radially symmetric function at each control point could result, though, in large registration errors away from the control points if the spacing between them varies greatly across the image domain.

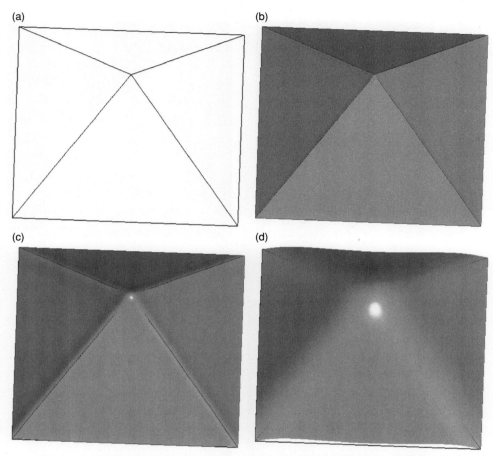

Figure 7.5. (a) Five control points producing four triangular regions; (b)–(d) surfaces approximating the triangle mesh using the weighted linear method with increasing values of smoothness parameter s of Eq. (7.17).

The smoothness parameter s in the weighted linear transformation makes it possible to change the global/local effect of the control points on the registration. If the scene is known to contain round and smooth surfaces, a larger smoothness parameter will be more suitable, while if the scene is known to contain sharp edges and corners, a smaller smoothness parameter will be preferred. For surface spline, information about scene content and spacing between the control points cannot be used to guide the registration. The stiffness parameter in surface spline can be used to control the smoothness of the overall surface, but because the basis functions are monotonically increasing, changes in the stiffness parameter will not make the transformation locally sensitive.

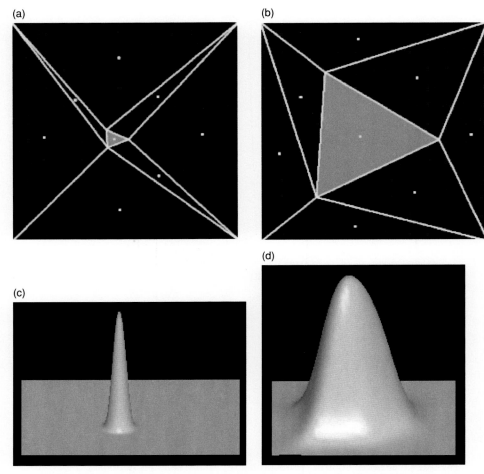

Figure 7.6. (a)–(b) Two different arrangements of control points in the reference image (the triangle centroids determined from the triangle vertices are used as the new control points); (c)–(d) weight functions associated, respectively, with the shaded triangles in (a) and (b). Weight functions adjust to the organization and density of the control points to produce a sum of 1 everywhere in the approximation domain.

Although it was theoretically shown that radial functions, such as surface splines and multiquadrics, can interpolate any arrangement of control points (Jackson, 1988), in practice, due to the global nature of the radial functions, depending on the organization of the control points and the local geometric difference between the images, the matrix of coefficients may become ill-conditioned. The weighted linear method requires the solution of only small systems of equations to find the parameters of the local linear functions. Therefore, the process is stable and is guaranteed to produce a solution even when a very large set of control points is given and the spacing between the control points varies greatly across the image domain.

7.6 Accuracy of weighted linear transformation

To evaluate the registration performance of the weighted linear transformation, we examine its approximation of 2D data. We will use control points in the reference image as the data sites and the X and Y components of the corresponding control points in the sensed image as the data values. The approximation of the weighted linear method will be compared to those obtained by the methods of surface spline (Goshtasby, 1988), modified Shepard (Franke and Nielson, 1980), and adaptive Shepard (Renka, 1988). Uniformly spaced control points, randomly spaced control points, control points with highly varying density, and control points along contours, as depicted in Figs. 7.7(a)–(d), will be used in the experiments.

7.6.1 Experiment 1

In this experiment, random samples from a number of known surfaces will be used to estimate the surface geometries. The samples will be at randomly spaced points in the approximation domain, as shown in Fig. 7.7(b), and the surface geometries will be those that are widely used in approximation theory to determine the accuracy of various approximation methods. The surfaces used in this experiment are given by the following:

$$
\begin{aligned}
F_1 = {} & .75 \exp[-((9x - 2)^2 + (9y - 2)^2)/4] \\
& + .75 \exp[-(9x + 1)^2 + (9y + 1)/10] \\
& + .5 \exp[-((9x - 7)^2 + (9y - 3)^2)/4] \\
& - .2 \exp[-(9x - 4)^2 - (9y - 7)^2],
\end{aligned}
\tag{7.22}
$$

$$
F_2 = [\tanh(9y - 9x) + 1]/9,
\tag{7.23}
$$

$$
F_3 = \left[1.25 + \cos(5.4y)\right]/\left[6 + 6(3x - 1)^2\right],
\tag{7.24}
$$

$$
F_4 = \exp\left\{-(81/16)\left[(x - .5)^2 + (y - .5)^2\right]\right\}/3,
\tag{7.25}
$$

$$
F_5 = \exp\left\{-(81/4)\left[(x - .5)^2 + (y - .5)^2\right]\right\}/3,
\tag{7.26}
$$

$$
F_6 = \left\{64 - 81\left[(x - .5)^2 + (y - .5)^2\right]\right\}^{1/2}/9 - .5.
\tag{7.27}
$$

The above surface geometries due to Franke (1982) are depicted in Figs. 7.8(a)–(f). Surfaces F_1 and F_5 have gentle as well as high-varying areas, F_2 has areas with very little variation separated by a sharply varying area, and F_3, F_4, and F_6 contain

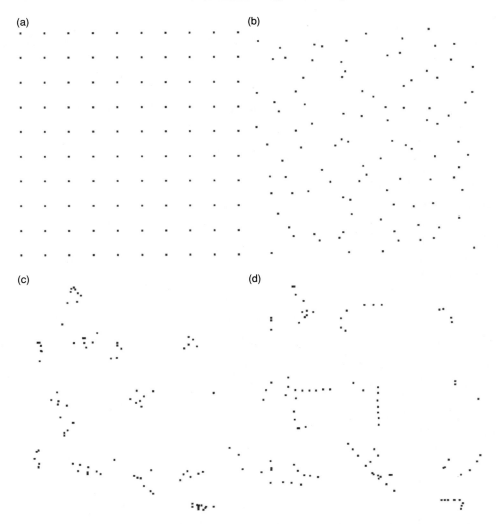

Figure 7.7. (a) Uniformly spaced control points, (b) randomly spaced control points, (c) control points with highly varying density, and (d) control points along contours.

low to moderate variations. Knowing the true surface values everywhere in the approximation domain, we will determine the accuracy of different approximation methods by estimating the above surface geometries for the random samples taken.

To compare the accuracies of different approximation methods, functional values at a grid of 33×33 uniformly spaced points across the approximation domain are determined and compared with the true surface values. Mean and maximum

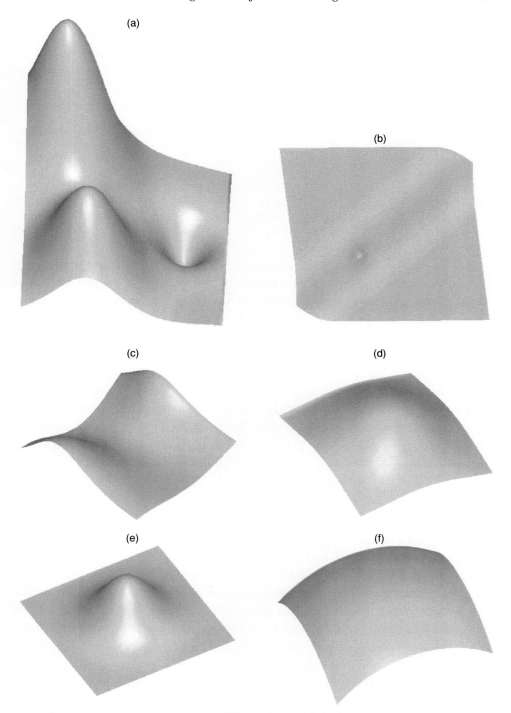

Figure 7.8. (a)–(f) The functions F_1–F_6 of Eqs. (7.22)–(7.27).

Table 7.2 *Mean and maximum approximation errors of the weighted linear approximation for the control points in Fig. 7.7(b) and the data values of the surfaces in Fig. 7.8 sampled at the control points; the results obtained by the modified and adaptive Shepard methods are also tabulated for comparison (best results appear in bold)*

Method	Error	F_1	F_2	F_3	F_4	F_5	F_6
Modified	Mean	.0078	.0026	.0011	.0006	.0018	.00026
Shepard	Max	.0573	.0468	.0125	.0039	.0218	.0036
Adaptive	Mean	**.0054**	.0020	.0009	.0004	.0012	.00024
Shepard	Max	**.0533**	.0249	.0124	.0032	.0099	.0039
Weighted	Mean	.0055	**.0015**	**.0003**	**.0002**	**.0006**	**.00022**
Linear	Max	.0587	**.0129**	**.0032**	**.0011**	**.0084**	**.0026**

differences between the true and estimated surface values are calculated and used as the error measures. Table 7.2 shows the mean and maximum approximation errors for the weighted linear approximation, as well as those errors for the methods of modified Shepard (Franke and Nielson, 1980) and adaptive Shepard (Renka, 1988).

From the results in Table 7.2, we observe that the use of gradient information greatly improves the approximation accuracy. Among the datasets tested, only the results from F_1 are worse than those obtained by the adaptive Shepard method (Renka, 1988), but only very slightly. The approximation errors of the weighted linear method for F_2–F_6 are better than those obtained by the modified Shepard and adaptive Shepard methods. (The best results obtained for each function appear in bold.)

7.6.2 Experiment 2

In the second experiment, the functions shown in Fig. 7.8 were sampled at the points depicted in Figs. 7.7(a)–(d) and used as data values. The approximation errors obtained by the weighted linear method are given in Table 7.3 along with the approximation errors obtained by surface spline for comparison. Since surface spline uses radially symmetric and monotonically increasing basis functions, it performs well on uniformly spaced data. The results are mixed, however. Surface spline works better than the weighted linear approximation on F_1, and has a lower mean error in four out of the six cases. However, the latter has a lower maximum error in five out of the six cases.

When randomly spaced data are used, since the rational functions of the weighted linear transformation adapt to the spacing surface between the control points, this

Table 7.3 *Mean and maximum errors due to surface spline and weight-linear approximations for the uniformly spaced points in Fig. 7.7(a) and the data values sampled at these points from the surfaces in Fig. 7.8 (best results appear in bold)*

Method	Error	F_1	F_2	F_3	F_4	F_5	F_6
Surface	Mean	**.0019**	.0012	**.0004**	**.0001**	.0003	**.0007**
Spline	Max	**.0205**	.0134	.0092	.0037	.0027	.0198
Weighted	Mean	.0023	**.0007**	.0010	.0009	**.0002**	.0014
Linear	Max	.0247	**.0057**	**.0062**	**.0036**	**.0021**	**.0039**

Table 7.4 *Same as Table 7.3, except for using the random points shown in Fig. 7.7(b)*

Method	Error	F_1	F_2	F_3	F_4	F_5	F_6
Surface	Mean	**.0053**	.0021	.0004	.00024	.0009	**.0004**
Spline	Max	**.0524**	.0273	.0052	.0016	.0182	**.0126**
Weighted	Mean	.0055	**.0015**	**.0003**	**.00022**	**.0006**	.0022
Linear	Max	.0587	**.0129**	**.0032**	**.0011**	**.0084**	.0142

approximation produces better results in four out of the six cases compared to surface spline, which is formulated by radially symmetric funtions. The errors obtained by the weighted linear and surface spline methods on the randomly spaced points in Fig. 7.7(b) are included in Table 7.4.

When points with highly varying spacing are used, the system of equations to be solved in surface spline becomes ill-conditioned, producing no solution. On the other hand, the weighted linear method does not require solving a large system of equations and it always returns a solution. The results for the points shown in Fig. 7.7(c) are summarized in Table 7.5. (Dashed lines indicate that surface spline approximation was not applicable.)

Interestingly, surface spline produced results for the contour dataset shown in Fig. 7.7(d). The results due to the surface spline and weighted linear methods are shown in Table 7.6. In the majority of cases, the weighted linear method produced results superior to those of surface spline.

Since surface heights over large areas in functions F_1–F_6 do not change sharply, it is possible for surface spline to perform relatively well. Multiview images of an urban scene could have sharp local geometric differences, which make surface spline an unsuitable transformation for their registration. This is demonstrated on some examples in the next section.

Table 7.5 *Same as Table 7.3, except for using control points with highly varying spacing as shown in Fig. 7.7(c) (dashed lines indicate that surface spline approximation was not applicable)*

Method	Error	F_1	F_2	F_3	F_4	F_5	F_6
Surface	Mean	–	–	–	–	–	–
Spline	Max	–	–	–	–	–	–
Weighted	Mean	**.0359**	**.0227**	**.0050**	**.0084**	**.0055**	**.0062**
Linear	Max	**.2249**	**.0681**	**.0488**	**.0284**	**.0421**	**.0849**

Table 7.6 *Same as Table 7.3, for the point set in Fig. 7.7(d)*

Method	Error	F_1	F_2	F_3	F_4	F_5	F_6
Surface	Mean	.0246	.0069	.0029	.0017	.0089	**.0025**
Spline	Max	.2087	**.0407**	.0312	**.0107**	.0465	.0484
Weighted	Mean	**.0141**	**.0068**	**.0025**	**.0013**	**.0027**	.0038
Linear	Max	**.0705**	.0503	**.0245**	.0162	**.0305**	.0354

7.7 Registration using weighted linear transformation

An example of image registration using the weighted linear transformation is given in Fig. 7.9. The reference and sensed images are shown in Fig. 7.9(a) and Fig. 7.9(b), respectively. A set of control points was selected in the reference image by the corner detector of Tomasi and Kanade (1992). Circular templates were centered at the control points in the reference image and searched for in the sensed image. The control points in the reference image were then triangulated, and based on the correspondence between the control points in the images, corresponding triangles were determined in the sensed image. From the coordinates of the vertices of the corresponding triangles, new control points were then generated at the centers of these triangles. The geometric gradient at each new control point was calculated by fitting a linear function to the vertices of the corresponding triangle containing that control point. This was done twice to determine both the geometric gradient of the X and Y components of the transformation. The local linear functions were then blended by using rational Gaussian weights to obtain the components of the transformation.

Resampling the sensed image with the transformation obtained provided the image shown in Fig. 7.9(c). (This is actually an overlay of the reference image and the resampled image, to demonstrate the quality of the registration.) For comparative purposes, the sensed image was also resampled using the global surface spline

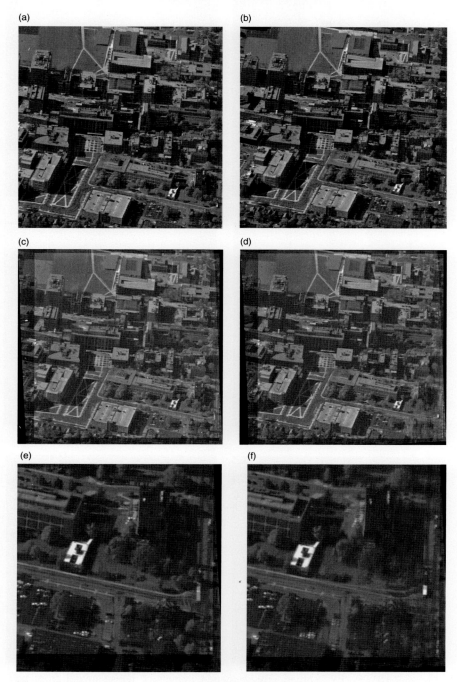

Figure 7.9. (a)–(b) Two images showing different views of an urban scene, (c) registration due to weighted linear transformation, (d) registration due to surface spline transformation, and (e)–(f) lower right corners of (c) and (d), respectively.

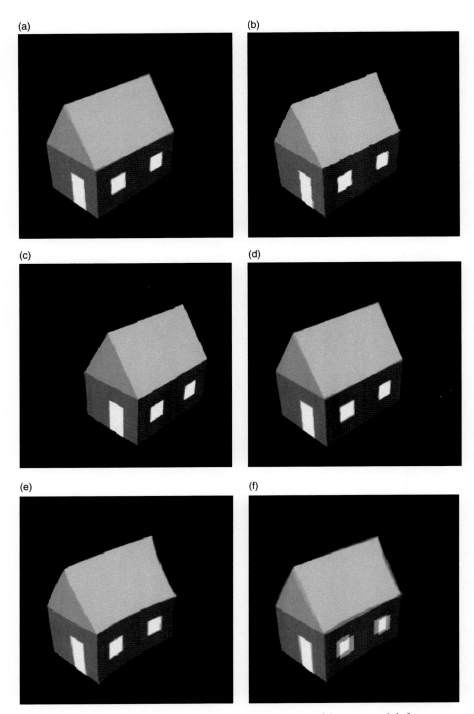

Figure 7.10. (a)–(b) Synthetically generated images of house model from two different view angles, (c) image (b) resampled to register with image (a) by the weighted linear transformation, (d) overlay of images (c) and (a), (e) image (b) resampled to register with image (a) by the surface spline transformation, and (f) overlay of images (e) and (a).

transformation. The overlay of this resampled image and the reference image is depicted in Fig. 7.9(d). Although the registration due to surface spline is acceptable in most areas in the images, the method excessively warped the lower right corner of the sensed image due to lack of control points there. Because of using monotonically increasing radial functions, surface spline tends to produce large errors away from the control points. When registering multiview urban scene images with significant local geometric differences, the errors due to global transformation functions can be large.

A second example is given in Fig. 7.10. Figures 7.10(a) and 7.10(b) are synthetically generated images of a house model. The images have perspective differences due to different viewing angles. House points closer to the camera baseline have larger disparities than points farther away. Corners of the house in the image in Fig. 7.10(a) were interactively selected as the control points. Then, circular templates centered at these corners were selected and searched for in the image of Fig. 7.10(b). Using the coordinates of the corresponding corners in the images, the weighted linear transformation was determined. Applying this transformation to the image of Fig. 7.10(b) followed by resampling results in the image shown in Fig. 7.10(c). The overlay of the images of Fig. 7.10(c) and 7.10(a) is shown in Fig. 7.10(d). Note that the weighted linear transformation is capable of compensating for sharp local geometric differences between the images.

To compare this with the results due to surface spline, the same control point correspondences were used to find the surface spline transformation. Applying the surface spline obtained to the image of Fig. 7.10(b), followed by resampling results in the image shown in Fig. 7.10(e). The overlay of the images of Figs. 7.10(e) and 7.10(a) is shown in Fig. 7.10(f). As expected, surface spline is incapable of capturing sharp geometric differences that exist between close-range images of 3D scenes. Although the registration is acceptable near the house corners, its performance away from the corners is poor, due to excessive warping.

7.8 Conclusions

Surface spline approximation is widely used for registration of satellite images, where local geometric differences between images are small or negligible. Due to the use of monotonically increasing and symmetric basis functions, surface spline can spread a local deformation to distant points, especially when the spacing between the control points varies greatly across the image domain. When there are local geometric differences between the images, a transformation function is needed that will keep these geometric differences local. To meet this objective, we introduced in this chapter a weighted linear transformation. This transformation captures the geometric difference between corresponding neighborhoods in images by blending a set of weighted linear functions. Specifically, we used asymmetric and monotonically decreasing weight functions. The asymmetric nature of the weight functions adapts the transformation to the spacing between the control points, and the monotonically decreasing nature of the weight functions ensures that geometric differences between corresponding local neighborhoods in the images do not propagate over the image domain.

Selecting the right transformation function is only one step in image registration. For a successful registration, one needs to find a sufficient number of appropriate control points. Determining a sufficient number of good control points is a major hurdle to overcome. Control points should be selected intelligently; rather than selecting control points that only satisfy a mathematical criterion, they should be selected to satisfy also the constraints imposed by the structures present in the scene and the viewing angles of the camera(s) that captures the images. In multiview images of urban scenes, there is a need to select control points that correspond to physical corners of buildings and other man-made structures.

Acknowledgements

This work was supported by the Air Force Research Laboratory (Award No. A8650-05-1-1914) and the State of Ohio. The author would like to thank the Robotics Institute, Carnegie-Mellon University, for the image in Fig. 7.10 and the Air Force Research Laboratory for the image in Fig. 7.9.

References

Alfeld, P. (1989). Scattered data interpolation in three or more variables. In T. Ly-che and L. L. Schumaker, eds., *Mathematical Methods in Computer Aided Geometric Design*. New York: Academic Press, pp. 1–33.

Barnhill, R. E. (1977). Representation and approximation of surfaces. In J. R. Rice, ed., *Mathematical Software III*. New York: Academic Press, pp. 69–120.

Chui, H. and Rangarajan, A. (2003). A new point pattern matching algorithm for nonrigid registration. *Computer Vision and Image Understanding*, **89**, 114–141.

Dyn, N. (1987). Interpolation of scattered data by radial functions. In C. K. Chui, L. L. Schumaker, and F. Utrerus, eds., *Topics in Multivariate Approximation*. New York: Academic Press, pp. 47–61.

Franke, R. (1982). Scattered data interpolation: Tests of some methods. *Mathematics of Computation*, **38**, 181–200.

Franke, R. (1987). Recent advances in the approximation of surfaces from scattered data. In C. K. Chui, L. L. Schumaker, and F. Utrerus, eds., *Topics in Multivariate Approximation*, New York: Academic Press, pp. 79–98.

Franke, R. and Nielson, G. (1980). Smooth interpolation of large sets of scattered data. *International Journal of Numerical Methods in Engineering*, **15**, 1691–1704.

Gordon, W. J. and Wixom, J. A. (1978). Shepard's method of 'metric interpolation' to bivariate and multivariate interpolation. *Mathematics of Computation*, **32**, 253–264.

Goshtasby, A. (1986). Piecewise linear mapping functions for image registration. *Pattern Recognition*, **19**, 459–466.

Goshtasby, A. (1987). Piecewise cubic mapping functions for image registration. *Pattern Recognition*, **20**, 525–533.

Goshtasby, A. (1988). Registration of images with geometric distortions. *IEEE Transactions on Geoscience and Remote Sensing*, **26**, 50–64.

Goshtasby, A. (2005). *2-D and 3-D Image Registration for Medical, Remote Sensing, and Industrial Applications*. New Hoboken, NJ: Wiley Press.

Grosse, E. (1990). A catalogue of algorithms for approximation. In J. Mason and M. Cox, eds., *Algorithms for Approximation II*. London: Chapman and Hall, pp. 479–514.

Harder, R. L. and Desmarais, R. N. (1972). Interpolation using surface spline. *Journal of Aircraft*, **9**, 189–191.

Hardy, R. L. (1990). Theory and applications of the multiquadric-biharmonic method: 20 years of discovery 1969–1988. *Computers and Mathematics with Applications*, **19**, 163–208.

Jackson, I. R. H. (1988). Convergence properties of radial basis functions. *Constructive Approximation*, **4**, 243–264.

Kanade, T. and Okutomi, M. (1994). A stereo matching algorithm with an adaptive window: Theory and experiments. *IEEE Transactions on Pattern Analysis and Machine Intelligence*, **16**, 920–932.

LeMéhauté, A., Schumaker, L. L., and Traversoni, L. (1996). Multivariate scattered data fitting. *Journal of Computational and Applied Mathematics*, **73**, 1–4.

Mount, D. M., Netanyahu, N. S., and Le Moigne, J. (1999). Efficient algorithms for robust feature matching. *Pattern Recognition*, **32**, 17–38.

Mundy, J. L. and Zisserman, A. (1992). *Geometric Invariance in Computer Vision*. Cambridge, MA: MIT Press.

Olson, C. F. (1997). Efficient pose clustering using a randomized algorithm. *International Journal of Computer Vision*, **23**, 131–147.

O'Neill, M. and Denos, M. (1995). Automated system for coarse-to-fine pyramidal area correlation stereo matching. *Image and Vision Computing*, **14**, 225–236.

Orrite, C. and Herrero, J. E. (2004). Shape matching of partially occluded curves invariant under projective transformation. *Computer Vision and Image Understanding*, **93**, 34–64.

Renka, R. J. (1988). Multivariate interpolation of large sets of scattered data. *ACM Transactions on Mathematical Software*, **14**, 139–148.

Sabin, M. A. (1980). Contouring: A review of methods for scattered data. In K. Brodlie, ed., *Mathematical Methods in Computer Graphics and Design*. New York: Academic Press, pp. 63–86.

Schumaker, L. L. (1976). Fitting surfaces to scattered data. In G. G. Lorentz, C. K. Chui, and L. L. Schumaker, eds., *Approximation Theory II*. New York: Academic Press, pp. 203–268.

Shepard, D. (1968). A two-dimensional interpolation function for irregularly spaced data. In *Proceedings of the Twenty Third National Conference on ACM*. New York, pp. 517–523.

Stockman, G., Kopstein, S., and Benett, S. (1982). Matching images to models for registration and object detection via clustering. *IEEE Transactions on Pattern Analysis and Machine Intelligence*, **4**, 229–241.

Tomasi, C. and Kanade, T. (1992). Shape and motion from image streams under orthography: A factorization method. *International Journal of Computer Vision*, **9**, 137–154.

Umeyama, S. (1993). Parametrized point pattern matching and its application to recognition of object families. *IEEE Transactions on Pattern Analysis and Machine Intelligence*, **15**, 136–144.

Zagorchev, L. and Goshtasby, A. (2006). A comparative study of transformation functions for image registration. *IEEE Transactions on Image Processing*, **15**, 529–538.

8

New approaches to robust, point-based image registration

DAVID M. MOUNT, NATHAN S. NETANYAHU,
AND SAN RATANASANYA

Abstract

We consider various algorithmic solutions to image registration based on the alignment of a set of feature points. We present a number of enhancements to a branch-and-bound algorithm introduced by Mount, Netanyahu, and Le Moigne (*Pattern Recognition*, Vol. 32, 1999, pp. 17–38), which presented a registration algorithm based on the partial Hausdorff distance. Our enhancements include a new distance measure, the *discrete Gaussian mismatch*, and a number of improvements and extensions to the above search algorithm. Both distance measures are robust to the presence of outliers, that is, data points from either set that do not match any point of the other set. We present experimental studies, which show that the new distance measure considered can provide significant improvements over the partial Hausdorff distance in instances where the number of outliers is not known in advance. These experiments also show that our other algorithmic improvements can offer tangible improvements. We demonstrate the algorithm's efficacy by considering images involving different sensors and different spectral bands, both in a traditional framework and in a multiresolution framework.

8.1 Introduction

Image registration involves the alignment of two images, called the *reference image* and the *input image*, taken of the same scene. The objective is to determine the transformation from some given geometric group that most nearly aligns the input image with the reference image. Our interest in this problem stems from its application in remote sensing, and in particular in the alignment of satellite images of the Earth taken possibly at different times, by different sensors, and from different spectral bands. The goal is to establish very close alignment, say to within a fraction of a pixel. The problem is complicated by issues such as

179

obscuration due to the presence of cloud cover, variations caused by time (such as coastline changes due to tidal effects or variations in shadows at different times of the day), and variations to surface features due to the differences in the sensors and spectral bands. Image registration and point pattern matching are closely related computational problems. We refer the reader to surveys by Ton and Jain (1989), Brown (1992), and Zitová and Flusser (2003), and Alt and Guibas (1999) for further information on the many approaches to these problems.

There are two common approaches to image registration. One approach makes direct use of the original image data and the other is based on matching discrete geometric feature points. Both approaches have their relative advantages and disadvantages. This chapter focuses on methods based on matching feature points. Features may be extracted by a number of methods. In our experiments, we have used a feature extraction process based on identifying relatively high-valued coefficients from a wavelet decomposition of the image (Le Moigne *et al.*, 2002). This approach has the advantage that it can produce feature points at multiple levels of resolution. This can be used to drive a progressive multiresolution registration algorithm, which registers images at increasing levels of accuracy (Cole-Rhodes *et al.*, 2003; Netanyahu *et al.*, 2004; Zavorin and Le Moigne, 2005). For the applications we have in mind, the transformations to be considered involve two-dimensional geometric similarities, that is, transformations resulting from the composition of rotation, translation, and uniform scaling.

Accurate image registration (whether point-based or image-based) is a computationally expensive task, especially when large images or point sets are involved and when the transformation space has many degrees of freedom. Hence, it is of interest to develop algorithms that are both accurate and efficient. The formulation that we shall consider in this chapter is based on the following characteristics, founded on the taxonomy proposed by Brown (1992):

Input space: The images are assumed to be presented as a discrete set of two-dimensional feature points. In our experiments, extracted feature points were based on the most significant coefficients of a wavelet decomposition of each of the images (Le Moigne *et al.*, 2002).

Search space: Our software system supports affine registration transformations (allowing for translation, rotation, uniform and nonunifom scaling, and shearing). All of our experiments were conducted on a subspace consisting of similarity transformations (allowing for translation, rotation, and uniform scaling). The user provides intervals limiting the maximum and minimum degree of translation, rotation, and scaling.

Search strategy: Our algorithm is based on a search of the transformation space for the optimal aligning transformation. Specifically, it employs a geometric variant of a branch-and-bound search. See Section 8.3 for a detailed description. This is an extension of the algorithm presented by Mount *et al.* (1999).

Distance metric: We used two different robust measures as the objective function of our search algorithm, the (directed) partial Hausdorff distance (PHD) (Huttenlocher and Rucklidge, 1993; Huttenlocher *et al.*, 1993b) and a new smoothed version of the symmetric difference distance called the *discrete Gaussian mismatch* (DGM). Both measures are robust in that they allow for missing, as well as spurious data points.

A large number of papers have been written on the point pattern matching problem in the fields of computer vision, pattern recognition, and computational geometry. It is beyond the scope of this chapter to survey the area in detail, and so we will focus on the most relevant results. Perhaps the simplest similarity among point sets involve the Hausdorff distance and its variants (Huttenlocher *et al.*, 1993a; Alt *et al.*, 1994; Chew *et al.*, 1997; Goodrich *et al.*, 1999). The standard notion of Hausdorff distance is not suitable for our application, since it requires that every point (from at least one set) have a nearby matching point in the other set. Computing the optimal alignment of two-point sets even under the relatively simple Hausdorff distance is computationally intensive. In an attempt to circumvent the high complexity of point pattern matching, some researchers have considered *alignment-based* algorithms. These algorithms use alignments between small subsets of points to generate potential aligning transformations, the best of which are then subjected to more detailed analysis. Examples of these approaches include work in the field of image processing (Stockman *et al.*, 1982; Goshtasby and Stockman 1985; Goshtasby *et al.*, 1986) and in the field of computational geometry (Heffernan and Schirra, 1994; Goodrich *et al.*, 1999; Gavrilov *et al.*, 2004; Cho and Mount, 2005; Choi and Goyal, 2006). Alignments can also be part of a more complex algorithm. For example, Kedem and Yarmovski (1996) presented a method for performing stereo matching under translation based on propagation of local matches for computing good global matches.

For our applications it will be important that the distance measure be robust, in the sense that it is insensitive to a significant number of feature points from either set that have no matching point in the other set. Examples of a robust distance measures include the partial Hausdorff distance (PHD) (Huttenlocher and Rucklidge, 1993; Huttenlocher *et al.*, 1993b) and symmetric and absolute differences (Alt *et al.*, 1996; Hagedoorn and Veltkamp, 1999). (See Section 8.2 for definitions.)

In this chapter we discuss a number of extensions to the prior work of Mount *et al.* (1999) on the problem of feature-based image registration. We have extended the software system of theirs to include the following new elements:

New distance measure: In addition to PHD, we introduce a new distance measure, called the discrete Gaussian mismatch (DGM), which offers a number of advantages over PHD.

New search algorithms: In addition to the two search algorithms introduced by Mount *et al.* (1999) (pure branch-and-bound and bounded alignment), we introduce a new search algorithm called *bounded least-squares alignment*. We demonstrate that in many cases this new algorithm exhibits significantly better performance. We have also added new variants in which the aforementioned algorithms perform their search, and showed that these variants were considerably more efficient than those of Mount *et al.* (1999).

More extensive experiments: We extend the experimental results of Mount *et al.* (1999) to consider registration of satellite images arising from different platforms and covering different spectral bands.

The rest of this chapter is organized as follows. In Section 8.2 we present the two distance functions that will be used by our algorithm. In Section 8.3, we present our registration algorithm. In Section 8.4 we discuss the results of our experiments on these algorithms.

8.2 Distance measures

The point-based registration can be defined abstractly as follows: We are given two point sets A and B. We refer to A as the *input set* and B as the *reference set*. We are given a space $\mathcal{T}$ of geometric transformations, including some a-priori limits on the range of transformations. (For example, we may limit the range of rotations to some interval of angles.) We are also given some distance function that measures the degree of dissimilarity of the two-point sets. The problem is to find the transformation $\tau \in \mathcal{T}$ that minimizes the distance between $\tau(A)$ and B.

There are two natural sources of error. The feature extraction process is subject to *noise*, that is, small errors in the coordinates due to sensing errors and digitization. The second source of error is the presence of *outliers*, that is, feature points from either image that are not present in the other image. As mentioned in the introduction, outliers can result from many different sources, and may constitute a relatively large (often unknown) fraction of the feature points. Following terminology from statistics (Rousseeuw and Leroy, 1987), we say that a distance measure is *robust* if it is insensitive to the presence of outliers.

In this chapter we consider two robust distance measures. The first is the *partial Hausdorff distance* (PHD) introduced by Huttenlocher and Rucklidge (1993) and Huttenlocher *et al.* (1993b). Consider the set of distances resulting from taking each point in one set, and finding the nearest point to it in the other set. Rather than taking the sum or the maximum of these distances, which may be affected by outliers, we consider the median or, in general, the kth smallest distance. More formally, given a set $S \subset \mathbb{R}$, and $1 \le k \le |S|$, let $\text{rank}_k S$ denote the kth smallest element of S. Given a point set B and a point a, let $\text{dist}(a, B)$ denote the distance from a to its closest point of B. Given two-point sets A and B, and an integer parameter $1 \leq k \leq |A|$, the *directed partial Hausdorff distance* of order k from A to B is defined to be

$$\text{PHD}_k(A, B) = \text{rank}_k\{\text{dist}(a, B) \mid \forall a \in A\}.$$

(Note that the standard directed Hausdorff distance arises as a special case when $k = |A|$.) To avoid the dependence on the size of A, we will replace the integer parameter k with a *quantile* q, where $0 < q \leq 1$ represents the fraction of inliers of A. We then set $k = \lceil q \cdot |A| \rceil$. The resulting measure is denoted by $\text{PHD}_q(A, B)$.

One shortcoming of the PHD is the need to estimate the value of k (or q), that is, the expected number of inliers. Since this quantity depends on characteristics of the images (such as the degree of cloud cover) that may be unknown at the time of registration, we introduce an alternative distance measure. This measure is motivated by the symmetric difference of two sets, that is, the number of points that are present in one set but not in the other. To allow for the presence of noise, we attach a weight to the existence of a match by a function that decreases with the distances between each point of A and its nearest neighbor of B. The user provides a positive real parameter σ, which intuitively represents the standard deviation of a Gaussian distribution. Each point $a \in A$ is assigned a weight based on a variant of the Gaussian distribution function applied to the distance to its nearest neighbor of B, such that,

$$w_\sigma(a) = \exp\left(-\frac{\text{dist}(a, B)^2}{2\sigma^2}\right).$$

Note that the weight is 1 if and only if a coincides with a point of B, and (depending on σ) decreases to zero as the distance increases. We define the *discrete Gaussian mismatch distance* (DGM) between A and B to be

$$\text{DGM}_\sigma(A, B) = 1 - \frac{\sum_{a \in A} w_\sigma(a)}{|A|}.$$

Observe that if every point $a \in A$ coincides with some point $b \in B$, then $\text{DGM}_\sigma(A, B) = 0$, and the distance increases to a maximum value of 1 as the degree dissimilarity between the two-point sets increases.

8.3 Framework of the registration algorithm

Our registration algorithms are based on a geometric branch-and-bound framework. This framework has been used by others including Huttenlocher and Rucklidge (1993), Rucklidge (1996), Rucklidge (1997), Hagedoorn and Veltkamp (1999), and Mount *et al.* (1999). Recall that we are given two-point sets, an input set A and a reference set B and a space of transformations $\mathcal{T}$. The problem is to find $\tau \in \mathcal{T}$

that minimizes the distance function, which is either PHD or DGM. The PHD is also parameterized by the *inlier quantile* $0 < q \leq 1$ and the DGM is parameterized by the *standard deviation* σ. Let us assume that A and B will be fixed for the remainder of the discussion, and let us define $\mathrm{PHD}_q(\tau)$ and $\mathrm{DGM}_\sigma(\tau)$ to be the respective distance measures between the transformed input set $\tau(A)$ and the reference set B. Let $\mathrm{PHD}_{\mathrm{opt}}$ and $\mathrm{DGM}_{\mathrm{opt}}$ denote the minimum distances under PHD and DGM, respectively, over all transformations $\tau \in \mathcal{T}$.

There are a number of different ways to represent a transformation of $\mathcal{T}$. Henceforth we assume that $\mathcal{T}$ consists of the space of geometric similarities (allowing for rotation, translation, and uniform scaling). We represent each such transformation by a four-element vector, whose entries are the rotation θ, the translation vector (t_x, t_y), and the scaling factor s.

There are also a number of ways to define the approximation error for each of our distance measures. We introduce four nonnegative error parameters:

- ε_{rm}: the relative metric error bound,
- ε_{am}: the absolute metric error bound,
- ε_{rq}: the relative quantile error bound,
- ε_{aq}: the absolute quantile error bound.

Only three of these parameters will be relevant to a particular distance measure. Intuitively, the metric error involves errors in the distance between points and quantile error involves errors in the number of points. First, for PHD, define $q^- = (1 - \varepsilon_{rq})q$. Note that since $q^- \leq q$, we have $\mathrm{PHD}_{q^-}(\tau) \leq \mathrm{PHD}_q(\tau)$, for any τ. We say that a transformation τ is *approximately optimal* for PHD relative to these parameters if either

$$\mathrm{PHD}_{q^-}(\tau) \leq (1 + \varepsilon_{rm})\,\mathrm{PHD}_{\mathrm{opt}} \qquad \text{or} \qquad \mathrm{PHD}_{q^-}(\tau) \leq \mathrm{PHD}_{\mathrm{opt}} + \varepsilon_{am}.$$

Thus, the approximate PHD solution is allowed to be less robust by a factor of $(1 - \varepsilon_{rq})$, and it may exceed the optimum distance by a relative error of ε_{rm} or an absolute error of ε_{am}.

For DGM, define $\sigma^+ = (1 + \varepsilon_{rm})\sigma$. Observe that $\mathrm{DGM}_{\sigma^+}(\tau) \leq \mathrm{DGM}_\sigma(\tau)$, for any τ. We say that a transformation τ is *approximately optimal* for DGM relative to these parameters if either

$$\mathrm{DGM}_{\sigma^+}(\tau) \leq (1 + \varepsilon_{rq})\,\mathrm{DGM}_{\mathrm{opt}} \qquad \text{or} \qquad \mathrm{DGM}_{\sigma^+}(\tau) \leq \mathrm{DGM}_{\mathrm{opt}} + \varepsilon_{aq}.$$

Thus, the approximate DGM solution is allowed to match points in a neighborhood that is larger by a factor of $(1 + \varepsilon_{rm})$, and it may exceed the optimum mismatch distance by a relative error of ε_{rq} or an absolute error of ε_{aq}. Note that the parameters ε_{rm} and ε_{rq} are used for both distance functions. Their meanings are slightly different but closely related.

Note that the earlier branch-and-bound algorithms of Huttenlocher and Rucklidge (1993) and Hagedoorn and Veltkamp (1999) implicitly provide ε_{am} but not the other two parameters. The algorithm of Mount *et al.* (1999) does allow for the same error settings as used for PHD above, but it does not include the DGM distance.

We begin by describing the general structure of the algorithm. We will not prove the algorithm's correctness formally, but it is a straightforward modification of the proof of Mount *et al.* (1999). As mentioned earlier, the algorithm is based on a geometric branch-and-bound search of transformation space. The algorithm implicitly generates a search tree, where each node of the tree is identified with an axis-parallel hyperrectangle in four-dimensional transformation space (for rotation, *x*-translation, *y*-translation, and scale). Each such rectangle, or *cell*, represents a subspace of possible transformations. The user provides initial limits on the subset of transformations to be considered, and the root of the search tree is associated with the associated cell.

The search processes each cell in a recursive manner, starting with the root cell. At any time there are a collection of *active cells* and a candidate transformation that is the best seen so far by the search. Let τ^* denote this best transformation, and let dist* denote the associated distance. When it can be determined that a cell does not contain a transformation whose distance is smaller than dist*, the algorithm *kills* the cell, that is, it eliminates it from further consideration. If a cell cannot be killed then it is processed as discussed below. The processing will generally involve hierarchically partitioning the cell into smaller cells, which will then be added to the list of active cells. The algorithm terminates when all cells have been killed, or when a user-supplied upper bound on the maximum number of cells to be processed has been exceeded. Upon termination, the best transformation encountered, τ^*, is returned.

Let us now present the algorithm in greater detail. For each cell T that we process, we are interested in the transformation of this cell, for which the distance measure is smallest. We compute an upper bound dist$^+(T)$ and a lower bound dist$^-(T)$ on this smallest distance (explained below). For each upper bound, there will be a specific transformation that serves as a witness to this upper bound.

Upper bound. To compute the upper bound, we may sample any transformation from within the cell. There are a few ways in which to do this. Our software implements three different approaches. The first two were introduced by Mount *et al.* (1999), and the third is new to this chapter.

Pure (PURE): The midpoint of the cell is selected as the candidate transformation to be used for the upper bound.

Bounded Alignment (BA): When the cell satisfies a given set of conditions (details given by Mount *et al.* (1999)) a small number of point pairs is sampled repeatedly

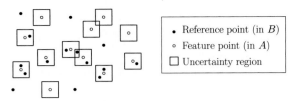

Figure 8.1. Uncertainty regions.

and at random from the subset of points of A that have a unique nearby point of B (after applying the cell's midpoint transformation). Each such pair $a_1, a_2 \in A$ is associated with its respective closest points $b_1, b_2 \in B$. There is a unique transformation τ mapping the pair (a_1, a_2) to (b_1, b_2). The distance (i.e., the similarity measure) of this transformation is computed. The transformation that produces the smallest distance is chosen to provide the upper bound for the cell. In our experiments, 10 samples were used.

Bounded Least-Squares Alignment (BLSA): Each point of A is associated with its closest point of B (after applying the cell's midpoint transformation). Given the resulting correspondences, the similarity transformation that minimizes the sum of squared distances between corresponding points is computed. This is done by an approach that first computes the transformation that aligns the centroids of the point sets, then computes the scale factor that aligns their spatial variances, and finally computes the rotation that minimizes the sum of squared distances. This is similar to an approach by Goshtasby *et al.* (1986) and Agarwal and Phillips (2006).

Nearest neighbors are computed by storing the points of B in a kd-tree data structure, and applying known efficient search techniques (Friedman *et al.*, 1977; Arya *et al.*, 1998; Arya and Mount, 2001). Let $\text{dist}^+(T)$ be the distance of the resulting sampled transformation.

Lower bound. To compute the lower bound, we use a technique presented by Mount *et al.* (1999), which is similar to that described by Huttenlocher and Rucklidge (1993) and Hagedoorn and Veltkamp (1999). Given any cell $T \subset \mathcal{T}$, and given any point $a \in A$, consider the image of a under *every* $\tau \in T$.

We compute a bounding rectangle enclosing this region, which we call the *uncertainty region* of a relative to T. In this way, each cell is associated with a collection of uncertainty regions, one for each point of A. (See Fig. 8.1.)

Define the distance between an uncertainty region and a point $b \in B$ to be the minimum distance between b and any part of the uncertainty region. (If b lies inside the uncertainty region, then the distance is zero.) To derive our lower bound for T, for each point $a \in A$, we compute the distance from the corresponding uncertainty region to its nearest neighbor of B. Observe that this distance is a lower bound on the distance from $\tau(a)$ to its nearest neighbor of B, for any $\tau \in T$. We then apply the distance computation to these lower bounds. For example, for the PHD we compute the qth smallest among these nearest neighbor distances, and for DGM we compute the match weights based on these distances. Let $\text{dist}^-(T)$ denote the result. Because it can never overestimate the distance from any point of A to its closest point of B, it is a lower bound on the actual PHD or DGM distance for any $\tau \in T$, and hence is indeed a lower bound for the cell. The nearest neighbor to an uncertainty region is computed by a straightforward generalization of the kd-tree-based nearest neighbor method described above.

Cell processing. As mentioned above, the algorithm operates by selecting an active cell T and processing it. Processing consists of the following steps. (A more detailed description is given by Mount *et al.*, 1999.) First, we compute the uncertainty regions for each point a and apply the aforementioned procedures to compute the upper and lower bound distances, $\text{dist}^+(T)$ and $\text{dist}^-(T)$. In the case of the PHD distance we kill the cell if either of the following two conditions hold:

$$\text{dist}^-(T) > \frac{\text{dist}^*}{1 + \varepsilon_{rm}} \qquad \text{or} \qquad \text{dist}^-(T) > \text{dist}^* - \varepsilon_{am}.$$

The rule for DGM is the same but replacing metric error parameters with their quantile counterparts. If $\text{dist}^+(T) < \text{dist}^*$, we set $\text{dist}^* \leftarrow \text{dist}^+(T)$ and save the associated transformation in τ^*. We then split the cell into two smaller subcells, T_1 and T_2, which replace T in the set of active cells.

There are three strategies that we implemented for selecting the next cell to be processed:

Maximum Uncertainty (MAXUN): The next cell is the active cell with the largest average diameter of its uncertainty regions.

Minimum Upper Bound (MINUB): The next cell is the one with the smallest upper bound.

Minimum Lower Bound (MINLB): The next cell is the one with the smallest lower bound.

We refer to the above choices as the *search priority*. The MAXUN method was used by Mount *et al.* (1999) and the other two are new. Based on the fact that it demonstrated the best performance in our preliminary analysis, we used MINLB in all of our experiments.

8.3.1 Multiresolution registration

A common approach for improving the efficiency of image registration is to apply a multiresolution framework. This framework has been considered extensively in the context of image registration for remotely sensed images. See, for example,

Cole-Rhodes *et al.* (2003), Netanyahu *et al.* (2004), and Zavorin and Le Moigne (2005). The approach involves representing the two images at a series of increasingly finer spatial resolutions. Feature points are extracted from each of these images. This is followed by progressive registration of the resulting point sets by applying the registration process from the coarsest level to the finest. As we proceed from one level to the next, the spatial resolution increases by a factor of two. Thus, the coarsest level involves the greatest degree of spatial uncertainty, but also involves the smallest number of feature points. The process begins by registering the images at the coarse level, using a significantly wider range of transformations. The transformation generated by our program at each stage is used as a center point for the transformation cell at the next stage. Thus, as we proceed level by level, the accuracy of the aligning transformation is expected to improve, while the running time increases, due to the increased number of feature points expected with higher-resolution images.

As observed by Zavorin and Le Moigne (2005), there are a number of advantages of using a multiresolution approach compared to working solely with the original images. It can reduce computation time by performing much of the work at coarse resolutions, leaving minor adjustments to later stages. Since this type of image decomposition usually involves low-frequency smoothing, this regularizes the registration problem thus yielding better convergence properties and improved accuracy of the search algorithm. Finally, if image scales differ significantly, decomposition could be used to bring the two images into similar scales, which may be advantageous for some registration algorithms.

8.4 Experimental studies

In order to assess the performance of our registration algorithms, we have implemented a number of variants and tested their performance on a combination of remotely-sensed satellite imagery. The algorithms have been implemented in C++ (g++ version 3.2.3), and all experiments were run on a PC with a 2.4-GHz processor running Linux 2.4. Nearest neighbor and range queries were performed using kd-trees as generated by the ANN library for approximate nearest neighbor searching (Arya *et al.*, 1998). In particular, we were interested in studying the relative performance of:

Distance function: DGM vs. PHD.
Search algorithm: PURE vs. BA vs. BLSA.

Our experiments involved satellite images that were taken from three distinct locations: *Konza* (Konza Prairie in the state of Kansas, July to August 2001), *Virginia* (Virginia's Hog Island Coast Reserve Area, October 2001), and *Cascades*

(Cascades Mountains, September 2000). In each case the images were taken from two satellite platforms (sensors), IKONOS (4 m per pixel) and ETM+ (30 m per pixel), and involved the red and near infrared (NIR) spectral bands. Thus, by considering all possible combinations, we have four images, denoted IKONOS-red (IR), IKONOS-NIR (IN), ETM+-red (ER), and ETM+-NIR (EN), for each location, resulting in six possible ways of pairing them for registration. Some examples are shown in Fig. 8.2.

We tested the performance of our algorithms for both the single-pair and multi-resolution frameworks. The results of the multiresolution experiments will be discussed in Subsection 8.4.3. In all of our experiments we considered matches under similarity transformations. Unless otherwise stated, the transformation width allows $4°$ of rotation, 4 pixels of x- and y-translation, and 20% of scaling. The initial cell of the search was centered at a random point whose maximum distance from the ground truth transformation was 25% of the transformation width. For the PHD we used an inlier quantile of $q = 0.5$ and for DGM we used a standard deviation of $\sigma = 1.0$. In all cases the program was allowed to execute for at most 10 000 cells, but it usually terminated well before then. We also used the following settings for the various error parameters:

Relative metric error: $\varepsilon_{rm} = 0.1$ (for both PHD and DGM),
Relative quantile error: $\varepsilon_{rq} = 0.2$ (for both PHD and DGM),
Absolute metric error: $\varepsilon_{am} = 0.4$ (for PHD),
Absolute quantile error: $\varepsilon_{aq} = 0.05$ (for DGM).

Most of our experiments involve computing two measures of performance. The first is running time, measured in CPU seconds. The second is a measure of accuracy, called the *transformation distance*. This is designed to measure how close the computed transformation is to our best estimate of "ground truth." We estimated the ground truth by visual inspection of the datasets and consensus of other image registration programs combined with prior analysis of these image (Le Moigne *et al.*, 2003). Examples of alignments produced for three of the data sets under the ground truth transformation are shown in Fig. 8.2. The transformation distance of a transformation τ is defined to be the average Euclidean distance of each point $p \in A$ of the input dataset from its image under this transformation, $\tau(p)$, and its image under the ground truth transformation.

8.4.1 Experiment 1: Comparison of distance functions

Our first experiment involves a comparison of the effectiveness of each of the distance functions. We used the simplest of the algorithms, namely pure branch-and-bound (PURE) together with the minimum lower bound (MINLB)

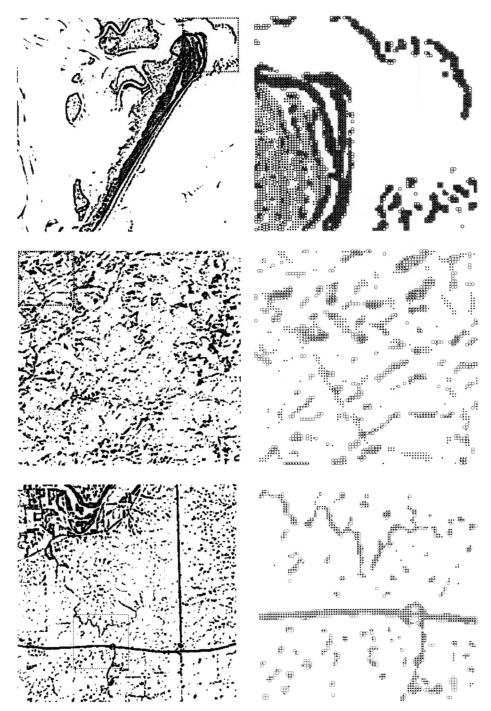

Figure 8.2. Feature-point sets for our experiments. Pixels of input and reference images are shown, respectively, as hollow and black points. Both sets are shown on the left, under our estimate of best aligning transformation. Detailed figures of highlighted subimages are shown on the right.

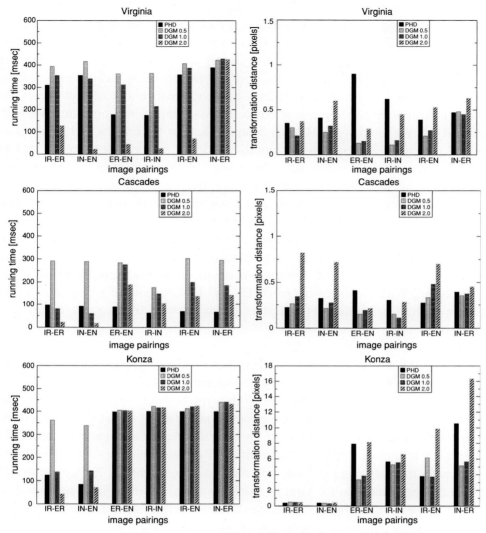

Figure 8.3. Results of Experiment 1 comparing the various distance functions. Relative performance, in terms of speed and accuracy, is shown as a function of distance measures.

search priority. We tested all six combinations of sensor-band pairs for each of the three images. For the discrete Gaussian mismatch, we tested values of $\sigma \in \{0.5, 1.0, 2.0\}$. For each experiment, we computed both execution time and transformation distance, and reported the average over five trials in each case. The results are presented in Fig. 8.3.

There are a number of conclusions that can be drawn from the experiments. First, increasing the standard deviation parameter σ in DGM tends to result in faster execution times, but also results in poorer performance with respect to transformation distance. (This is most clearly seen in the cases of Virginia and Cascades.) This is because increasing σ has the effect of making each feature point "fuzzier," which in turn makes it easier to localize matches but makes the algorithm less sensitive to minor errors in placement. If we compare PHD with DGM 0.5, we see that DGM takes comparable time to compute and achieves as good or better transformation distances than PHD. In one case (Virginia IR-IN) the difference is quite dramatic. The problem with this dataset is evident from Fig. 8.2. This dataset has a high number of distinctive features that match very well, and it has a much lower number of outliers that do not match at all. The DGM measure seeks to match as many points as possible, while PHD is satisfied once it has matched the given quantile of points (which was $q = 0.5$ in this case). We feel that its greater degree of sensitivity to the actual number of outliers is the principal strength of DGM. Note that many of the Konza registrations were not successful. With the exception of IR-ER and IN-EN (both of which involve the same spectral band) the sets of feature points between the two images are very different.

A deeper understanding of the nature of distance functions is illustrated in Fig. 8.4. We first computed the value of the distance function with respect to ground truth. We then applied an additional horizontal shift of the input set A to both the left and right and reevaluated the cost function. One would expect the distance function to achieve a minimum at an offset of 0 and then to increase on either side. The ideal shape of an objective function is one that gradually descends towards a single, well-concentrated global minimum (at 0). The figure illustrates the challenges of doing this with the existing images. First, observe that both distance functions for Virginia exhibit not one, but many local minima. The PHD cost function is worse (note the different vertical scales), since it exhibits multiple local minima with identical distance values for $q = 0.25$ and $q = 0.5$. Thus, it is not surprising that the algorithm does not distinguish among these minima and produces an erroneous transformation. In the case of DGM, as the value of σ increases, the local minima are smoothed out (which explains the faster execution times) but the accuracy decreases as well. In the case of Konza, the objective function shows a single global minimum, but the objective function is not well concentrated for all parameter settings. Setting the σ value too small is problematic (as seen in the case of Konza for $\sigma = 0.1$) since there may be no trans-formation under which a significant number of feature points match within the σ bound.

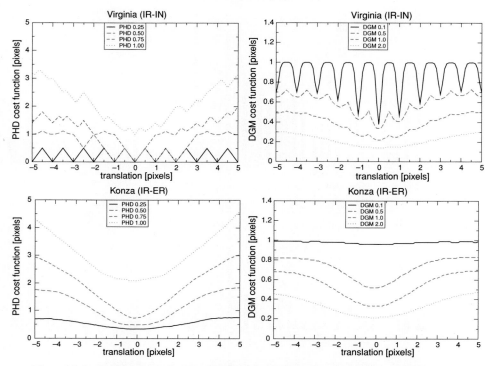

Figure 8.4. Comparison of objective functions subject to a horizontal shift relative to the ground truth transformation (at 0).

8.4.2 Experiment 2: Different sensors and bands

Given the relatively large number of possible experimental combinations (six pairs for registration from each of the three principal locations, for a total of 18), in our next experiment we compared the performance of the algorithms in all instances, in order to identify a relatively small set of representative cases. The hypothesis on which this experiment is based, is that the greater degree of commonality among the input images, the easier the registration should be in terms of running time and accuracy. Because some features are apparent only at certain spectral bands, we have noted that in the images tested, differences in the spectral band seem to be more significant the differences in the sensor. To test these effects, we grouped registration pairings into the following three groups:

Case 1: Images of the same spectral band but from different sensors (IR-ER, IN-EN).
Case 2: Images from the same sensor but from different spectral bands (ER-EN, IR-IN).
Case 3: Images from different spectral bands and different sensors (IR-EN, IN-ER).

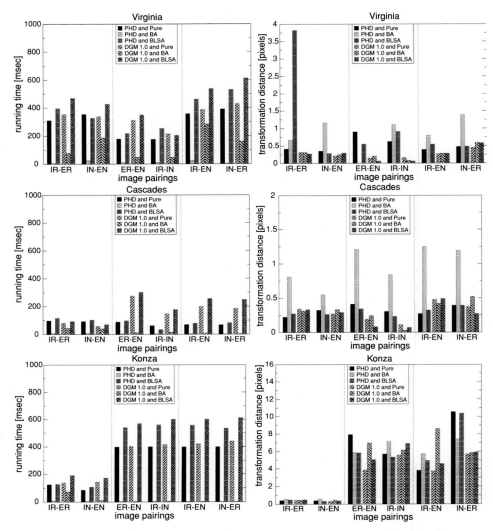

Figure 8.5. Results of Experiment 2 comparing the performance of the algorithms for various choices of sensor and spectral bands. See color plates section.

For each of the three locations, we ran all six combinations of image pairings. The interpretation of the labels is given below. Based on the results of Experiment 1, we set $\sigma = 0.5$ for our experiments involving DGM, since this almost always produced the most accurate results. Throughout we considered matching under similarity transformations and used MINLB as the search priority. We tested all three search algorithms. As before, we measured execution time and transformation distance averaged over five trials. The results are presented in Fig. 8.5. Each plot is split into three groups; the leftmost group corresponds to Case 1, the middle group to Case 2, and the right group to Case 3.

The experiments show that there are similarities among the various cases. This is most dramatically true for Konza and Cascades, where the patterns of execution times and transformation distances are notably similar within each group and dissimilar between groups. Among the different search algorithms, PURE and BLSA demonstrated generally steady and predictable performance. In contrast, BA was almost always the fastest of the algorithms, but it demonstrated the highest degree of variation in the quality of the results. In some instances (e.g., Virginia ER-EN), the BA algorithm convincingly outperformed the optimal transformation. However, in other instances (e.g., Virginia IR-EN and IN-ER), its performance was significantly worse. Inspection of the individual trials showed that in two out of five trials it found the optimal transformation, and in three cases it was off by a full pixel.

8.4.3 Experiment 3: Multiresolution framework

For this experiment we considered the performance of the algorithm in a multiresolution framework as described in Subsection 8.3.1. The algorithm was applied to four different resolution levels, and in each case the output from one level was used as the starting transformation for the next level. At the coarsest level of resolution we used a relatively high range of transformations, allowing for $16°$ of rotation, 32 pixels of x- and y-translation, and 30% of scaling. At all the other levels the transformation width allows $6°$ of rotation, 6 pixels of x- and y-translation, and 30% of scaling. Subsequent levels used the more restrictive transformation ranges described at the start of Section 8.4. Otherwise, we used the same parameter settings as in Experiment 2. As always, the results were averaged over five trials, each with a different random starting transformation. The other parameter settings were the same as in Subsection 8.4.2. We tested three cases, Virginia IR-IN (Case 2), Cascades IN-ER (Case 2), and Konza IR-ER (Case 1).

To determine the relative performance of our algorithms in the multiresolution framework, we measured execution time and transformation distance. The results are shown in Fig. 8.6. (Note that plots are on a logarithmic scale, and values less than 0.01 have been rounded up to 0.01.) A number of trends are apparent from the plots. First, as expected, the running times of the algorithm increase roughly exponentially with each subsequent level, since the image sizes and, hence, number of feature points increase similarly. Also, as expected, the transformation distances tend to decrease monotonically, since the accuracy of the feature points is increasing. There are two notable exceptions. In the case of PHD for Virginia IR-IN, the accuracy either exhibits very little change or actually gets worse (in the case of BLSA). Virginia is known to be a hard case for PHD, and this anomalous behavior reflects

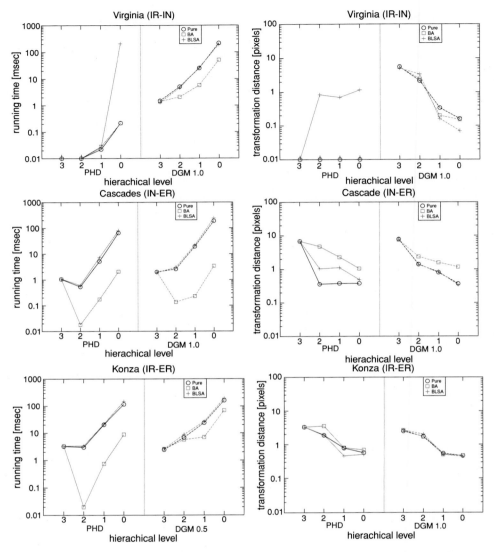

Figure 8.6. Results of Experiment 3 on multiresolution registration. The units for transformation distance are pixels at the highest level of resolution.

this fact. In contrast, DGM does quite well in this case. Other than this anomaly, both distance functions and all algorithms tended to perform quite similarly. As in the previous experiment, BA is the fastest of the methods. In terms of accuracy, it tends to be slightly worse in some instances than others.

Conclusions

In this chapter we have presented a number of enhancements to a feature-based registration algorithm introduced by Mount *et al.* (1999). In particular, we have considered a new distance measure, the discrete Gaussian mismatch (DGM), new search algorithms based on a new method (BLSA) for computing the upper bound associated with a cell, and variants for selecting the next active cell to be processed, based on the cell's lower or upper bounds (MINLB and MINUB, respectively). Our experimental studies show that DGM is almost always as good as PHD, providing significant improvement in a few difficult cases. The other innovations offer tangible improvements, but these improvements are less pervasive and of lesser significance. Finally, we have demonstrated the algorithm's efficacy in significantly more general instances than reported by Mount *et al.* (1999), by considering images from different sensors and covering various spectral bands. We have further demonstrated the algorithm's efficacy in both a traditional single-pair framework and in a multiresolution framework.

Acknowledgements

The work of David Mount was partially supported by the National Science Foundation under grant CCR-0635099 and the Office of Naval Research under grant N00014-08-1-1015. The work of San Ratanasanya was supported in part by the National Science Foundation under grant CCR-0635099 while the author was visiting in the Department of Computer Science at the University of Maryland, College Park.

References

Agarwal, P. K. and Phillips, J. M. (2006). On bipartite matching under the RMS distance. In *Proceedings of the Eighteenth Canadian Conference on Computational Geometry*, Kingston, Canada, pp. 143–146.

Alt, H. and Guibas, L. J. (1999). Discrete geometric shapes: Matching, interpolation, and approximation. In J.-R. Sack and J. Urrutia, eds., *Handbook of Computational Geometry*. Amsterdam: Elsevier Science B.V., North-Holland, pp. 121–153.

Alt, H., Aichholzer, O., and Rote, G. (1994). Matching shapes with a reference point. In *Proceedings of the Tenth Annual ACM Symposium on Computational Geometry*, Stony Brook, NY, pp. 85–91.

Alt, H., Fuchs, U., Rote, G., and Weber, G. (1996). Matching convex shapes with respect to the symmetric difference. In *Proceedings of the Fourth Annual European Symposium on Algorithms*. London: Springer-Verlag, pp. 320–333.

Arya, S. and Mount, D. M. (2001). Approximate range searching. *Computational Geometry: Theory and Applications*, **17**, 135–163.

Arya, S., Mount, D. M., Netanyahu, N. S., Silverman, R., and Wu, A. (1998). An optimal algorithm for approximate nearest neighbor searching. *Journal of the ACM*, **45**, 891–923.

Brown, L. G. (1992). A survey of image registration techniques. *ACM Computing Surveys*, **24**, 325–376.

Chew, L. P., Goodrich, M. T., Huttenlocher, D. P., Kedem, K., Kleinberg, J. M., and Kravets, D. (1997). Geometric pattern matching under Euclidean motion. *Computational Geometry Theory and Applications*, **7**, 113–124.

Cho, M. and Mount, D. M. (2005). Improved approximation bounds for planar point pattern matching. In *Proceedings of the Ninth Workshop on Algorithms and Data Structures*, Waterloo, Canada; *Lecture Notes in Computer Science*, Springer-Verlag, Vol. 3608, pp. 432–443.

Choi, V. and Goyal, N. (2006). An efficient approximation algorithm for point pattern matching under noise. In *Proceedings of the Seventh Latin American Theoretical Informatics Symposium*, Valdivia, Chile; *Lecture Notes in Computer Science*, Springer-Verlag, Vol. 3887, pp. 298–310.

Cole-Rhodes, A. A., Johnson, K. L., Le Moigne, J., and Zavorin, I. (2003). Multiresolution registration of remote sensing imagery by optimization of mutual information using a stochastic gradient. *IEEE Transactions on Image Processing*, **12**, 1495–1511.

Friedman, J. H., Bentley, J. L., and Finkel, R. A. (1977). An algorithm for finding best matches in logarithmic expected time. *ACM Transactions on Mathematical Software*, **3**, 209–226.

Gavrilov, M., Indyk, P., Motwani, R., and Venkatasubramanian, S. (2004). Combinatorial and experimental methods for approximate point pattern matching. *Algorithmica*, **38**, 59–90.

Goodrich, M. T., Mitchell, J. S. B., and Orletsky, M. W. (1999). Approximate geometric pattern matching under rigid motions. *IEEE Transactions on Pattern Analysis and Machine Intelligence*, **21**, 371–379.

Goshtasby, A. and Stockman, G. C. (1985). Point pattern matching using convex hull edges. *IEEE Transactions on Systems, Man, and Cybernetics*, **15**, 631–637.

Goshtasby, A., Stockman, G. C., and Page, C. V. (1986). A region-based approach to digital image registration with subpixel accuracy. *IEEE Transactions on Geoscience and Remote Sensing*, **24**, 390–399.

Hagedoorn, M. and Veltkamp, R. C. (1999). Reliable and efficient pattern matching using an affine invariant metric. *International Journal of Computer Vision*, **31**, 103–115.

Heffernan, P. J. and Schirra, S. (1994). Approximate decision algorithms for point set congruence. *Computational Geometry Theory and Applications*, **4**, 137–156.

Huttenlocher, D. P. and Rucklidge, W. J. (1993). A multi-resolution technique for comparing images using the Hausdorff distance. In *Proceedings of the IEEE Conference on Computer Vision and Pattern Recognition*, New York, pp. 705–706.

Huttenlocher, D. P., Kedem, K., and Sharir, M. (1993a). The upper envelope of Voronoi surfaces and its applications. *Discrete and Computational Geometry*, **9**, 267–291.

Huttenlocher, D. P., Klanderman, G. A., and Rucklidge, W. J. (1993b). Comparing images using the Hausdorff distance. *IEEE Transactions on Pattern Analysis and Machine Intelligence*, **15**, 850–863.

Kedem, K. and Yarmovski, Y. (1996). Curve based stereo matching using the minimum Hausdorff distance. In *Proceedings of the Twelfth Annual ACM Symposium on Computational Geometry*, Philadelphia, Pennsylvania, pp. C15–C18.

Le Moigne, J., Campbell, W. J., and Cromp, R. F. (2002). An automated parallel image registration technique of multiple source remote sensing data. *IEEE Transactions on Geoscience and Remote Sensing*, **40**, 1849–1864.

Le Moigne, J., Morisette, J., Cole-Rhodes, A., Netanyahu, N. S., Eastman, R., and Stone, H. (2003). Earth science imagery registration. In *Proceedings of the IEEE Geoscience and Remote Sensing Symposium*, Toulouse, France, pp. 161–163.

Mount, D. M., Netanyahu, N. S., and Le Moigne, J. (1999). Efficient algorithms for robust point pattern matching. *Pattern Recognition*, **32**, 17–38.

Netanyahu, N. S., Le Moigne, J., and Masek, J. G. (2004). Georegistration of Landsat data via robust matching of multiresolution features. *IEEE Transactions on Geoscience and Remote Sensing*, **42**, 1586–1600.

Rousseeuw, P. J. and Leroy, A. M. (1987). *Robust Regression and Outlier Detection*. New York: Wiley.

Rucklidge, W. J. (1996). Efficient visual recognition using the Hausdorff distance. *Lecture Notes in Computer Science*, Vol. 1173. Berlin: Springer-Verlag.

Rucklidge, W. J. (1997). Efficiently locating objects using the Hausdorff distance. *International Journal of Computer Vision*, **24**, 251–270.

Stockman, G. C., Kopstein, S., and Benett, S. (1982). Matching images to models for registration and object detection via clustering. *IEEE Transactions on Pattern Analysis and Machine Intelligence*, **4**, 229–241.

Ton, J. and Jain, A. K. (1989). Registering Landsat images by point matching. *IEEE Transactions on Geoscience and Remote Sensing*, **27**, 642–651.

Zavorin, I. and Le Moigne, J. (2005). Use of multiresolution wavelet feature pyramids for automatic registration of multisensor imagery. *IEEE Transactions on Image Processing*, **14**, 770–782.

Zitová, B. and Flusser, J. (2003). Image registration methods: A survey. *Image and Vision Computing*, **21**, 977–1000.

9

Condition theory for image registration and post-registration error estimation

C. S. KENNEY, B. S. MANJUNATH, M. ZULIANI, AND K. SOLANKI

Abstract

We present in this chapter applications of condition theory for image registration problems in a general framework that is easily adapted to a variety of image processing tasks. After summarizing the history and foundations of condition theory, a short analysis is given of computational sensitivity for point correspondence between images with respect to translation, rotation-scale-translation (RST), and affine pixel transforms. Several surprising results follow from this analysis, including the principal result that increasing transform complexity is mirrored by increasing computational sensitivity, i.e., $K_{Trans} \leq K_{RST} \leq K_{Affine}$. The utility of condition-based corner detectors is also seen in the demonstrated equivalence between the translational condition number and the commonly used Shi-Tomasi corner function. These results are supplemented by a short discussion of sensitivity estimation for the computed transform parameters and any resulting registration misalignment.

9.1 Introduction

The central issue in image registration is the problem of establishing correspondence between image features, whether they are point features (e.g., corner locations) or extended features (e.g., level sets). Corresponding features then act as input for the process of computing a low-dimensional pixel map between images. The success of this approach depends on the accuracy of the feature correspondence, in the sense that mismatched features can lead to completely erroneous transform estimates. Mismatches may be due to computational constraints that limit the complexity of the feature-matching algorithm or may be intrinsic to the image pair as is the case with identical local features, such as windows in an office building. Poorly matched or completely mismatched features must be detected and eliminated before the image transform parameters are computed. This can be accomplished by utilizing condition theory, which provides a sound theoretical methodology for addressing this problem.

The goal of condition theory is to quantitatively evaluate the numerical sensitivity of computed results with regard to errors in the input information (Rice, 1966). In Kenney *et al.* (2003) the sensitivity of point matching objective functions with respect to translation, rotation-scale-translation (RST), and affine transforms was investigated with several surprising results.

First, it was established that for objective functions based on translational window matching, the resulting condition number is identical to the commonly used Shi-Tomasi corner function (Shi and Tomasi, 1994). Second, it was shown that the condition numbers for translation, RST, and affine transforms satisfy a nesting inequality with regard to increasing transformation complexity, i.e., $K_{Trans} \leq K_{RST} \leq K_{Affine}$.

More recent work, for example, by Zuliani *et al.* (2004) and Kenney *et al.* (2005) posited a group of desirable properties, such as rotational invariance, that one might expect in a reasonable corner detector. Analysis of these requirements for local corner detectors showed that only condition-based detectors satisfied all of the given properties for general images (e.g., color, multispectral, tomographic, etc.).

Over the past 30 years the problem of selecting image points for reliably determining optical flow, image registration parameters and 3D reconstruction has been extensively studied, and many good schemes for selecting feature points have been advanced. Kearney *et al.* (1987) showed that the normal matrix associated with locally constant optical flow is critical in determining the accuracy of the computed flow. This matrix has the form

$$A^T A \equiv \begin{bmatrix} \sum g_x g_x & \sum g_x g_y \\ \sum g_x g_y & \sum g_y g_y \end{bmatrix}, \tag{9.1}$$

where $g = g(x, y)$ refers to the image intensity, subscripts indicate differentiation, and the summation is over a window about the point of interest. Kearney *et al.* also reported that ill-conditioning in the matrix $A^T A$ and large residual error in solving the equations for optical flow could result in inaccurate flow estimates. This was supported by the work of Barron *et al.* (1994) who looked at the performance of different optical flow methods (see, for example, Beauchemin and Barron, 1996).

More recently, Shi and Tomasi (1994) presented a technique for measuring the quality of local windows for the purpose of determining image transform parameters (translational or affine). For local translation they argued that in order to overcome errors introduced by noise and ill-conditioning, the smaller eigenvalue of the normal matrix $A^T A$ must be above a certain threshold, i.e., $\lambda \leq \min(\lambda_1, \lambda_2)$, where λ is the prescribed threshold and λ_1, λ_2 are the eigenvalues of $A^T A$. When this condition is met, the point of interest has good features for tracking (Tommasini *et al.*, 1998). The latter paper builds on the work of Shi and Tomasi (1994) by examining the statistics of the residual difference between a window and a computed backtransform of the corresponding window in a second image with the goal of deriving conditions for rejecting a putative match. The papers (Brooks *et al.*, 2001) and (Kanazawa and Kanatani, 2001) give further consideration to the importance of the $A^T A$ in estimating vision parameters.

In Schmidt *et al.* (2000) the authors presented results for the problem of evaluating interest point detectors from the standpoint of repeatability (i.e., whether the point is repeatedly detected in a series or sequence of images) and information content (which measures the distinctiveness of the interest point by means of the likelihood of the local gray value descriptor). The above paper provides also a good survey of interest point detectors through contour-based methods (Zheng and Chellappa, 1993; Li *et al.*, 1995), corner detection (Harris and Stephens, 1988), intensity-based methods and parametric model approaches. The evaluation of computed video georegistration for aerial video was considered in Wildes *et al.* (2001).

Regardless of which procedure is used to identify and match corners, one still faces the problem of extracting transform parameters from a set of candidate matched points that may consist of more than 50% outliers, that is, the majority of the points are essentially mismatched. Because of this difficulty, all registration algorithms should employ (either explicitly or implicitly) a culling procedure that detects and eliminates mismatches from the subset of corresponding features that is used in computing the transform between images. The random sample consensus procedure (RANSAC) developed by Fischler and Bolles (1981) deals with the problem of fitting data to a given model class. It was developed specifically for handling the large number of mismatches that typically occur in feature-based image registration. Unlike least-squares approximation, RANSAC exhibits good performance when the data consist of a number of outliers, and has become the standard for culling mismatched candidate pairs before estimating the transform parameters.

Subsequent to parameter estimation one must face the problem of assessing the accuracy of the computed results. The condition-based assessment approach presented in this chapter can be extended to both the RANSAC-generated set of matched points and the pixel offset error in the registered images.

9.2 Point matching sensitivity

Matching points in two or more images is generally accomplished by first defining a scalar feature function that measures the suitability of a point or region for matching. This yields a suitability surface over the image with maxima corresponding to the best points for matching purposes. We illustrate this for window matching. Given two images g and $\hat{g}$ and a point (x, y), we define the point-matching objective function for a transformation T as

$$f(T) = \frac{1}{2} \sum \left[g(T(x', y')) - \hat{g}(x', y') \right]^2, \tag{9.2}$$

where the summation is over (x', y') in a window about (x, y).

We consider three types of transformations, each associated with a parameter vector p, and seek the optimal parameter vector that minimizes the objective function (9.2). The transformation types considered are:

(1) Translation, for which we seek the best shift $p = (a, b)$ that minimizes the objective function (9.2), where T is given by

$$T(x', y') = \begin{bmatrix} x' + a \\ y' + b \end{bmatrix}. \tag{9.3}$$

(2) Rotation-scale-translation (RST), for which we search for the best rotation-scale-translation vector $p = (\theta, r, a, b)$ that minimizes the objective function (9.2), where T is given by

$$T(x', y') = r \begin{bmatrix} \cos \theta & \sin \theta \\ -\sin \theta & \cos \theta \end{bmatrix} \begin{bmatrix} x' - x \\ y' - y \end{bmatrix} + \begin{bmatrix} a \\ b \end{bmatrix}. \tag{9.4}$$

(3) Affine, for which we need to find the best affine parameter vector $p = (m_{11}, m_{12}, m_{21}, m_{22}, a, b)$ that minimizes the objective function (9.2), where T is given by

$$T(x', y') = \begin{bmatrix} m_{11} & m_{12} \\ m_{21} & m_{22} \end{bmatrix} \begin{bmatrix} x' - x \\ y' - y \end{bmatrix} + \begin{bmatrix} a \\ b \end{bmatrix}. \tag{9.5}$$

To measure the sensitivity of the minimizing solutions we use the following definition.

Definition: Let $p = p(x, y, g, \hat{g}, T)$ denote the minimizer to the objective function (9.2). The condition number K_T, which measures the sensitivity of p to perturbations $(\Delta g, \Delta \hat{g})$, is defined by

$$K_T \equiv \lim_{\delta \to 0} \max_{\|(\Delta g, \Delta \hat{g})\| \leq \delta} \frac{\|\Delta p\|}{\|(\Delta g, \Delta \hat{g})\|}, \tag{9.6}$$

where $\|(\Delta g, \Delta \hat{g})\| = (\|\Delta g\|^2 + \|\Delta \hat{g}\|^2)^{1/2}$, and where Δp denotes the change in the parameter vector p due to the perturbation $(\Delta g, \Delta \hat{g})$.

In the next section we present the background and standard theory of condition measurement, which lead to Thm. (9.2) and the main result of Kenney *et al.* (2003), i.e., that $K_{Trans} \leq K_{RST} \leq K_{Affine}$. This result is very intuitive, in the sense that minimizing the three types of objective functions is analogous to trying to extract more information from a fixed dataset and should give results that are increasingly uncertain.

9.3 Condition theory

9.3.1 Historical background

Condition theory grew out of the derivation of the computerized procedure of solving large systems of linear equations and eigenproblems. The question facing investigators at that time was whether such problems could be solved reliably. The work of Wilkinson (1961), Wilkinson (1965) and Rice (1966) in the early 1960s established a general theory of condition for computed functions. In the 1970s this led to power method condition estimates for matrix inversion by Cline *et al.* (1979), as well as condition estimates for the exponential matrix function by Ward (1977) and Moler and Loan (1978), invariant subspaces associated with eigenvalues by Stewart (1973), and other problems. The 1980s saw the extension of this work (especially the ideas of Cline *et al.*, 1979) to the matrix square root function by Björck and Hammarling (1983), the Lyapunov and Riccati equations by (Hewer and Kenney, 1988; Kenney and Hewer, 1990), and the distance to the nearest unstable matrix by Loan (1985) and Hinrichsen and Pritchard (1986). At the same time, the work of Kenney and Laub (1989, 1994) provided computational procedures for estimating the sensitivity of general matrix functions, and Demmel *et al.* (Demmel, 1987, 1988; Demmel *et al.*, 2001) developed a global rather than local theory of sensitivity based on the distance to the nearest ill-posed problem. The latter work represents a step towards more realistic condition results since it goes beyond the assumptions of sensitivity based on smoothness and differentiability. This is especially appropriate for image processing since many perturbation effects, such as quantized gray level intensity changes, are discrete rather than continuous.

9.3.2 General condition measures

The solution of a system of equations can be viewed as a mapping from the input data $D \in R^n$ to the solution or output $X = X(D) \in R^m$. If a small change in D produces a large change in $X(D)$, then X is ill-conditioned at D. Following Rice (1966), we define the δ-condition number of X at D by

$$K_\delta = K_\delta(X, D) \equiv \max_{\|\Delta D\| \leq \delta} \frac{\|X(D + \Delta D) - X(D)\|}{\|\Delta D\|}, \tag{9.7}$$

where $\| \cdot \|$ denotes the vector 2-norm, i.e., $\|D\|^2 = \Sigma |D_i|^2$. For any perturbation ΔD with $\|\Delta D\| \leq \delta$, the perturbation in the solution satisfies $\|X(D + \Delta D) - X(D)\| \leq \delta K_\delta$. The condition number K_δ inherits any nonlinearity in the function X and consequently is usually impossible to compute. For this reason the standard procedure is to take the limit as $\delta \to 0$. If X is differentiable at D, then we can define the (local or differential) condition number $K = K(X, D) \equiv \lim_{\delta \to 0} K_\delta(X, D)$. Under suitable smoothness assumptions, a first-order Taylor expansion gives $X(D + \Delta D) = X(D) + X_D \Delta D + O(\|\Delta D\|^2)$, where X_D is the $m \times n$ gradient matrix with entries $(X_D)_{ij} = \dfrac{\partial X_i}{\partial D_j}$. This expansion shows that the local condition number is just the norm of the gradient matrix, i.e., $K(X, D) = \|X_D\|$.

9.3.3 Conditioning of objective function

The general theory of condition as applied to the sensitivity of the minimizer of the objective function (9.2) gives the definition of Eq. (9.6). This condition number incorporates all possible perturbations to an arbitrary pair of images $(g, \hat{g})$. Unfortunately, this generality makes the analysis difficult. To overcome this, we consider the condition number for a more reasonable class of problems, where g is related to $\hat{g}$ via the transform T and $\hat{g}$ is perturbed by noise. This leads to the following result (Kenney *et al.*, 2003):

Theorem 9.1 *Define the "transform plus noise" problem as* $\hat{g}(x', y') = g(T(x', y'))$ *with perturbations* $\Delta \hat{g}(x', y') = \hat{\eta}(x', y')$, *where* $\|\hat{\eta}\| \leq \delta$ *and* (x', y') *varies over the window centered at* (x, y). *In the limit, as* $\delta \to 0$ *we find that*

$$K_T^2 = \|(A^T A)^{-1}\|, \quad \text{where} \quad A = \begin{bmatrix} v^1 \\ \vdots \\ v^n \end{bmatrix} \qquad (9.8)$$

with row vectors v^i *depending on the type of transformation T. The row vectors are given by*

(1) $v^i = (\hat{g}_x^i, \hat{g}_y^i)$,
(2) $v^i = (\hat{g}_x^i, \hat{g}_y^i, \hat{g}_x^i(x^i - x) + \hat{g}_y^i(y^i - y), \hat{g}_x^i(y^i - y) - \hat{g}_y^i(x^i - x))$, *and*
(3) $v^i = (\hat{g}_x^i, \hat{g}_y^i, \hat{g}_x^i(x^i - x), \hat{g}_x(y^i - y), \hat{g}_y^i(x^i - x), \hat{g}_y^i(y^i - y))$,

for translation, RST, and affine transformations, respectively, where (x^i, y^i) *is the ith point in the window centered at* (x, y), $\hat{g}_x^i = \hat{g}_x(x^i, y^i)$, $\hat{g}_y^i = \hat{g}_y(x^i, y^i)$, *and subscripts denote differentiation.*

Note that the condition numbers are given in terms of $\hat{g}_x$ and $\hat{g}_y$ and do not require knowledge of the minimizing parameters of the transformation. For example, the

2×2 matrix $A^T A$ associated with the condition number for translation, reduces to the matrix given by Eq. (9.1).

The connection between conditioning and the matrix $A^T A$ is not surprising considering that the eigenvalues of this matrix have been commonly used in the analysis of local image structure to distinguish between flat, linear and corner-like structures (Harris and Stephens, 1988; Trucco and Verri, 1998). Further motivation for this connection is given (in the following subsection) in the discussion on optical flow accuracy.

A numerical evaluation of the translational condition number can be done efficiently by noting that $A^T A$ is a 2×2 matrix whose inverse can be written explicitly. As a technical point, it is numerically easier to deal with the modified condition number $\tilde{K}^2_{Trans} \equiv \|(A^T A + \epsilon I)^{-1}\|$, where ε is some small number (see Golub and Loan, 1989, for the relationship with the pseudo-inverse). This formulation avoids problems associated with inverting singular or nearly singular matrices. In our experiments we used $\varepsilon = 10^{-8}$.

To simplify the computation, we considered the Schatten 1-norm defined by $\|M\|_S = \Sigma \sigma_i$, where $\sigma_1, \ldots, \sigma_n$ are the singular values of M (see Horn and Johnson, 1991, p. 199), which yields the closed-form expression

$$K^2_{Trans,Schatten} \equiv \|(A^T A + \epsilon I)^{-1}\|_S = \frac{2\epsilon + \Sigma \hat{g}_x^2 + \Sigma \hat{g}_y^2}{\left(\Sigma \hat{g}_x^2 + \epsilon\right)\left(\Sigma \hat{g}_y^2 + \epsilon\right) - \left(\Sigma \hat{g}_x \hat{g}_y\right)^2}.$$

Note that the Schatten norm is essentially equivalent to the regular 2-norm because the 2-norm of a matrix M is equal to the largest singular value, i.e., $\|M\|_2 = \sigma_2$. Thus, for a 2×2 matrix M, we have $\|M\|_2 \le \|M\|_S \le 2\|M\|_1$. We also note that

$$\|(A^T A + \epsilon I)^{-1}\|_S = \frac{\text{trace}(A^T A + \epsilon I)}{\det(A^T A + \epsilon I)} = \frac{\mu_1 + \mu_2}{\mu_1 \mu_2} \ge \frac{1}{\mu_2},$$

where here, μ_1 and μ_2 are the eigenvalues of the matrix $A^T A + \epsilon I$, and $0 \le \mu_2 \le \mu_1$. This bodes well with the Shi-Tomasi-Kanade feature tracker (e.g., Shi and Tomasi, 1994), where a point is a feature candidate for tracking if μ_2 is large or equivalently if $1/\mu_2$ is small. The connection arises from the inequality $\frac{1}{\mu_2} \le \frac{\mu_1 + \mu_2}{\mu_1 \mu_2} \le \frac{2}{\mu_2}$ for $0 \le \mu_2 \le \mu_1$. That is,

$$\frac{1}{\mu_2} \le K^2_{Trans,Schatten} = \|(A^T A + \epsilon I)^{-1}\|_S = \frac{\mu_1 + \mu_2}{\mu_1 \mu_2} \le \frac{2}{\mu_2}.$$

Moreover, working with the 2-norm of the matrix, we can write $K^2_{Trans} = \|(A^T A)^{-1}\| = \frac{1}{\min(\lambda_1, \lambda_2)}$, where λ_1, λ_2 are the eigenvalues of $A^T A$. Thus the Shi-Tomasi requirement that $\lambda < \min(\lambda_1, \lambda_2)$ for some threshold value λ is equivalent

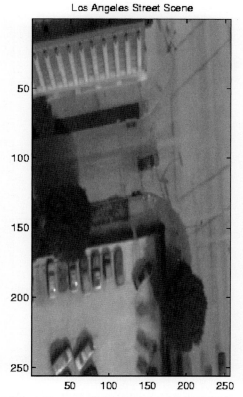

Los Angeles Street Scene

Figure 9.1. Image selected for illustration of point-matching condition numbers for translation, RST, and affine transformations. (Source: Kenney *et al.*, 2003, © IEEE, reprinted with permission.)

to requiring that $K_{Trans} \leq \frac{1}{\sqrt{\lambda}}$, i.e., the Shi-Tomasi feature condition is equivalent to specifying a maximum condition value for translation.

Returning to the problem of conditioning with respect to the hierarchy of the transforms in question, it was shown by Kenney *et al.* (2003) that point matching sensitivity increases as the complexity of the transform increases. This is formulated by the following result:

Theorem 9.2 *Under the assumptions of* Thm. *9.1 we have* $K_{Trans} \leq K_{RST} \leq K_{Affine}$.

Figure 9.1 shows a noisy infrared (IR) image containing an urban street scene that we use to demonstrate the condition number for translation, RST, and affine matching. Figure 9.2 shows the condition surfaces for these three transformation types. In the images shown, dark points have good conditioning. The striking feature of Fig. 9.2 is the overall similarity of the condition numbers for translation,

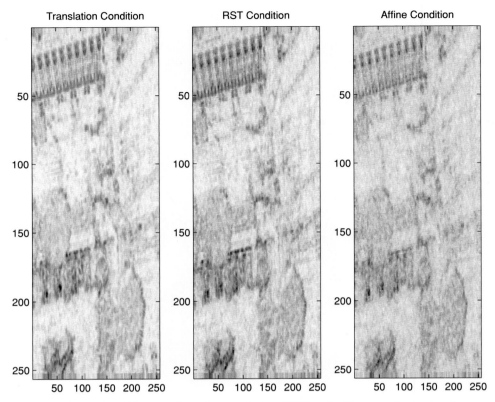

Figure 9.2. Condition numbers for translation, RST, and affine transformation for the LA street scene; dark is better conditioned. (Source: Kenney *et al*., 2003, © IEEE, reprinted with permission.)

RST, and affine point matching for the urban street image. In part, this similarity stems from Thm. 9.2, in the sense that well-conditioned points comprise only a small percentage of the overall image pixels and they form a nested series of subsets for translation, RST, and affine transformations. This works to our advantage as the computation of the condition number for translation requires merely the inversion of a 2×2 matrix, which can be easily computed. In contrast, the condition numbers for both the RST and affine transformations are much more expensive to compute, as they involve inverting a 4×4 matrix and a 6×6 matrix, respectively, at each point in the image, in accordance with Thm. 9.1.

9.3.4 Optical flow

The area of optical flow is closely related to the problem of determining the best local translation for matching points between images. Unfortunately, many existing strategies for evaluating the reliability of optical flow computations can lead to unstable estimates.

Let $g = g(x, y, t)$ be the intensity of a time-varying image. If an image point $(x, y) = (x(t), y(t))$ maintains constant brightness with respect to time, i.e., $g(x(t), y(t), t) = c$, then its time derivative is zero (Horn, 1986), i.e., $0 = g_x x_t + g_y y_t + g_t$, where subscripts denote differentiation. We would like to know (x_t, y_t) in order to track the approximate motion of the point $(x(t), y(t)) \approx (x(0), y(0)) + t\,(x_t, y_t)$. If we identify g_t with the difference between the first and second images, i.e., $g_t = \hat{g} - g$, and estimate g_x and g_y via finite differences, then we have one equation for the two unknowns x_t and y_t at each point in the image. The vector (x_t, y_t) is referred to as the optical flow at (x, y). Several approaches have been advocated for overcoming the underdetermined nature of optical flow equations. Horn and Schunck (1981) imposed a smoothness constraint on the optical flow by casting the optical flow as a minimization problem over the entire image. This results in a large system of elliptic linear equations which is usually solved iteratively starting with the *normal flow* $(x_t, y_t)_{nor} = (-g_t g_x, -g_t g_y)/(g_x^2 + g_y^2)^{1/2}$ as an initial guess. Unfortunately, this approach is computationally costly and has problems at points in the image where the optical flow is discontinuous as a result of, for example, the motion of occluding objects (Irani *et al.*, 1994).

To avoid these problems, various alternatives have been suggested. For example, Hildreth (1984) proposed an optical flow computation of edges in an image sequence, which is a 1D version of Horn and Schunck's algorithm. Sundareswaran and Mallat (1992) combined Hildreth's approach with multiscale wavelet information to regularize the optical flow computation. (See also Hewer *et al.*, 1994, for a detailed review.)

One of the more successful alternative methods assumes that the optical flow is constant in a local neighborhood about the point (x, y) (Lucas and Kanade, 1981; Tanner and Mead, 1989; Bergen *et al.*, 1992). Under this assumption, we obtain the following system of equations for the two unknowns (x_t, y_t):

$$A \begin{bmatrix} x_t \\ y_t \end{bmatrix} = v, \qquad (9.9)$$

where

$$A = \begin{bmatrix} g_x^1 & g_y^1 \\ \vdots & \\ g_x^n & g_y^n \end{bmatrix}, \qquad v = \begin{bmatrix} -g_t^1 \\ \vdots \\ -g_t^n \end{bmatrix}, \qquad (9.10)$$

and each superscript denotes the point (out of the n points) in the window centered at (x, y) at which the functions in question are evaluated. Assuming that A is full rank, the least-squares solution of the above system is given by

$$\begin{bmatrix} x_t \\ y_t \end{bmatrix} = (A^T A)^{-1} A^T v. \tag{9.11}$$

The assumption that the optical flow is constant over a window is used to overcome the underdetermined nature of optical flow equations. Unfortunately, this approach may fail. For example, if the image consists of a linear edge moving from left to right then the least-squares system of equations is rank deficient and does not have a unique solution. This well-known problem leads to a loss of accuracy in the estimate of the optical flow vector. A variety of related effects can also cause loss of accuracy in the computed optical flow as was noted by Kearney *et al.* (1987). This has prompted some researchers (e.g., Irani *et al.*, 1994) to assign a reliability measure to the flow estimates based on K_L, that is, the condition number with respect to inversion of the least-squares system. In general, the condition number with respect to inversion of a square matrix L is $K_L = \|L\| \|L^{-1}\|$. For the optical flow problem we have $L = A^T A$, as in Eq. (9.1). K_L values near 1 indicate that L is well conditioned with respect to inversion. However, K_L values near 1 do not necessarily ensure accurate optical flow computation, as can be seen by the following example. Consider a flat background point in the image. The least-squares matrix L in this case is doubly rank deficient, that is, it has rank zero. Now add a slight amount of noise to the image. Since K_L is the ratio of the largest singular value of L to the smallest singular value of L, we see that K_L can be close to 1 under the addition of even arbitrarily small noise.

An example: Starting with an image of constant intensity, we added at each point random Gaussian noise of mean zero and standard deviation $\sigma = 10^{-8}$. Using a 3×3 window we then formed the least-squares matrix $L = A^T A$ (where the summation extends over the 3×3 window) and evaluated both K_L (the condition number of L with respect to inversion) and $K_{Trans} = \|L^{-1}\|^{1/2}$ (the condition number with respect to matching via translation; see Thm. 9.1). Having run this test 100 times we found that the condition number for inversion stayed in the range $1.2 \leq K_L \leq 7.4$. This suggests excellent conditioning with respect to inversion for each of the 100 test samples. At the same time, the translation matching condition number was approximately 10^4, and in general, $K_{Trans} \approx 1/\sqrt{\sigma}$ for this example. Clearly, using K_L as a condition measure for optical flow for this problem is wrong, as K_L stays roughly 1 even though L is doubly rank deficient for the underlying problem (i.e., the one without noise).

9.3.5 A thought experiment

We can assess which points are likely to give bad optical flow estimates by a simple ansatz. Suppose that the scene is static, so that the true optical flow is zero, i.e.,

$\mathbf{v}_{exact} = \mathbf{0}$. Also assume that the images of the scene vary only by additive noise η (i.e., the difference between frames). The error in the optical flow estimate is given by $\mathbf{e} = \mathbf{v}_{exact} - \mathbf{v}_{computed}$, namely

$$\begin{aligned}
\|\mathbf{e}\| &= \|\mathbf{v}_{exact} - \mathbf{v}_{computed}\| \\
&= \|\mathbf{0} - (A^T A)^{-1} A^T \eta\| \\
&= \|(A^T A)^{-1} A^T \eta\| \\
&\leq \|(A^T A)^{-1} A^T\| \, \|\eta\|.
\end{aligned}$$

Thus, the term $\|(A^T A)^{-1} A^T\|$ controls the error multiplication factor, namely the factor by which the input error (i.e., the noise η) is multiplied to get the output error (i.e., the error in the optical flow estimate). Large values of $\|(A^T A)^{-1} A^T\|$ correspond to points in the image where we cannot estimate the optical flow accurately in the presence of noise, at least for the static image case. If we use, however, the 2-norm together with the result that for any matrix M $\|M\|_2^2 = \lambda_{max}(MM^T)$, where $\lambda_{max}(MM^T)$ is the largest eigenvalue of MM^T, we obtain

$$\begin{aligned}
\|(A^T A)^{-1} A^T\|_2^2 &= \lambda_{max}((A^T A)^{-1} A^T A (A^T A)^{-1}) \\
&= \lambda_{max}((A^T A)^{-1}) \\
&= \frac{1}{\lambda_{min}(A^T A)}.
\end{aligned}$$

We conclude that the error multiplication factor for the 2-norm in the optical estimate for the static noise case is equal to $\frac{1}{\lambda_{min}(A^T A)}$. This motivates the use of the gradient normal matrix in feature detection, since the ability to accurately determine optical flow at a point is intimately related to its suitability for establishing point correspondence between images (i.e., whether it is a good corner).

9.4 Conclusions

Although condition theory has demonstrated usefulness in the field of numerical linear algebra and control theory, it has seen until recently limited application in image processing, in particular image registration. The work of Kenney *et al.* (2003) changed this state dramatically with the introduction of condition numbers for point-matching for translation, rotation-scaling-translation, and affine window matching between images. Based on first principles, this work produced some surprising results: (1) The translational condition number was shown to be equivalent to the commonly used Shi-Tomasi corner function, and (2) as the complexity of the transform increases, the sensitivity of the computed transform parameters also increases. This initial work then led to a comparative study of corner detectors with respect to a set of desirable prerequisites such as rotational invariance. This study demonstrated (Zuliani *et al.*, 2004; Kenney *et al.*, 2005) that only condition-based detectors were entirely compliant for general images (color, multispectral, tomographic, etc.). Current research in this area has focused on post-processing condition estimates for registration misalignment, and has produced very promising results on an extensive series of test problems.

Acknowledgements

This research was supported by the Office of Naval Research under ONR Grant Number N00014-02-1-0318.

References

Barron, J. L., Fleet, D. J., and Beauchemin, S. (1994). Performance of optical flow techniques. *International Journal of Computer Vision*, **12**, 43–77.

Beauchemin, S. S. and Barron, J. L. (1996). The computation of optical flow. *ACM Computing Surveys*, **27**, 433–467.

Bergen, J., Anandan, P., Hanna, K., and Hingorani, R. (1992). Hierarchical model-based motion estimation. In *Proceedings of the Second European Conference on Computer Vision*. New York: Springer-Verlag, pp. 237–252.

Björck, A. and Hammarling, S. (1983). A Schur method for the square root of a matrix. *Linear Algebra and its Applications*, **52–53**, 127–140.

Brooks, M., Chojnacki, W., Gaeley, D., and van den Hengel, A. (2001). What value covariance information in estimating vision parameters? In *Proceedings of the Eighth IEEE International Conference on Computer Vision*, Vancouver, Canada, pp. 302–308.

Cline, A., Moler, C., Stewart, G., and Wilkinson, J. (1979). An estimate for the condition number of a matrix. *SIAM Journal on Numerical Analysis*, **16**, 368–375.

Demmel, J. (1987). On condition numbers and the distance to the nearest ill-posed problem. *Numerische Mathematik*, **51**, 251–289.

Demmel, J. (1988). The probability that a numerical analysis problem is difficult. *Mathematics of Computation*, **50**, 449–480.

Demmel, J., Diament, B., and Malajovich, G. (2001). On the complexity of computing error bounds. *Foundations of Computational Mathematics*, **1**, 101–125.

Fischler, M. A. and Bolles, R. C. (1981). Random sample consensus: A paradigm for model fitting with applications to image analysis and automated cartography. *Communications of the ACM*, **24**, 381–395.

Golub, G. H. and Loan, C. F. V. (1989). *Matrix Computations*. Baltimore, MD: John Hopkins University Press.

Harris, C. and Stephens, M. (1988). A combined corner and edge detector. In *Proceedings of the Fourth Alvey Vision Conference*, Manchester, UK, pp. 147–151.

Hewer, G. A. and Kenney, C. S. (1988). The sensitivity of the stable Lyapunov equation. *SIAM Journal on Control and Optimization*, **26**, 321–344.

Hewer, G. A., Kenney, C. S., and Kuo, W. (1994). A survey of optical flow methods for tracking problems. In *Proceedings of the SPIE Conference on Wavelet Applications*, Orlando, FL, Vol. 2242, pp. 561–572.

Hildreth, E. C. (1984). Computations underlying the measurement of visual motion. *Artificial Intelligence*, **23**, 309–354.

Hinrichsen, D. and Pritchard, A. J. (1986). Stability radii of linear systems. *Systems and Control Letters*, **7**, 1–10.

Horn, B. K. P. (1986). *Robot Vision*. Cambridge, MA: MIT Press.

Horn, B. K. P. and Schunck, B. (1981). Determining optical flow. *Artificial Intelligence*, **17**, 185–203.

Horn, R. A. and Johnson, C. R. (1991). *Topics in Matrix Analysis*. Cambridge: Cambridge University Press.

Irani, M., Rousso, B., and Peleg, S. (1994). Computing occluding and transparent motion. *International Journal of Computer Vision*, **12**, 5–16.

Kanazawa, Y. and Kanatani, K. (2001). Do we really have to consider covariance matrices for image features? In *Proceedings of the Eighth IEEE International Conference on Computer Vision*, Vancouver, Canada, pp. 301–306.

Kearney, J. K., Thompson, W. B., and Boley, D. L. (1987). Optical flow estimation: An error analysis of gradient-based methods with local optimization. *IEEE Transactions on Pattern Analysis and Machine Intelligence*, **9**, 229–244.

Kenney, C. S. and Hewer, G. A. (1990). The sensitivity of the algebraic and differential Riccati equations. *SIAM Journal on Control and Optimization*, **28**, 50–69.

Kenney, C. S. and Laub, A. J. (1989). Condition estimates for matrix functions. *Matrix Analysis and Applications*, **10**, 191–209.

Kenney, C. S. and Laub, A. J. (1994). Small-sample statistical condition estimates for general matrix functions. *SIAM Journal on Scientific Computing*, **15**, 36–61.

Kenney, C. S., Manjunath, B. S., Zuliani, M., Hewer, G. A., and Nevel, A. V. (2003). A condition number for point matching with application to registration and post-registration error estimation. *IEEE Transactions on Pattern Analysis and Machine Intelligence*, **25**, 1437–14541.

Kenney, C. S., Zuliani, M., and Manjunath, B. (2005). An axiomatic approach to corner detection. In *Proceedings of the IEEE Conference on Computer Vision and Pattern Recognition*, San Diego, CA, pp. 191–197.

Li, H., Manjunath, B., and Mitra, S. (1995). A contour based approach to multisensor image registration. *IEEE Transactions on Image Processing*, **4**, 320–334.

Loan, C. V. (1985). How near is a stable matrix to an unstable matrix? *Contemporary Mathematics*, **47**, 465–477.

Lucas, B. and Kanade, T. (1981). An iterative image registration technique with an application to stereo vision. In *Proceedings of the DARPA Image Understanding Workshop*, Washington, DC, pp. 121–130.

Moler, C. B. and Loan, C. F. V. (1978). Nineteen dubious ways to compute the exponential of a matrix. *SIAM Review*, **20**, 801–836.

Rice, J. R. (1966). A theory of condition. *SIAM Journal on Numerical Analysis*, **3**, 287–310.

Schmidt, C., Mohr, R., and Baukhage, C. (2000). Evaluation of interest point detectors. *International Journal of Computer Vision*, **27**, 151–172.

Shi, J. and Tomasi, C. (1994). Good features to track. In *Proceedings of the IEEE Conference on Computer Vision and Pattern Recognition*, Seattle, WA, pp. 593–600.

Stewart, G. W. (1973). Error and perturbation bounds for subspaces associated with certain eigenvalue problems. *SIAM Review*, **15**, 727–764.

Sundareswaran, V. and Mallat, S. (1992). Multiscale optical flow computation with wavelets. Preprint, Courant Institute of Mathematical Sciences, New York University.

Tanner, J. and Mead, C. (1989). Optical motion sensor. In C. Mead, ed., *Analog VLSI and Neural Systems*. Reading, MA: Addision-Wesley, pp. 229–255.

Tommasini, T., Fusiello, A., Trucco, E., and Roberto, V. (1998). Making good features track better. In *Proceedings of the IEEE Conference on Computer Vision and Pattern Recognition*, Santa Barbara, CA, pp. 178–183.

Trucco, E. and Verri, A. (1998). *Introductory Techniques for 3-D Computer Vision*. New Jersey: Prentice Hall.

Ward, R. C. (1977). Numerical computation of the matrix exponential with accuracy estimate. *SIAM Journal on Numerical Analysis*, **14**, 600–614.

Wildes, R., Horvonen, D., Hsu, S., Kumar, R., Lehman, W., Matei, B., and Zhao, W. (2001). Video georegistration: Algorithm and quantitative evaluation. In *Proceedings of the Eighth IEEE International Conference on Computer Vision*, Vancouver, Canada, pp. 343–350.

Wilkinson, J. H. (1961). Error analysis of direct methods of matrix inversion. *Journal of the ACM*, **8**, 281–330.

Wilkinson, J. H. (1965). *The Algebraic Eigenvalue Problem*. Oxford: Clarendon Press.

Zheng, Q. and Chellappa, R. (1993). A computational vision approach to image registration. *IEEE Transactions on Image Processing*, **2**, 311–326.

Zuliani, M., Kenney, C. S., and Manjunath, B. S. (2004). A mathematical comparison of point detectors. In *Proceedings of the Second IEEE Workshop on Image and Video Registration*, Washington, DC, p. 172.

10

Feature-based image to image registration

VENU MADHAV GOVINDU AND RAMA CHELLAPPA

Abstract

Recent advances in computer vision address the problem of registration of multiple images or entire video sequences. Such registration methods have a wide variety of application in constructing mosaics, video summarization, site modeling and as preprocessing for tasks such as object tracking and recognition. In this chapter we present a variety of registration techniques that utilize image features such as points and contours. Computational issues such as robustness to data outliers and recent developments in accurate feature extraction are discussed. A correspondenceless method that works on multimodal images is outlined. We also present approaches that efficiently utilize the information redundancy in a sequence of images to solve the problem of image registration. All of these methods are illustrated with appropriate examples.

10.1 Introduction

The underlying geometry of image formation has been well studied over the recent years in the discipline of computer vision (Hartley and Zisserman, 2004). This understanding of the image geometry has been accompanied by increasingly sophisticated computational models that can be solved on modern hardware. Many methods and ideas developed for solving various aspects of the motion estimation problem in computer vision are applicable to problems relating to image registration. In particular, image features like points, edges, and contours have been used in a range of applications like the construction of mosaics from video sequences, shape estimation, object tracking and recognition, etc. In this chapter, we describe a variety of methods dedicated to utilizing image features for solving the problem of registration. We present methods based on the most commonly used features, namely, contours extracted from high-contrast image edges and feature points

215

characterized by properties of local extrema in the image space. While the problem is relatively simpler in the case when such features are matched, finding the correspondences is often difficult, especially for images taken from multiple sensors or when images are acquired under greatly varying illumination conditions. Consequently, we also describe approaches that avoid the use of explicit correspondences and work on aggregated information that encodes the required geometric transformation. Also, in the scenario where we do use matched features, the estimation process has to account for erroneous correspondences obtained from the aforementioned matching process. Therefore, robustness to deviations from underlying assumptions is essential. A well-established computer vision technique known as RANSAC (Fischler and Bolles, 1981) and its variants are briefly described, and their application to the problem of robustness in image registration is also detailed. A relative advantage with images typically acquired from a platform that is distant from the scene is that the geometry of the scene can be assumed to be planar. Throughout this chapter we assume that we want to solve for either an affine or a homography transformation to represent the relative geometry between the images to be registered. While there are some strong similarities between tracking and image registration, the vast variety of approaches to tracking of rigid or deformable models is outside the scope of this chapter.

The rest of this chapter is organized as follows. In Section 10.2 we describe our correspondenceless approaches to image registration using contours and their discrete counterparts represented by points and lines. While this approach starts with extracting contour information from high-contrast edges in images, Section 10.3 considers issues surrounding the more common approach of using corresponding feature points to solve for the registration geometry. In this section we describe a variety of low-level approaches to the detection and matching of such feature points. Introducing robustness into the estimation process, via RANSAC, is also described. Section 10.4 describes different methods of image registration using feature points. This includes a *consistency framework* that efficiently and accurately exploits the large amounts of information redundancy in a video sequence to solve the global registration problem. This approach puts all the information contained in a video sequence into a common frame of reference. Section 10.5 provides some concluding remarks.

10.2 Contours

In this section we describe our correspondenceless approach to image registration using the geometric information in image contours and their discrete counterparts, i.e., points and lines.

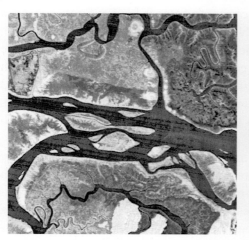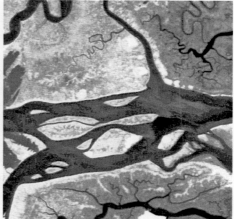

Figure 10.1. Landsat TM (left) and SPOT (right) images of a river. Note the intensity differences in structures observed by the different sensors. The scales of the two images are also quite different. (Source: Govindu and Shekhar, 1999, © IEEE, reprinted with permission.)

Both area- and feature-based approaches rely on a given notion of *similarity* to establish matches between areas or features across images. This is done either implicitly, in the form of a maximizable cost function for area-based approaches, or explicitly for matching features (e.g., image points). A significant advantage of the latter approach is that once correspondence is established, solving for the required transformation for registration is easy. For example, in the case of an affine model for registration, all point correspondences can be stacked into a system of equations that can be solved linearly to give a least-squares solution. However, the advantage of easy computation of the image transformation is somewhat countervailed by the fact that establishing correspondences of feature points across images is not necessarily an easy task. In the overwhelming majority of cases, feature matching is achieved by assuming identical feature representations of the matched features. (A feature representation could be, for example, a vector of oriented filter responses based on a local region of support around the feature point.)

This assumption is not valid in certain important scenarios, for example, for images acquired from different sensors or under significantly different illumination conditions. Figure 10.1 shows two satellite images of the same region acquired by the Landsat TM and SPOT modalities. As can be noted, apart from significant changes in the scale of the images, the geographical features themselves appear differently in the two images since the wavelengths of the image acquisition bands are different. Given these radiometric differences, establishing point feature correspondences is not feasible for such multisensor data.

However, while establishing feature correspondence is hard, we notice that the region boundaries are defined in a stable manner in both images. For instance, despite the different intensity levels, the boundaries that define the banks of the river in both images exhibit the same geometric properties. Therefore, for such cross-modality registration, contours extracted from image edges are the preferred features. Thus, while the image intensities may vary according to the sensor used, the localization of the edges present in the images is relatively invariant to such intensity or radiometric variations.

In Li *et al.* (1995), the authors used image contours for such multisensor image registration as follows. Zero crossings are first detected to define edges from which chain-coded contour representations are extracted. Then, the chain-coded representations of the contours in the two images are matched, and once correspondences are established, the image transformation is estimated. Since there could be significant geometric differences between the contours, as the two images are acquired from different sensors (optical and Synthetic Aperture Radar (SAR) imagery in this case), the authors further developed an elastic contour matching technique to overcome these variations and establish explicit correspondence between contours. While this works well in most scenarios, one can do away entirely with the need for any explicit correspondence. This is achieved by identifying specific geometric properties of contours that encode the image transformation information in a systematic manner, and by aggregating such information for the entire image. This method was developed in Govindu and Shekhar (1999) and is summarized below.

The key observation for this approach is that we identify certain differential geometric features of contours that depend on the camera transformation parameters in a simple and systematic manner. These contour measurements are then aggregated into distributions that encode the transformation parameters to be estimated. This estimation is carried out in a simple optimization that can be efficiently solved using cross-correlation of the distributions. Our approach can be best illustrated by a simple transformation model, say an image rotation. Consider the contours in Fig. 10.2, where one contour is a rotated version of the other for some rotation angle θ. Let $\mathbf{p}_1$ be a point that lies on a contour in the first image. We define the slope angle of its tangent as $\psi(\mathbf{p}_1)$. Since the contours are related by a rotation of angle θ, the slope angle for the tangent at the corresponding point $\mathbf{p}_2$ in the second image is $\psi(\mathbf{p}_2) = \psi(\mathbf{p}_1) + \theta$. Since we do not have explicit knowledge of the corresponding points in the two contours, we cannot directly utilize this relationship to solve for the rotation angle.

Nevertheless, the problem can be solved without explicit correspondences by using the distributions of the slope angles of both contours. If we examine our model, we notice that for every point in the first contour with a given slope angle

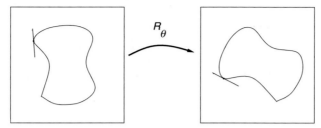

Figure 10.2. The contour on the right is a rotated version of the contour on the left. Tangent lines at corresponding points on the two curves are indicated. (Source: Govindu and Shekhar, 1999, © IEEE, reprinted with permission.)

there is a point on the second contour whose slope angle is shifted by θ. This translates to the relationship that the histograms of the slope angles for the two contours are identical to each other up to a shift of the rotation angle θ. This is illustrated in Fig. 10.3. Thus the required rotation angle can be easily estimated by a one-dimensional correlation function whose maximum occurs at the correct rotation angle (see Fig. 10.3(c)). The advantage of such an approach is twofold. Firstly, we completely do away with the need to establish correspondences and work instead on geometric properties of contours that can be reliably estimated from images. Secondly, the one-dimensional search for the transformation parameter can be performed both robustly and efficiently. The principle as elucidated above extends to multiple contours. Thus, we can apply this approach, based on global distributions of local geometric properties, to general images. Moreover, we do not require here that every contour point used in one image be present in the other image. Rather, as long as there is a sufficient number of common contour points, the alignment can be estimated.

In general, we are interested in geometric transformation models that are richer than rotations. For 2D planes, the affine transformation consisting of six parameters is often sufficient to capture the geometric relationship between images. To be able to exploit the approach outlined above we need to break down the transformation into a series of one-dimensional searches which makes the computational search process very efficient. The individual parameters are then estimated in a given sequence to solve for the full transformation. For example, while the scaling of a transformation can be computed independently of the other parameters, the image translation can only be estimated after all the other parameters are solved and the image contours are transformed up to the unknown translation parameter. In a general form, we need to find a geometric descriptor D that encodes the transformation parameter g in a known form, i.e., $D(\mathbf{p}_1) = D(\mathbf{p}_2, g)$, for all corresponding points $\mathbf{p}_1$, $\mathbf{p}_2$. In the case of rotation described above, we have $g = \theta$ and $D = \psi$. Given this descriptor relationship for contour points on two images, we can also relate to

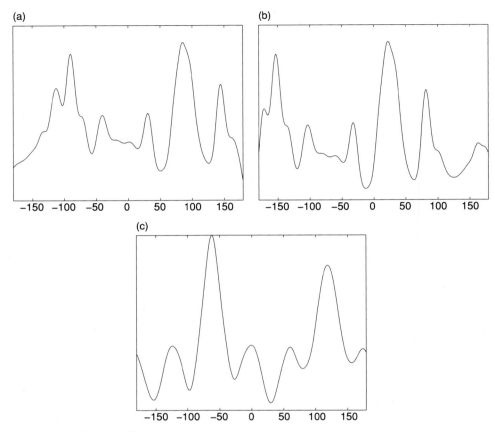

Figure 10.3. (a), (b) Distributions of slope angles for the two contours, and (c) the cross-correlation between the distributions. The peak occurs at the correct rotation angle. (Source: Govindu and Shekhar, 1999, © IEEE, reprinted with permission.)

their distributions. In Govindu and Shekhar (1999) we showed that the maximum likelihood estimation of the parameter g obtained from the two distributions of the descriptors is given by

$$\arg \min_{g} ||\mathcal{P}(D(\mathbf{p}_1)) - \mathcal{P}(D(\mathbf{p}_2)|g)||, \qquad (10.1)$$

where $\mathcal{P}$ denotes the probability distribution of the descriptor values on the image contours. To clarify this representation, we further illustrate our approach by describing the estimation of the image scale factor. For corresponding points in the first and second images, denoted as $\mathbf{p}_1$ and $\mathbf{p}_2$, we have the relationship

$$\mathbf{p}_2 = \mathbf{A}\mathbf{p}_1 + \mathbf{t}, \qquad (10.2)$$

where $\mathbf{A}$ and $\mathbf{t}$ are the parameters of the affine transformation. We define a geometric descriptor $D(.)$ of a point on a contour by taking derivatives of the (x- and y-coordinates of the) point with respect to a specific parameterization of the contour's curve. Specifically,

$$D(\mathbf{p}_1) = \begin{bmatrix} \dot{\mathbf{p}}_1 & \ddot{\mathbf{p}}_1 \end{bmatrix} = \begin{bmatrix} \dot{x}_1 & \ddot{x}_1 \\ \dot{y}_1 & \ddot{y}_1 \end{bmatrix},$$

$$D(\mathbf{p}_2) = \begin{bmatrix} \dot{\mathbf{p}}_2 & \ddot{\mathbf{p}}_2 \end{bmatrix} = \begin{bmatrix} \dot{x}_2 & \ddot{x}_2 \\ \dot{y}_2 & \ddot{y}_2 \end{bmatrix}. \tag{10.3}$$

Letting D_1 and D_2 denote the geometric descriptors at corresponding points $\mathbf{p}_1$ and $\mathbf{p}_2$, respectively, and letting s_1 and s_2 denote the parameterizations representing corresponding contours extracted from the two images, respectively, it can be shown according to Govindu and Shekhar (1999) that

$$D_2 = \mathbf{A} D_1 \frac{ds_1}{ds_2},$$

which implies that $|D_2| = |\mathbf{A}||D_1|$, i.e.,

$$\ln|D_2| = \ln|\mathbf{A}| + \ln|D_1|, \tag{10.4}$$

provided that $ds_1/ds_2 = 1$. (See Govindu and Shekhar, 1999, for further details.) Thus, based on the distributions of $\ln|D_1|$ and $\ln|D_2|$, we can obtain an estimate of $|\mathbf{A}|$, i.e., of the transformation magnitude.

This approach is very robust to a variety of deviations from the idealized assumptions. Since we use contours, our method is successful in the presence of radiometric variations, as can be seen in Fig. 10.4. This figure illustrates the registration obtained for the two multisensor images of Fig. 10.1. As can be noted, the registration is accurate despite the significant difference in scale between the two images. It may be noted that the specific forms of our representation of the geometric properties allow us to break down the transformation into a series of manageable probability representations that encode the parameters to be estimated in a simple manner.

The well-known method based on *mutual information* (Viola and Wells, 1997) is somewhat related to this underlying principle. In the mutual information approach, the cost function is based on estimating the mutual information between an image and the corresponding registered version. Optimization is typically achieved using nonlinear methods. While the mutual information approach has been shown to be very accurate and useful, especially in the area of medical image registration, the issue of convergence remains as with most nonlinear optimization approaches. Therefore, mutual information can only be used when the images to be aligned are close to each other in the space of relative transformations. In contrast, since we

Figure 10.4. The registered SPOT and Landsat TM images of the river. (Source: Govindu and Shekhar, 1999, © IEEE, reprinted with permission.)

break down the search (or optimization) procedure into a series of one-dimensional minimizations, our approach is applicable for any transformation magnitude, provided that the descriptors themselves are bounded in magnitude.

The underlying notion of correlating distributions of descriptors in our approach can also be interpreted as a mechanism for *feature consensus*. In general, given two corresponding features α_1 and α_2 from two images, and assuming that their values are related by a relationship $\alpha_2 = g_\theta(\alpha_1)$, then the parameter θ can be considered to be *observable* if it can be written in a parametric form $\theta = h_\theta(\alpha_1, \alpha_2)$. This formulation is utilized in the discrete case of points and lines in Shekhar *et al.* (1999). In a general sense, it allows us to write a *consensus function C_θ*, where

$$C_\theta(\theta) = \sum_{i,j} \delta(\theta - h_\theta(\alpha_1, \alpha_2)) \tag{10.5}$$

for features α_1 and α_2 in the two images. Given this representation, the desired image transformation θ can be estimated as the location of the maximum of the consensus function C_θ which serves as a voting mechanism. In Fig. 10.5(a)–(b) we show individual visible and SAR images to be registered. Figure 10.5(c) shows the

(a) (b)

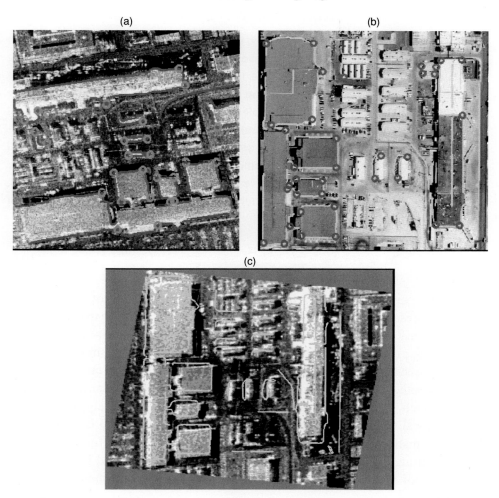

(c)

Figure 10.5. (a) Radar, (b) visual images to be registered (feature points are indicated by circles), and (c) resulting transformation represented by overlaying edges onto the radar image. See color plates section. (Source: Shekhar *et al.*, 1999, p. 49, © Elsevier Science, reprinted with permission.)

results obtained using the feature consensus approach applied to discrete features (points and lines) extracted from the individual images. The papers by Govindu and Shekhar (1999) and Shekhar *et al.* (1999) provide continuous and discrete implementations of the same underlying approach of utilizing one-dimensional distributions of local geometric features that encode specific parameters of the underlying image transformation. Both techniques share similar properties and implementational issues that can be obtained from the above mentioned papers.

10.3 Detection and matching of feature points

By far, points are the most commonly used features for image registration. In this section we discuss some recent developments in the detection and matching of feature points across images. While there are many approaches of matching feature points for image registration, the detection and localization of the feature points themselves is a significant problem in low-level image processing. Most feature-point detectors are based on identifying points with corner-like characteristics. Almost all of these approaches are based on computing a weighted local differential structure of the image around a point. In particular, letting I_x and I_y denote, respectively, the image derivatives in the x and y direction at a point (x, y), the local image structure can be characterized by the following moment matrix of image derivatives:

$$\mathbf{M} = g(\sigma_I) * \begin{bmatrix} \sum I_x(\sigma_d)I_x(\sigma_d) & \sum I_x(\sigma_d)I_y(\sigma_d) \\ \sum I_x(\sigma_d)I_y(\sigma_d) & \sum I_y(\sigma_d)I_y(\sigma_d) \end{bmatrix}, \tag{10.6}$$

where each summation is weighted over a local neighborhood of (x, y) and the asterisk represents a convolution operation. More specifically, note that the moment matrix estimate is defined by two scales, σ_d and σ_I. The first scale defines an isotropic Gaussian kernel that is often used to obtain a locally smoothed image. The subsequent computation of the derivative filters is carried out with respect to the smoothed image. The second scale defines another Gaussian kernel (i.e., $g(\sigma_I)$), which serves as a weight function for computing the weighted averages over the local neighborhood of a given image point, according to Eq. (10.6). The local information captured by the moment matrix $\mathbf{M}$ is useful for feature-point assessment and detection. This information can be characterized by the relative magnitude of the two eigenvalues of $\mathbf{M}$. In turn, many algorithms exploit this property.

An evaluation of the various feature-point detection schemes is presented in Schmid *et al.* (2000). For example, the well-known Harris detector (Harris and Stephens, 1988) computes the local derivative structure that captures the information of image gradients in all directions about a point. This is achieved by selecting image points that satisfy the relationship

$$\det(\mathbf{M}) - \alpha \, \text{trace}^2(\mathbf{M}) > T, \tag{10.7}$$

where T is a given threshold parameter and α is a fixed scalar value. Points that are stable with respect to this measure are selected as features to be matched. Further processing is required to weed out false detections and a variety of approaches are used to match such features across images. However, since the computation of the Harris detector is done at a predetermined scale, there is no guarantee that the local measure of the differential image structure around a feature point would be preserved if the image were to undergo a significant change of scale. In fact, the localization of the corner feature is significantly affected by the choice of the two scale parameters σ_I and σ_d. To overcome this problem, a variety of approaches have been developed in recent years. For example, in a companion chapter in this volume, Kenney *et al.* present an approach based on measuring condition numbers.

In the domain of computer vision, the problem of matching feature points across images in a stable manner could become particularly difficult in many circumstances, as one encounters significant image scale variations. A moving vehicle whose size in the images (or video sequence) increases significantly as it approaches the camera location is an example of such a scenario. Another significant challenge is due to the fact that the local image structure itself could be drastically altered. This situation is best imagined by considering a plane that rotates about an axis that is parallel to the camera's imaging plane. As the rotation approaches 90°, i.e., the viewing direction is aligned to lie almost in the object plane, we can expect to see a drastic variation in the image structure due to the acuteness of the perspective effect. Even a milder scenario of significant skew poses serious problems to any low-level mechanism that desires to extract the intrinsic image information.

Indeed, the problem of low-level feature extraction in a manner that is invariant to some of the above effects has attracted much attention. Recent developments in this research area are epitomized by Lowe (2004) and Mikolajczyk and Schmid (2003), which provide significant improvement to the problem of feature-matching across images with widely varying scales and viewpoints. Specifically, the *scale invariant feature transform* (SIFT) of Lowe (2004) provides a successful approach of detecting and matching features across a wide range of image scales. (An image pyramid representation is used to search for stable features across a range of image scales.) Once local extrema are detected in this scale-space, a vector of local image descriptors is computed for each of these extrema. These descriptor vectors could be used, for example, for image registration (by matching across images), object recognition (by matching against databases), etc.

Once such stable features are extracted, the problem of matching them across images still needs to be addressed. Most approaches use simple methods of feature comparison to generate matches. But this is seldom sufficient to establish a good set of correspondences. This has led to the development of specific approaches that utilize the underlying geometric structure of the problem to guide the matching process. Some of these approaches also address specifically the problem of data outliers present in the detected feature-point sets.

In Zhang *et al.* (1995) and Gold *et al.* (1998) examples of different approaches to the problem of matching feature points are presented. The work in Zhang *et al.* (1995) focuses on a robust estimate of the epipolar geometry between images. A given set of initial matches is used to solve for the epipolar geometry between uncalibrated images. First, a robust *least median of squares* estimator (Rousseeuw and Leroy, 1987) is applied to yield a solution that is reasonably accurate. Drawing on this estimate, a larger set of correspondences is then generated to accurately recover the camera geometry. This approach uses a correlation window around a point to make the matching process stable.

However, to account for significant scaling effects due to geometric transformations, many authors have introduced a variety of methods to estimate locally an affine invariant neighborhood around a feature point. The reader should consult Mikolajczyk and Schmid (2003) and references therein for more details. The approach in Gold *et al.* (1998) deals with the matching problem in a more direct fashion. Given two sets of points that are to be matched (with possibly a different number of points including outliers), a combination of deterministic annealing and an assignment method was used to gradually tune the search space for determining matches between the two-point sets and identifying outliers.

10.3.1 Robustness in registration via RANSAC

Despite accurate modeling and representation of the image features, the matching process is often erroneous and results in incorrect matches. Consequently, any estimation that utilizes these feature matches has to incorporate robustness to alleviate such errors.

A well-known approach for incorporating robustness in computer vision is the *random sampling consensus* (RANSAC) method (Fischler and Bolles, 1981). This randomized scheme has been shown to have desirable statistical properties in that it can effectively identify data *outliers* that do not satisfy a given geometric model. The idea behind RANSAC is illustrated in Fig. 10.6, where most of the points lie on a straight line, along with some outlying data. If we were to seek the least-squares fit for the full set of data points, the resulting line fit would be grossly incorrect as it would average over the correct points and the outliers.

The RANSAC approach for detecting outliers works by generating solutions that use the minimal number of data points. Since a line can be defined by two nonidentical points, we randomly select a pair of points and use the line passing through them as a hypothesis. All points that fall within a prespecified distance from this hypothesized line are declared to fit the line. In Fig. 10.6 this range is indicated by the two dotted lines around the true line. For each trial, we count the

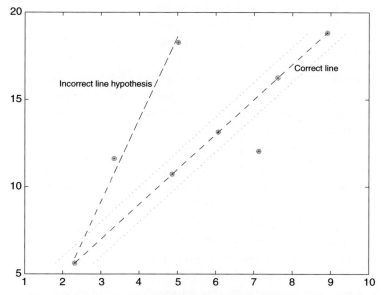

Figure 10.6. Illustration of the RANSAC approach. Most of the points fall on a line, while some of them are outliers that corrupt the dataset.

number of points that fall within this bounding region. For a given number of trials, the hypothesis with the maximum number of points within the bounding region is selected and all points within the bounding region are declared as *inliers*, while those outside this region are classified as outliers. The ultimate line estimate is now obtained by least-squares fitting to all the inliers. Figure 10.6 illustrates a labeled line hypothesis (containing outliers) versus a correct line estimate. The score for the former is obviously smaller than that for the correct hypothesis, implying robustness to as many as 50% of data outliers.

Since the entire set of hypothesis is combinatorially large, we need to determine the number of samples required to ensure that there is a high probability of correctly identifying the outliers. Denote the probability of obtaining a good sample (i.e., a sample that leads to a good fit) by p (e.g., $p = 95\%$), and assume that the proportion of outliers in the original dataset is ε. If we draw k elements at a time for a single hypothesis, then using elementary probability, it can be shown (Rousseeuw and Leroy, 1987) that the number of samples required to be drawn (by any randomized algorithm, including RANSAC) is given by

$$N_{samples} = \frac{\ln(1 - p)}{\ln(1 - (1 - \epsilon)^k)}. \tag{10.8}$$

In general, we draw a larger number of samples than $N_{samples}$ to avoid degeneracies. In an obvious sense, to solve the problem with a given degree of confidence, the number of samples required increases rapidly if the degree of contamination is increased.

The RANSAC approach has proved very successful in a variety of computer vision problems over the years. Its basic premise of sampling a solution space has been modified to yield a more accurate information representation (Torr and Zisserman, 2000) and provide greater computational efficiency (Chum and Matas, 2005). While RANSAC uses the number of inliers to score a hypothesis, the approach of Torr and Zisserman (2000) modifies the scoring function to represent a maximum likelihood estimate (MLESAC). RANSAC and MLESAC both use an unbiased sampler to generate the hypothesis sets of minimal data. In contrast, the approach outlined by Chum and Matas (2005) progressively biases the search space. This method makes use of the idea that intermediate samples drawn from the search space contain useful information of the probable correspondences, i.e., a hypothesis that leads to a fit of good quality is likely to contain correct correspondences. Thus, by ordering the search space with this information, a great deal of computational savings can be achieved.

10.4 Registration using feature points

The most common approach to image registration is based on matching of feature points. Despite the considerable number of issues and computational considerations described in Section 10.3, matched feature points offer a great degree of flexibility and can be efficiently exploited to solve for the camera geometry required for image registration. The literature on this subject is vast. Other than the survey chapter of this book and the specific chapters on feature matching, we point also to the well-known survey of Brown (1992) and the more recent overview of Zitová and Flusser (2003), which refer to numerous sources on image registration using feature matching.

Matching image feature points that differ by a substantial transformation is a significant challenge. While the use of affine invariant patch descriptors around feature points addresses this issue (Mikolajczyk and Schmid, 2003), a considerable simplification can be obtained in the case of planar imagery. As mentioned earlier, for a large number of registration scenarios the scene can be assumed to be planar and hence the camera geometry can be greatly simplified. However, in the presence of a significant transformation between images, say a large rotation, the matching of localized feature points remains a problem. In Zheng and Chellappa (1993), the authors used specific knowledge of the imaging conditions to solve this problem. Their paper addressed the problem of registering a pair of images taken from an unmanned balloon hovering above the imaged surface. Since the balloon itself could be rotated between the two images, the angle of the camera rotation could be large. Instead of trying to match the feature points directly under this scenario, the authors used illumination information to solve for the camera rotation. Assuming that the source of illumination (in this case, the Sun) is stationary, the azimuth angles of the illumination direction can be directly estimated from the individual images. The difference of these two estimates is a good approximation for the camera rotation. Once this estimate is used for unrotating the images, the matching of feature points can be carried out in a conventional manner. Using this approach the authors developed a robust technique for registering images that are significantly rotated with respect to each other.

If a 3D model of a site is available, the problem of registration becomes one of aligning an image with the site model, i.e., of solving for the camera viewing geometry. In an obvious sense, due to perspective projection, the local coordinates of features in an image are now related in a nonlinear manner to their physical counterparts in the 3D site model. Assuming that the intrinsic parameters of the camera are known and that it is calibrated accordingly, and representing the physical coordinates of a point on the site model as $\mathbf{P} = [X, Y, Z]^T$ and its location in the image as $\mathbf{p} = [x, y]^T$, we have the relationship

$$
\begin{bmatrix} x \\ y \end{bmatrix} = \mathcal{P} \left(\mathbf{R} \begin{bmatrix} X \\ Y \\ Z \end{bmatrix} + \begin{bmatrix} t_X \\ t_Y \\ t_Z \end{bmatrix} \right), \tag{10.9}
$$

where $\mathbf{R}$ and $\mathbf{T} = [t_X, t_Y, t_Z]^T$ represent, respectively, the 3D rotation and translation between the camera and the coordinate frame attached to the site model, and where $\mathcal{P}$ represents the projection operation, i.e.,

$$
\mathcal{P} \left(\begin{bmatrix} X \\ Y \\ Z \end{bmatrix} \right) = \begin{bmatrix} \frac{X}{Z} \\ \frac{Y}{Z} \end{bmatrix}. \tag{10.10}
$$

The problem of registering images to site models and that of registering multimodal images was addressed by Chellappa *et al.* (1997). The authors utilized a single image registered to a set of 3D control points to provide a basis for registering other images. A series of multiresolution searches for matches followed by checks for consistency are used to derive sets of correspondences with control points. Finally, the external orientations of the camera are reestimated by means of resectioning of the cameras. The registration of images to 3D site models has several applications to image exploitation tasks, ranging from detection and counting of objects (e.g., vehicles) to object grouping to registration and fusion of multisource datasets.

Since the camera geometry of SAR images is different from that of regular images, a method of registering SAR images to site models was also presented in Chellappa *et al.* (1997).

10.4.1 A consistency framework for registration

While there are many robust and efficient image registration methods, recent work in computer vision has focused on the development of image analysis and exploitation of video sequences, i.e., a large number of images. Thus, various approaches have been developed for constructing image mosaics (see Szeliski, 2006, and many references therein), as well as extracting 3D information, for example, camera trajectories and object shape (Hartley and Zisserman, 2004). A significant issue in registering large sets of images is the ability to handle the computational complexity associated with this task in an efficient and principled manner. In the case of extended sequences, most methods that work on adjacent or closeby image frames suffer from the problem of error accumulation. This accumulation of error results in a progressively degraded solution, as the registration error with respect to the frame of reference tends to increase beyond acceptable limits (as the length of the sequence increases). Our more recent work (Govindu, 2004) has focused on computationally efficient solutions that exploit the *inherent redundancy* of information in an image sequence to efficiently solve for the problem of registration or motion estimation for long sequences. The rest of this section is devoted to motivating and developing our approach.

Specifically, the key idea of our approach is that there is a highly redundant set of relative camera motions that can be estimated in a sequence. This in turn can be efficiently averaged to solve for global camera motions. Note that while a sequence of N images can be registered by $N - 1$ camera transformations, we can estimate many more transformations between images in the sequence. Since there are $\frac{N(N-1)}{2}$ distinct pairs of images for which we could compute the relative motions (assuming a sufficient number of correspondences), we have a highly redundant set of observations from which we can efficiently derive the global camera motions.

The formulation was described along with a linear solution for two- and three-dimensional camera motion estimation in Govindu (2001). Since Lie groups provide an adequate geometric framework for Euclidean motion, the approach was further refined in Govindu (2004), using motion averaging in a Lie-algebraic scheme. However, for registering images taken from a distance, the scene can be effectively modeled as a planar region and hence a simple affine camera model is sufficient to capture the image geometry. This leads to a significantly simpler formulation since the affine group is a linear group and hence the additional requirements of the

Lie-group can be dispensed with. For further details on the Lie-algebraic averaging approach, see Govindu (2004). In the following discussion we shall assume that the motion is represented by an affine model.

To register N images in a common reference frame, we assume that the registration geometry can be described by $N - 1$ motions.[*] Since the motion model is affine, we have the following relationship between corresponding feature points in image i and image j:

$$\begin{bmatrix} x_j \\ y_j \end{bmatrix} = \mathbf{A}_{ij} \begin{bmatrix} x_i \\ y_i \end{bmatrix} + \mathbf{t}_{ij}, \tag{10.11}$$

where (x_i, y_i) and (x_j, y_j) are the coordinates of the corresponding feature points in the ith image and jth image, respectively, and the affine transformation is given by the matrix $\mathbf{A}_{ij}$ and the translation vector $\mathbf{t}_{ij}$. If we use a homogeneous coordinate representation for the feature points, we can subsume this into a single matrix representation for the motion, i.e.,

$$\begin{bmatrix} x_j \\ y_j \\ 1 \end{bmatrix} = \underbrace{\begin{bmatrix} \mathbf{A}_{ij} & \mathbf{t}_{ij} \\ 0 & 1 \end{bmatrix}}_{=\mathbf{M}_{ij}} \begin{bmatrix} x_i \\ y_i \\ 1 \end{bmatrix}, \tag{10.12}$$

where the 3×3 matrix $\mathbf{M}_{ij}$ represents the affine motion that registers the jth image to the ith image. Thus, the matrix $\mathbf{M}_{ij}$ has six degrees of freedom as does the affine transformation of the image plane. If we represent the terms of the affine matrix $\mathbf{M}_{ij}$ as a vector $\mathbf{m}_{ij}$ of six unknowns, we have

$$\begin{bmatrix} x_j \\ y_j \end{bmatrix} = \begin{bmatrix} x_i & y_i & 1 & 0 & 0 & 0 \\ 0 & 0 & 0 & x_i & y_i & 1 \end{bmatrix} \mathbf{m}_{ij}, \tag{10.13}$$

i.e., each point correspondence gives us two relationships (in the above equation), and the affine transformation $\mathbf{M}_{ij}$ can thus be estimated in a least-squares manner given at least three-point correspondences.

If we denote the motion between frame i and the reference frame as $\mathbf{M}_i$, the relative motion between the two frames i and j can be described as $\mathbf{M}_{ij} = \mathbf{M}_j \mathbf{M}_i^{-1}$. This is the basis of our notion of *motion consistency*. Intuitively, this implies a loop of transformations, i.e., a series of transformations that starts and ends with the same frame, which is effectively equivalent to the identity transformation. For

---

[*] We need only $N - 1$ motions since the frame of reference can be attached to any image, for example, the first image in this case. Also, our solution has the desirable property that it is invariant to the choice of reference frame.

example, in the simple case of translation, this amounts to the requirement that the vectorial summation of the sides of any closed polygon be zero.

Since the original data (i.e., the feature point locations) are always noisy, the estimated camera transformations will not be consistent. However, we can rewrite the given relationship as a constraint on the global motion model $\mathbf{M}_g = \{\mathbf{M}_2, \ldots, \mathbf{M}_N\}$ which completely describes the motion.* In general, since there are up to $\frac{N(N-1)}{2}$ such constraints, we have an overdetermined system of equations,

$$\mathbf{M}_j \mathbf{M}_i^{-1} = \mathbf{M}_{ij}, \forall i \neq j, \tag{10.14}$$

where the variables on the left-hand side are unknowns to be estimated in terms of the observed data $\mathbf{M}_{ij}$. Intuitively, we want to estimate a global motion model $\mathbf{M}_g$ that is most consistent with the measurements $\{\mathbf{M}_{ij}\}$ derived from the data. By reorganizing the terms, we can rewrite Eq. (10.14) in the linear form

$$\overbrace{[\ldots \quad \mathbf{M}_{ij} \quad \ldots \quad -\mathbf{I} \quad \ldots]}^{\mathbf{C}_{ij}} \underbrace{\begin{bmatrix} \vdots \\ \mathbf{M}_i \\ \vdots \\ \mathbf{M}_j \\ \vdots \end{bmatrix}}_{\mathbf{M}_g} = \mathbf{0}, \tag{10.15}$$

which offers a linear constraint on the solution space of the global motion model $\mathbf{M}_g$ with $\mathbf{I}$ denoting an identity matrix. Note that Eq. (10.15) represents the constraint on $\mathbf{M}_g$ due to a *single* relative motion $\mathbf{M}_{ij}$ which can be written as $\mathbf{C}_{ij}\mathbf{M}_g = \mathbf{0}$. With a sufficient number of such constraints, the global motion can be solved linearly by stacking all the constraints as rows into a single system of equations. In particular, the unknowns of all the affine matrices can be stacked onto one large vector, say, $\mathbf{m}_g = [\mathbf{m}_2, \ldots, \mathbf{m}_N, 1]^T$ which is the homogeneous form of the stacked affine terms. Therefore, for an N-image sequence, the vector $\mathbf{m}_g$ is of size $6(N-1)+1$. Given this form of the unknown terms, Eq. (10.14) can be rewritten as $\mathbf{V}_{ij}^T \mathbf{m}_g = \mathbf{0}$ for the constraint between image i and image j, where $\mathbf{V}_{ij}$ represents terms from the observed relative transformations $\mathbf{M}_{ij}$. In turn, by stacking all these terms we get the form $\mathfrak{V}^T \mathbf{m}_g = \mathbf{0}$, where $\mathfrak{V}$ contains all the $\mathbf{V}_{ij}$ matrices. The least-squares solution of this form is obtained by the minimization of $\mathbf{m}_g^T \mathfrak{V}\mathfrak{V}^T \mathbf{m}_g$, which can be found due to the well-known linear algebra technique of *singular value decomposition* (SVD).

---

* Taking the first image as the reference frame, $\mathbf{M}_1$ is the identity transformation.

The advantage of using such a unified approach to averaging all the possible relative affine estimates is that the errors in the individual estimates of $\mathbf{M}_{ij}$ are cancelled out, which is very useful for long or closed loops where the solution often drifts off due to error accumulation. The presence of multiple constraints between different images results in a tight tethering of the solution, thus preventing it from drifting away as would happen if only adjacent frames were used in the registration process. We also note that for a solution of the global motion model, we are not required to use every pairwise constraint. For extended sequences, there is seldom any overlap between frames that are well separated in time. Therefore, their relative two-frame motions cannot be estimated. However, we can still get a consistent solution as long as we have at least $N - 1$ relative motions. This is a distinct advantage over methods that use a rank-constraint condition, which requires the computation of the entire set of relative motions (see, for example, Malis and Cipolla, 2000).

Figure 10.7 shows the result of registering a large set of images in a common reference frame. The mosaic represents an image region of a scene containing a road and a bridge, which was acquired from a video sequence of aerial images. As can be seen, the images contain individual vehicles driving along the road (the first vehicle pulling into the area on the left). The video sequence contains a total of more than 1500 images.

For correspondences between images we used the publicly available KLT tracker (Shi and Tomasi, 1994) to generate a set of 200 matched points in every pair of adjacent images. When the same point is persistent across the views they result in tracks of matched points. As the field of view changes, some of these tracks disappear and new ones are created. Having tracked the features across the entire sequence, we decimated the image sequence to 74 images which were used in the final mosaic created. For every pair of images in the sequence that are at most seven frames apart, we estimated the relative affine transformation. This resulted in a total of 484 relative affine motions over the entire sequence. These relative motions were finally averaged to obtain the global registration using the approach outlined above.

Since the video sequence contains independently moving vehicles, we could not estimate directly the affine transformations between image pairs. A direct solution would be grossly erroneous since it would be corrupted by the presence of independently moving objects. Instead, we used the RANSAC method described in Subsection 10.3.1 for every set of correspondences between image pairs to obtain a robust estimate of the affine transformation that ignores the independently moving objects and aligns the image to the ground plane.

Once the global transformation was solved, we stacked the transformed images and created a global mosaic of the 74 images by taking the median value for each

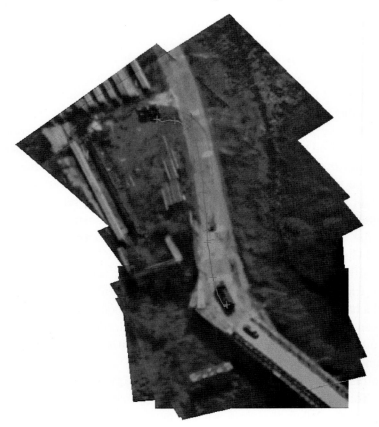

Figure 10.7. Mosaic created out of a long sequence with independent motions; trajectory of one independently moving feature point is shown. See color plates section.

pixel. This mosaic, which can be seen to be correct from Fig. 10.7, provides a succinct summarization of the information in the entire video sequence. Furthermore, given the accurate ground plane estimation, we can use the registration information to reconstruct the tracks of the independently moving objects. One such feature track (on the large truck shown in Fig. 10.7) clearly indicates the motion of the truck in the video sequence.

Finally, we demonstrate the usefulness of our notion of consistency in constructing a mosaic using *image homographies*. Figures 10.8(a)–(b) show mosaics constructed with the same set of 24 images. Here, images were acquired along four horizontal strips with six overlapping images along each strip. Unlike our previous examples, where a 2D affine transformation was sufficient to capture the relative geometry between images, in this case, since the images exhibit significant

(a)

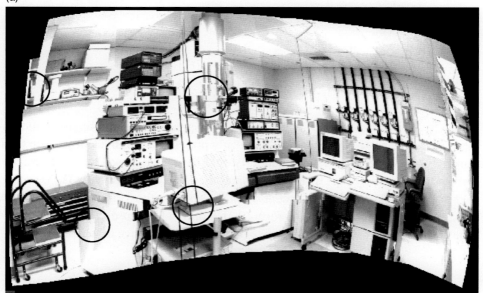

(b)

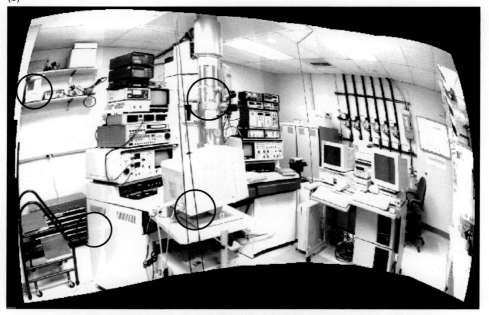

Figure 10.8. (a) Mosaic created by aligning images along horizontal strips; gross vertical errors (indicated by circles) can be seen at the boundaries of some strips, and (b) mosaic created by our method that computes a consistent set of transformations; note the good alignment throughout the image in this case.

perspective effects, we used a homography to represent the geometric transformation. A homography is a linear transformation on the projective representation of image points, i.e.,

$$\begin{bmatrix} x_j \\ y_j \\ 1 \end{bmatrix} = \lambda_{ij} \mathbf{H}_{ij} \begin{bmatrix} x_i \\ y_i \\ 1 \end{bmatrix}, \tag{10.16}$$

where λ_{ij} is an unknown scale factor and $\mathbf{H}_{ij}$ represents the interimage homography. Since the relative homographies are only projectively equivalent (i.e., up to an unknown scale factor), we have to modify our consistency-based motion averaging method expressed by Eq. (10.14). Also, since the 3×3 transformation $\mathbf{H}_{ij}$ is projective, λ_{ij} is an unknown scalar, except when $\mathbf{H}_{ij}$ takes on the form of an affine matrix (i.e., when its third row is of the form $(0, 0, 1)$), in which case $\lambda_{ij} = 1$. Hence the linear system that solves for a consistent system of projective transformations is

$$\mathbf{H}_{ij}\mathbf{H}_i - \lambda_{ij}\mathbf{H}_j = \mathbf{0}, \tag{10.17}$$

where $\mathbf{H}_i$ and $\mathbf{H}_j$ denote, respectively, the homographies with respect to the reference frame.

In the case of projective transformations, the scaling of different elements is not the same. The third column consists of the translation components that have a much larger scale than the rest of the entries (these values can be on the order of the size of the image itself, say 256). On the other hand, the third row contains elements that are much smaller compared to the rest of the elements of the transformation (the third row is typically close to $(0, 0, 1)$). This unequal scaling implies that it is important to apply a transformation to the $\mathbf{H}_{ij}$ estimates before solving for a consistent solution. This transformation has to be chosen in such a way as to approximately *whiten* the $\mathbf{H}_{ij}$ estimates. Indeed, our method solves for the different relative transformations $\mathbf{H}_{ij}$, and applies an approximately whitening transformation to give equal weights to the different terms.

Since we also need to incorporate the estimation of the unknown scaling factors λ_{ij}, we use the following iterative scheme to solve for a consistent set of transformations:

(1) Compute the average value s_k of the elements of the kth row ($k = 1, 2, 3$) of all the matrices $\mathbf{H}_{ij}$.
(2) Compute the approximate whitening transformation $S = diag([\frac{1}{s_1}, \frac{1}{s_2}, \frac{1}{s_3}])$.
(3) Apply the transformation to individual estimates as $\mathbf{H}'_{ij} \leftarrow S\mathbf{H}_{ij}S^{-1}$. Now apply the following iterative scheme:
  (a) Set all $\lambda_{ij} = 1$;
  (b) Solve the linear system $\mathbf{H}'_{ij}\mathbf{H}'_i - \lambda_{ij}\mathbf{H}'_j = \mathbf{0}$;

(c) Update $\lambda_{ij} = \frac{\|\mathbf{H}'_{ij}\mathbf{H}'_i\|}{\|\mathbf{H}'_j\|}$;

(d) Repeat until convergence.

(4) 'Unwhiten' the individual global motion models as $\mathbf{H}_i \leftarrow S^{-1}\mathbf{H}'_i S$.

In Fig. 10.8(a), the mosaic is constructed by computing pairwise homography transformations between adjacent images along the horizontal direction, and then aligning images in each horizontal strip with respect to the reference frame by computing the transformation between the rightmost image in each strip and the reference frame. Thus, for each image we can compute a product of projective transformations that will align it with respect to the reference frame. As a result, there is good alignment along a horizontal strip.

However, as can be clearly seen, there is gross misalignment between different strips. Indeed, moving from right to left along a strip, the errors in individual horizontal strips accumulate. And since these errors are independent, this results in gross misalignment in the vertical direction, as indicated by the circles in Fig. 10.8(a). In addition to the relative transformations used in Fig. 10.8(a), Fig. 10.8(b) illustrates the resulting mosaic due to our method, which first computes some of the relative transformations between adjacent pairs along the vertical direction, and then constructs a consistent set of transformations as described above. As can be observed by comparing the circled regions in the two images, our method exhibits good alignment over the entire mosaic and none of the anomalies present in Fig. 10.8(a).

10.5 Conclusions

In this chapter we have described a variety of approaches to image registration based on features such as contours and points extracted from images. Low-level processing issues like detection and matching of features were discussed and an approach to robust estimation was outlined. The current body of literature pertaining to motion estimation in computer vision contains many interesting ideas that can find potential application in the domain of image registration. While the geometry of linearized representations is well understood, more accurate estimators can be derived by considering the statistical properties of features and their distributions in the transformation space that represents camera geometry. Another issue of interest is the application of known techniques to large datasets of images which would involve significant questions pertaining to robustness and computational efficiency.

Acknowledgements

The work of Rama Chellappa was partially supported by a MURI from the Army Research Office under Grant W911NF0410176.

References

Brown, L. G. (1992). A survey of image registration techniques. *ACM Computing Surveys*, **24**(4), 325–376.

Chellappa, R., Zheng, Q., Burlina, P., Shekhar, C., and Eom, K. B. (1997). On the positioning of multisensor imagery for exploitation and target recognition. *Proceedings of the IEEE*, **85**(1), 120–138.

Chum, O. and Matas, J. (2005). Matching with PROSAC – progressive sample consensus. In *Proceedings of the IEEE Conference on Computer Vision and Pattern Recognition*, San Diego, CA, Vol. 1, pp. 220–226.

Fischler, M. A. and Bolles, R. C. (1981). Random sample consensus: A paradigm for model fitting with applications to image analysis and automated cartography. *Communications of the ACM*, **24**(6), 381–395.

Gold, S., Rangarajan, A., Lu, C. P., Pappu, S., and Mjolsness, E. (1998). New algorithms for 2D and 3D point matching: Pose estimation and correspondence. *Pattern Recognition*, **31**(8), 1019–1031.

Govindu, V. M. (2001). Combining two-view constraints for motion estimation. In *Proceedings of the IEEE Conference on Computer Vision and Pattern Recognition*, Hawaii, pp. 218–225.

Govindu, V. M. (2004). Lie-algebraic averaging for globally consistent motion estimation. In *Proceedings of IEEE Conference on Computer Vision and Pattern Recognition*, Washington, DC, pp. 684–691.

Govindu, V. and Shekhar, C. (1999). Alignment using distributions of local geometric properties. *IEEE Transactions on Pattern Analysis and Machine Intelligence*, **21**(2), 1031–1043.

Harris, C. and Stephens, M. (1988). A combined corner and edge detector. In *Proceedings of the Fourth Alvey Vision Conference*, Manchester, UK, pp. 147–151.

Hartley, R. and Zisserman, A. (2004). *Multiple View Geometry in Computer Vision*, 2nd edn. Cambridge: Cambridge University Press.

Li, H., Manjunath, B. S., and Mitra, S. (1995). A contour-based approach to multi-sensor image registration. *IEEE Transactions on Image Processing*, **4**, 320–334.

Lowe, D. G. (2004). Distinctive image features from scale-invariant keypoints. *International Journal of Computer Vision*, **60**(2), 91–110.

Malis, E. and Cipolla, R. (2000). Multi-view constraints between collineations: Application to self-calibration from unknown planar structures. In *Proceedings of the Sixth European Conference on Computer Vision*, Dublin, Ireland; *Computer Vision – ECCV 2000*, D. Vernon, ed. Berlin: Springer, Vol. 2, pp. 610–624.

Mikolajczyk, K. and Schmid, C. (2003). A performance evaluation of local descriptors. In *Proceedings of the IEEE Conference on Computer Vision and Pattern Recognition*, Madison, Wisconsin, Vol. 2, pp. 257–263.

Rousseeuw, P. J. and Leroy, A. M. (1987). *Robust Regression and Outlier Detection*. New York: Wiley.

Schmid, C., Mohr, R., and Baukhage, C. (2000). Evaluation of interest point detectors. *International Journal of Computer Vision*, **37**(2), 151–172.

Shekhar, C., Govindu, V., and Chellappa, R. (1999). Multisensor image registration by feature consensus. *Pattern Recognition*, **32**(1), 39–52.

Shi, J. and Tomasi, C. (1994). Good features to track. In *Proceedings of the IEEE Conference on Computer Vision and Pattern Recognition*, Seattle, Washington, pp. 593–600.

Szeliski, R. (2006). Image alignment and stitching. In N. Paragios, Y. Chen, and O. Faugeras, eds., *Mathematical Models in Computer Vision: The Handbook*. New York: Springer, pp. 273–292.

Torr, P. and Zisserman, A. (2000). MLESAC: A new robust estimator with application to estimating image geometry. *Computer Vision and Image Understanding*, **78**(1), 138–156.

Viola, P. and Wells, W. (1997). Alignment by maximization of mutual information. *International Journal of Computer Vision*, **24**(2), 137–154.

Zhang, Z., Deriche, R., Faugeras, O., and Luong, Q. T. (1995). A robust technique for matching two uncalibrated images through the recovery of the unknown epipolar geometry. *Artificial Intelligence*, **78**(1–2), 87–119.

Zheng, Q. and Chellappa, R. (1993). A computational vision approach to image registration. *IEEE Transactions on Image Processing*, **2**(3), 311–326.

Zitová, B. and Flusser, J. (2003). Image registration methods: A survey. *Image and Vision Computing*, **21**(11), 977–1000.

11

On the use of wavelets for image registration

JACQUELINE LE MOIGNE, ILYA ZAVORIN, AND HAROLD STONE

Abstract

Wavelets provide a multiresolution description of images according to a well-chosen division of the space-frequency plane. This description provides information about various features present in the images that can be utilized to perform registration of remotely sensed images. In the last few years, many wavelet filters have been proposed for applications such as compression; in this chapter, we review the general principle of wavelet decomposition and the many filters that have been proposed for wavelet transforms, as they apply to image registration. In particular, we consider orthogonal wavelets, spline wavelets, and two pyramids obtained from a steerable decomposition. These different filters are studied and compared using synthetic datasets generated from a Landsat-Thematic Mapper (TM) scene.

11.1 Introduction

The main thrust of this chapter is to describe image registration methods that focus on computational speed and on the ability of handling multisensor data. As was described in Chapter 1 and in Brown (1992), any image registration method can be described by a feature space, a search space, a search strategy, and a similarity metric. Utilizing wavelets for image registration not only defines the type of features that will be matched, but it also enables the matching process to follow a multiresolution search strategy. Such an iterative matching at multiple scales represents one of the main factors that will define the accuracy of such methods.

A more detailed review of wavelet and wavelet-like representations will be provided in Section 11.2, but as a general description, a wavelet decomposition of a two-dimensional (2D) image signal is the process that provides a complete representation of the signal according to a well-chosen division of the space-frequency plane. Through iterative filtering by low-pass and high-pass filters, this process provides information about low and high frequencies of the signal at successive scales. Following this description, using wavelets for the registration of remotely sensed image data is therefore justified by the following:

- Multiresolution wavelets, such as those utilized for the compression standard JPEG-2000 (Schelkens *et al.*, 2009), can bring multisensor/multiresolution data to the same spatial resolution without losing significant information and without blurring the higher-resolution data.
- At the lower resolutions, the process preserves important global features such as rivers, lakes, and mountain ridges, as well as roads and other man-made structures, while at the same time eliminating weak higher-resolution features often considered as "noise" or "spurious pixels." Of course, some of the important finer features will be lost at the lowest resolutions but, if needed, they can be retrieved in the iterative higher levels of decomposition.
- The multiresolution iterative search focuses progressively toward the final transformation with a decreasing search interval and an increasing accuracy at each iteration. Hence, this type of strategy achieves higher accuracies with higher speeds than a full search at the full resolution.

Additionally, wavelet decomposition and multiresolution iterative search are very well suited for fine-grained parallelization, thus speeding up the computations even more; see El-Ghazawi and Le Moigne (2005) and Chan *et al.* (1995).

This chapter first provides a brief review of wavelets and wavelet-like transforms (Section 11.2), then Section 11.3 describes some experiments dealing with the use of orthonormal wavelets for image registration. Section 11.4 compares the use of orthonormal wavelets to an overcomplete representation, such as the one provided by Simoncelli's steerable pyramid framework, showing how such representation can lead to better accuracy, consistency, and reliability. Finally, conclusions will be presented in Section 11.5.

11.2 Brief review of wavelet and wavelet-like transforms

This section does not constitute an exhaustive survey of wavelet theory but rather intends to give the minimal amount of information necessary to understand the experiments described in Sections 11.3 and 11.4. For more complete information on the wavelet theory, the reader is referred to general works in this area (Daubechies, 1988; Strang, 1989; Chui, 1992; Daubechies, 1992; Heil and Walnut, 2006).

11.2.1 Wavelet theory and orthogonal/orthonormal wavelets

Similarly to a Fourier transform, wavelet transforms provide space-frequency representations of a 2D signal, which can be inverted for later reconstruction.

Although it is straightforward to reconstruct a signal from its Fourier representation, either the original signal or the Fourier representation must be infinite in extent. Both cannot be finite. For practical calculations, both the signal and its transformation should have finite extent, that is, should be "local." To alleviate this problem, windowed Fourier transforms, and as a special case, Gabor transforms, were introduced (Heil and Walnut, 2006). The signal is analyzed after filtering the original signal by a fixed window, so these transforms have the spatial localization property that traditional Fourier transforms lack. But with the envelope of the signal being the same for all frequencies, Gabor transforms do not provide optimal space details for high frequencies, where more sample points are needed. Wavelets were then introduced to allow for a better spatial localization as well as a better division of the space-frequency plane than a Fourier or even than a windowed Fourier transform (for more details, see Chui, 1992, and Daubechies, 1988).

In a wavelet representation, the original signal is filtered by the translations and the dilations of a basic function, called the *mother wavelet*. Equation (11.1) shows the general continuous form of a wavelet transform of an image I,

$$Wav(I)(a, b) = \frac{1}{\sqrt{a}} \int_{-\infty}^{\infty} \int_{-\infty}^{\infty} I(u, v) \cdot W\left(\frac{u - b_1}{a}, \frac{v - b_2}{a}\right) du\, dv, \quad (11.1)$$

where W represents the mother wavelet, $b = (b_1, b_2)$ is the translation factor, and a is the dilation factor. All the dilations and translations of the mother wavelet form an orthonormal basis in which the function image is uniquely represented, and therefore the transformation can be inverted to produce the original image from the unique representation.

A similar equation in the discrete domain is given by

$$Wav(I)(a, b) = \sum_{m} \sum_{n} a^{-m/2} I(x, y) \cdot W(a^{-m}x - nb_1, a^{-m}y - nb_2). \quad (11.2)$$

Typical values for a and b are $a = 2$ and $b = 1$.

As pointed out in Daubechies (1992), "wavelets are a fairly simple mathematical tool with a great variety of possible applications." For these various applications, different types of wavelets have been proposed. The simplest wavelet is the Haar wavelet, while some of the most popular wavelets are the ones proposed by Daubechies, in particular the compactly supported wavelets of different sizes (Daubechies, 1988). Other wavelet types are the Meyer wavelet, Morlet's wavelet, and the Mexican hat wavelet (Chui, 1992; Djamdji *et al.*, 1993; Heil and Walnut, 2006). For the remaining of this chapter, "Daubechies wavelets" or "orthonormal"

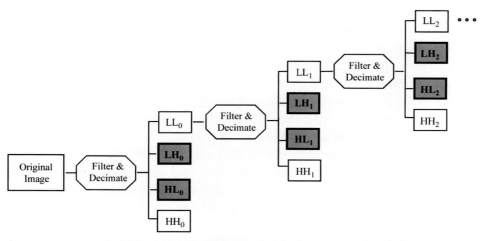

Figure 11.1. Multiresolution wavelet decomposition described in Mallat (1989). (Source: Le Moigne and Zavorin 2000, © SPIE, reprinted with permission.)

or "orthogonal" wavelets will be referring to the group of compactly supported wavelets that were originally developed by Daubechies.

Mallat (1989) demonstrated the connection between filter banks and wavelet basis functions. His results became very important for the application of wavelet theory to image processing applications. Specifically, for a 2D image, this work shows that wavelet analysis can be implemented in a separable fashion by filtering the original image by a high-pass and a low-pass filter, iteratively in a multiresolution fashion, and separately in rows (vertical filter) and in columns (horizontal filter). At each level of decomposition, four new images are computed. Each of these images has one quarter of the number of pixels of the original image at the previous level, and it represents the low-frequency or high-frequency information of the image in the horizontal or/and the vertical directions; images LL (Low/Low), LH (Low/High), HL (High/Low), and HH (High/High) of Fig. 11.1. The Low/Low image appears to be the same as the original image, except that it has a lower resolution. The other three images contain edge information, which when added to the Low/Low image can recreate the original image at full resolution. Because the Low/Low image has one fourth the number of pixels of the original image, and because it has the same appearance, the Low/Low image is a compressed version of the original image. An illustration of these observations can be found in Le Moigne (1994, Figures 1 and 2). Starting from the "compressed" image (or image representing the low-frequency information), the process can be iterated, thus building a hierarchy of lower and lower resolution images. Since, at each iteration, the size of the compressed image decreases, by leaving the size of the filter unchanged, the area being filtered in the compressed image corresponds to a larger and larger portion of the original image. Figure 11.1 summarizes the multiresolution decomposition.

The features provided through this framework of wavelet decomposition are of two different types: (1) Low-pass features, which provide a compressed version of the original data and some texture information, and (2) high-pass features, which provide detailed information very similar to edge features.

Since they are orthogonal, Daubechies wavelets are computationally efficient and are, in fact, very well suited to fine-grained parallelization (El-Ghazawi and Le Moigne, 2005); but for the same reason, they also have poor invariance properties. As an image is shifted, energy shifts both within and across subbands (Simoncelli *et al.*, 1992). The energy shift within a subband is illustrated by a change in appearance of, for example, the High/High subband depending on shift position of the original image. In the same fashion, the energy shift across subbands may be illustrated by the energy in the Low/High subband reducing and the energy in the High/High subband increasing as an image is shifted; an example of this phenomenon is shown in Simoncelli *et al.* (1992, Figure 1). However, it was shown (Stone *et al.*, 1999) that, when combined with correlation as a similarity measure, orthogonal wavelets can still provide fairly accurate registration results (that study involved only translation; see Subsection 11.3.1).

This shift- and rotation-invariance is particularly important for our registration application. Ideally, a search strategy would provide a candidate spatial transformation point at which features would be extracted by applying the wavelet transform to the image warped according to that candidate spatial transform. But since it would be prohibitively expensive to recompute features for each transform, we prefer to compute the wavelet transform once and then apply the geometric spatial transformation. Further studies comparing Daubechies with Simoncelli-steerable pyramids are presented in Section 11.4.

11.2.2 Shift-invariant wavelets and Simoncelli steerable decomposition

Several approaches have been proposed to overcome the deficiencies of orthogonal wavelets, and with the large variety of wavelets that have been developed for image processing applications we define the following requirements that are essential for a registration application (Zavorin and Le Moigne, 2005):

(1) The order of decomposition and warping for both reference and input images can be interchanged without altering the result of the registration or its accuracy. By warping, we mean the geometric transformation that is necessary to match the input image to the reference image. For this chapter, images to be registered are assumed to undergo rigid geometric warping transformations known as *RST transforms*, which are sequences of

rotations, scale changes, and translations. Because our search space for image registration transformations is limited to these RST transforms, the wavelet pyramids we employ should be, ideally, invariant to shifts and rotations. In other words, the wavelet transform of a shifted and rotated image should be the same as the shifted and rotated wavelet transform of the original image.

(2) A pyramid should be implemented efficiently, i.e., the corresponding image representation must be computationally inexpensive to generate.

(3) The resulting representation should also be efficient when used in the registration phase.

Among the wavelet schemes that satisfy these requirements is an efficient shift-invariant pyramid image representation based on polynomial splines that was designed by Unser and his colleagues (Unser *et al.*, 1993a; 1993b). This scheme has several attractive properties. *First*, it minimizes least-squares differences between successive image representations in the pyramid. In terms of image registration, such optimality ensures that accurate approximations to the final geometric transformation can be recovered at coarse pyramid levels. *Second*, the pyramid is implemented efficiently using simple linear filters derived from the recursive B-splines. Starting with the original image, a pyramid $\{S_1, S_2, \ldots\}$ is produced by recursive anti-aliasing prefiltering, followed by decimation by a factor of two. *Third*, it can be tied neatly to the cubic spline interpolation used at a given pyramid level during registration, for geometric warping, and exact computation of derivatives of the objective function (Thévenaz *et al.*, 1998). Fourth, it is based on the spline theory, the theoretical and practical aspects of which have been thoroughly developed since the original work of Schoenberg (1969). This spline-based representation has the shift and rotational invariance property required by our application.

Other approaches that satisfy our three basic requirements are based on the use of *overcomplete wavelets* (or frames). Since for applications such as image registration, a minimal representation of the 2D signal is not required, it is possible to consider wavelet frames or overcomplete representations of the signal. This overcompleteness has been proposed by various authors to overcome the non-invariance issue discussed in conclusion of Subsection 11.2.1 (Simoncelli *et al.*, 1992; Coifman and Donoho, 1995; Simoncelli and Freeman, 1995; Karasaridis and Simoncelli, 1996; Magarey and Kinsgbury, 1997). In Coifman and Donoho (1995), the signal is transformed using a range of shifts and the resulting shifts are averaged, thus suppressing artifacts. The method was used for denoising and a similar method could be devised for image registration, but it would considerably increase the computational complexity of the process and might affect the accuracy of the registration. In Magarey and Kinsgbury (1997), the efficient discrete wavelet transform is combined with complex-valued Gabor-like filters that have nearly optimal localization, thus producing a pyramid with approximate shift invariance.

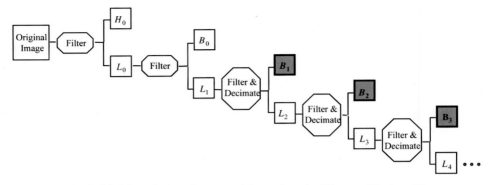

Figure 11.2. Multiresolution decomposition using the Simoncelli steerable pyramid. (Source: Le Moigne and Zavorin 2000, © SPIE, reprinted with permission.)

At each level of the pyramid, six complex-valued subimages are produced from the original image using equally spaced oriented filters. Although this scheme might be more reliable than most other wavelets from a translation-invariant point of view, it might also significantly increase the computation time.

On the other hand, the steerable pyramid proposed by Simoncelli *et al.* (1992) builds translation- and rotation-invariant filters by relaxing the critical sampling condition of the wavelet transforms. The overcomplete invertible wavelet representation generated by Simoncelli's steerable pyramid shown in Fig. 11.2 satisfies the three criteria defined above. In this scheme, the original image is first preprocessed by a high-pass prefilter and a low-pass prefilter to create H_0 and L_0. The resulting two images are of the same resolution as the original image. Then, the low-pass image is further processed by a band-pass filter to create a band-pass component, B_0, and by a low-pass filter to create a low-pass component, L_0. Note that although in Fig. 11.2, only a single band-pass filter is shown, it is possible to have several such filters, each emphasizing image features oriented in a specific rotational angle. If there are k band-pass filters, then the pyramid is $4k/3$ overcomplete, with each band $j = 1, \ldots, k$ oriented at the angle $\pi j / k$. Although in certain cases, it may be beneficial to use more than one oriented filter (Liu *et al.*, 2002), our study is limited to $k = 1$ to reduce computational and memory requirements. In order to ensure shift invariance, the outputs of the high-pass prefilter and of the band-pass filter(s) are not subsampled. As a result, the steerable pyramid produces a representation of an image composed of two multiresolution series of components, the low-pass series $\{L_0, L_1, \ldots\}$ and the band-pass series $\{B_0, B_1, \ldots\}$, where both $L_{n-1}{}^*$ and B_n are 2^{-n} times the original size. In our experiments, downsampling of the steerable pyramid was slightly modified to remove *shift bias* caused by filter nonsymmetries (Zavorin *et al.*, 2002).

---

* With L_0 being the same as the original.

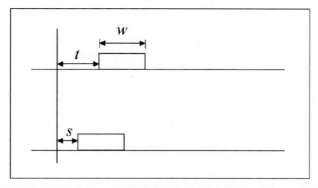

Figure 11.3. Pulse model function used for registration studies. (Source: Stone *et al.*, 1999, © IEEE, reprinted with permission.)

In summary, the three schemes that we investigated within the framework of image registration were:

(1) Mallat's multiresolution analysis using orthonormal Daubechies filters.
(2) The Thévenaz-Unser spline wavelets scheme.
(3) Simoncelli's steerable pyramid.

Results of these investigations are summarized below in Sections 11.3 and 11.4.

11.3 Orthogonal wavelet experiments

11.3.1 Translation sensitivity experiments

As described in Stone *et al.* (1999), we performed experiments to examine the sensitivity of wavelet subband correlations to translation. For these experiments, the Haar wavelet and the Daubechies wavelet of size 4 (referred to as "Daubechies-4") were used.

We first conducted some pulse-model experiments, as shown in Fig. 11.3. The one-dimensional model consists of a pulse of unit height and width w offset from the origin by an amount t. It is correlated against a pulse of identical width offset by an amount s. Therefore, in the absence of noise, the correlation peak occurs with the offset $|t - s|$. The model uses a baseline of length 128 and varied pulse widths from 1 to 64. The study covered all distinct values of $|t - s| \bmod B$, where B is the filter size, e.g., 2 and 4 in our experiments, for the Haar filter and the Daubechies-4 compactly supported filter, respectively. In 2D, a similar two-dimensional pulse-model is also built. Results of the experiments found in Stone *et al.* (1999) show that the height of the correlation peak is a function of the pulse width for both

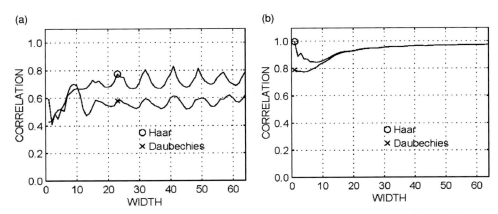

Figure 11.4. Average height of correlation peak as a function of pulse width in 2D: (a) Level-3 high-pass subbands of level-2 low-pass subbands, and (b) level-3 low-pass subbands. (Source: Stone *et al.*, 1999, © IEEE, reprinted with permission.)

low-pass and high-pass filters, for 1D and 2D pulses. Both Haar and Daubechies-4 and all decimation levels were being considered, and results were similar in each case. They show that the low-pass filters produce very high correlation peaks in the worst case, and on the average for features greater than $2B$ (e.g., 4 or 8) in size. The low-pass correlation peak is as low as 0.8 in 1D and 0.7 in 2D for features twice the filter size or more. Overall, the average height of the low-pass correlation peak exceeds 0.9 for both the 1D and 2D models. This proves that the low-pass subband is essentially insensitive to translation for feature sizes at least twice the size of the wavelet filter (note that this is consistent with the *sampling theorem* (Shannon, 1949)). On the other hand, as shown in Stone *et al.* (1999), the high-pass correlation peak is much more sensitive to translation, and can fall below 0.3 in 1D and below 0.1 in 2D for some translation phases. Nevertheless, the average correlation peak is 0.5 or higher for both 1D and 2D, which indicates that high-pass correlation can be used, but with care, for example, by carrying out multiple transformation candidates. As an illustration, Fig. 11.4 shows the average height of correlation peak as a function of pulse width for the 2D case and level-3 low-pass and high-pass subbands.

Additional experiments were performed using Gaussian white-noise distributions and show essentially the same results. See Stone *et al.* (1999) for more details on these experiments.

11.3.2 Orthogonal wavelets for image registration

Based on the results shown in Subsection 11.3.1, registration was performed using the Mallat multiresolution wavelet decomposition and Daubechies-4 wavelets (Le

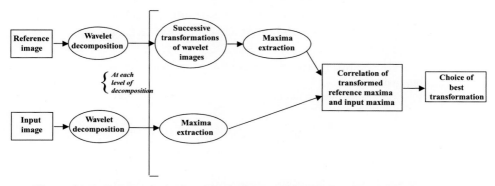

Figure 11.5. Summary of exhaustive search strategy for wavelet-correlation image registration method. (Source: Le Moigne and Zavorin, 2000, © SPIE, reprinted with permission.)

Moigne, 1994; Le Moigne *et al.*, 2002a). After performing the wavelet decomposition of both reference and input images according to Fig. 11.1, the histograms of the LH and HL images were computed for all levels of decomposition. Then, only those points whose intensities belonged to the top $x\%$ of the histograms were kept. These points or "maxima of the wavelet coefficients" represent the features that were utilized in the registration process. Although this process does guarantee that these points belong to edges of strong image features, it does not guarantee that they belong to features larger than at least twice the size of the wavelet filter. But by averaging the independent correlations (LH–LH) and (HL–HL), and by refining these results at each decomposition level, the algorithm provides accurate results. Another way to make the algorithm more robust is to choose the parameter x adaptively; at the lowest level of resolution, the best transformation is computed successively for the four values $x = 5\%$, 10%, 15%, and 20%. A correlation measure was associated with each of these four values, and the parameter x was then chosen as the average of these values weighted by the corresponding correlations.

The experiments described in Le Moigne *et al.* (2002a) used a search space composed of 2D rotations and translations, and the search follows the multiresolution approach provided by the wavelet decomposition. Figure 11.5 and Table 11.1 give a summary of this search strategy. If only looking for rotations at each level (Table 11.1), an optimal rotation is computed and the exhaustive search is refined at the next level, searching around this optimal rotation more accurately, in a smaller search interval.

In summary, since we showed earlier how wavelets are not rotationally invariant, our search had to generate all of the rotations at one level to find the "optimal" rotation. At successive levels, the search was limited to a region around the optimal rotation, but the search still had to be done because of the lack of rotational

Table 11.1 *Exhaustive search strategy summary for rotations and no initial guess (source: Le Moigne et al., 2002a, © IEEE, reprinted with permission)*

Decomposition level	Search interval	Accuracy	Result rotation
n	$[0, \pi/2]$	$2^{n-1}d$	R_n
$n-1$	$[R_n - 2^{n-1}d, R_n + 2^{n-1}d]$	$2^{n-2}d$	R_{n-1}
$n-2$	$[R_{n-1} - 2^{n-2}d, R_{n-1} + 2^{n-2}d]$	$2^{n-3}d$	R_{n-2}
...	...	...	...
2	$[R_3 - 2^2d, R_3 + 2^2d]$	$2d$	R_2
1	$[R_2 - 2d, R_2 + 2d]$	d	R_1

invariance. The drawback of this method is that if the original estimate of the optimal rotation is wrong, then the registration fails. Also, the extra computations required to search for the rotation reduces the efficiency of this method (although as shown below, the method can be very easily implemented on a parallel computer). The method also lacks translation invariance in general, but, as pointed out in Subsection 11.3.1, the Low/Low subbands at the various levels are largely translation-invariant. This explains why the results shown below are still quite acceptable, especially when transformations are mainly limited to translations.

In these experiments, the algorithm was tested on Landsat-Thematic Mapper (TM) and on the Advanced Very High Resolution Radiometer (AVHRR) images, and the results show that the average optimum correlations range from 0.2 to 0.55 at the lowest level but may increase to 0.6 or more at the highest level. Overall, the accuracies obtained during these experiments range from 0.17 pixels for the Landsat images to 0.425 pixels for the AVHRR images after applying a cloud mask. The method was implemented on several high-performance computers, including a Single Instruction Multiple Data (SIMD) architecture, a Cray T3D, an Intel Paragon, a Convex SPP, and a Beowulf cluster of workstations. As an example, the registration of a 256×256 image on the Beowulf architecture requires about 0.545 seconds when 16 processors are utilized. More details can be found in Le Moigne *et al.* (2001, 2002a) and in El-Ghazawi *et al.* (1997).

11.4 Steerable pyramid experiments

This section focuses on Simoncelli's steerable filters, using first the same exhaustive search as the one described in Subsection 11.3.2 to compare these filters to orthogonal wavelets. The study is then generalized utilizing a systematic test dataset, by replacing exhaustive search with an optimization scheme, and extending the comparison to include also spline wavelets.

11.4.1 Preliminary experiments

Following the experiments described in Section 11.3, a study involving a few well-chosen test images was performed, using the same exhaustive search algorithm and comparing registration accuracies obtained with features from a Simoncelli steerable pyramid decomposition (from Subsection 11.2.2) to those obtained with orthogonal wavelet features. As the LH and HL features were utilized for the orthogonal wavelet features, steerable pyramid features were extracted from the band-pass subbands B (see Fig. 11.2). Results reported in Le Moigne and Zavorin (1999) show that the two types of filters give similar results for ideal or low-noise images but, as expected, as the noise level increases, the registration accuracy is much more stable with Simoncelli's filters. Overall, the translation accuracy obtained with these filters stays around one pixel, and even reaches subpixel accuracy for some of the datasets, while the accuracy obtained using Daubechies' orthogonal filters greatly varies depending on the content of the images. The results were consistent for all levels of thresholds chosen in the maxima selection, even when the noise level increases. These results, obtained within the framework of an exhaustive search, verify that the Simoncelli steerable filters possess the translational and rotational invariance property lacked by orthogonal filters and, therefore, provide more reliable and stable results.

A more systematic study was then conducted using synthethic datasets (presented in Subsection 11.4.3). Additionally, recognizing that an exhaustive search involving multiple cross-correlations is computationally expensive, we focused subsequent work on a systematic comparison using optimization as a search strategy and tested using carefully designed test data. Additional work was also done looking at different subbands provided by the Simoncelli filters and comparing them to low-pass spline wavelet subbands. The experimental setup is described in Subsection 11.4.2 and the results are given in Subsection 11.4.3.

11.4.2 Systematic experiments setup

To continue with the systematic study of steerable filters, it is essential to have accurate ground truth information. Therefore, we used synthetic test data generated from a single 1024×1024 Landsat-7 ETM+ (Enhanced Thematic Mapper) scene; this control process provides known ground truth while emulating real data. In the process, the 512×512 center of the image is extracted and utilized as the "reference image." Then, three types of modifications are applied to the "source" image in various combinations to create the test dataset: (1) Geometric

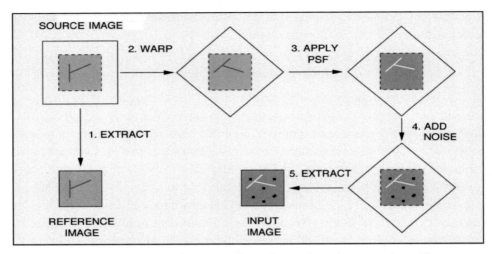

Figure 11.6. Synthetic image generation. See color plates section. (Source: Zavorin and Le Moigne, 2005, © IEEE, reprinted with permission.)

warping, (2) radiometric variations, and (3) addition of noise. *Geometric warping* is introduced by applying an RST transformation; *radiometric variations* are introduced by convolving the image with a *point spread function* (PSF) chosen to mimic how an instrument would process a scene; and *Gaussian noise* is added to the image to emulate imperfections of optical systems and models used in preprocessing the satellite data.

Figure 11.6 summarizes the process by which these synthetic data, the reference image, and a set of input images are generated for a variety of geometric transformations and varying levels of noise. More details about the generation of these synthetic data are given in Zavorin and Le Moigne (2002).

11.4.3 *Experimental results*

11.4.3.1 *Orthogonal and steerable filters study*

The first set of experiments deals with the comparison of Daubechies' orthogonal filters with Simoncelli's band-pass filters. In this study (Le Moigne and Zavorin, 2000),

- Translation parameters were varied in the horizontal and vertical directions by amounts of 0 to 50 pixels.
- Rotation parameters were varied with angles going from 0 to 50 degrees.
- Noise was added by a Gaussian function with noise variance varying from 0 to 0.16.

Note that for this experiment, no radiometric transformation was applied. Results of the study are shown in Fig. 11.7 to 11.10. Figures 11.7 and 11.8 show the results of the study when translation and rotation parameters are considered. Figure 11.7 is a contour plot of the shift errors and rotational errors in the registrations for Daubechies-based filters as a function of the relative shift and rotation of the images. Figure 11.8 shows shift errors in the Simoncelli-based registrations as a function of relative shift in the x direction for various image pairs. As expected, both Daubechies- and Simoncelli-based registrations exhibit some periodicity patterns of the shift errors. With no noise and for small rotations ($0°–18°$), Daubechies-based registration exhibits a shift error of period 2 (i.e., half the size of the filter). But for large rotations ($20°–50°$), periodicity disappears and the shift and rotation errors of the Daubechies/orthogonal-based registration increase with the size of the rotation angle (see Figs. 11.7(a) and 11.7(b), respectively). Identical tests were run with Simoncelli filters and showed that for any size rotation ($0°–50°$), the shift error varies from 0 to 2 with a periodicity of 4 (half the size of the filter) and is not affected by the size of the rotation angle. The rotation error obtained with a Simoncelli-based registration is always equal to 0. Additional tests were run for the Simoncelli band-pass filters in which the horizontal-shift error t_x was studied, with vertical shift t_y, rotation angle θ, and noise level fixed. The results show that whether t_y is small or large, the periodicity of 4 is almost always present, while the shift error varies between 0 and 2 when at least two of the three parameters are small, and may vary up to 10 pixels when all parameters are large; see Fig. 11.8 for an illustration of these observations when t_y is large.

Figures 11.9 and 11.10 show how rotation and translation errors vary as a function of noise. While noise variance varies from 0 to 0.16, three cases are considered for rotation and translation:

(1) Rotation angle and both translation parameters are small ($\theta = 0°$, $t_x = 4$, $t_y = 2$).
(2) Rotation angle larger and both translation parameters small ($\theta = 10°$, $t_x = 1$, $t_y = 2$).
(3) Rotation angle larger and both translation parameter large ($\theta = 10°$, $t_x = 20$, $t_y = 30$).

In all cases, Figs. 11.9 and 11.10 show that the Simoncelli-based registration is much more robust under noise conditions, with rotation and translation errors constant until the noise variance reaches 0.12 for small translations, and 0.10 for large translations.

11.4.3.2 Spline and Simoncelli filters study

Further experiments were then conducted using an L_2-based gradient least-squares optimization as the search strategy. This scheme, which we denote by the acronym *TRU*, was developed by Thévenaz *et al.* (1998). The algorithm is a multiresolution approach based on a modified version of the Levenberg-Marquardt

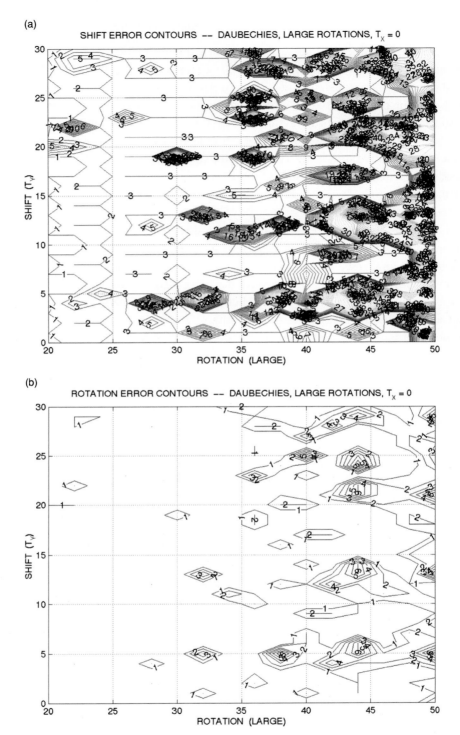

Figure 11.7. Errors obtained by Daubechies-based registration: (a) Shift errors for large rotations, and (b) rotation errors for large rotations. (Source: Le Moigne and Zavorin, 2000, © SPIE, reprinted with permission.)

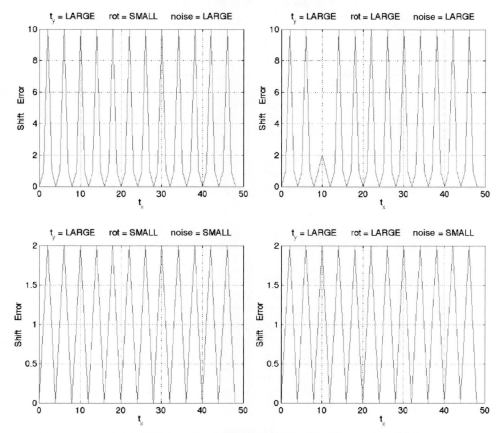

Figure 11.8. Periodicity of shift error: Simoncelli-based registration for large vertical translations (t_y). (Source: Le Moigne and Zavorin, 2000, © SPIE, reprinted with permission.)

(LM) algorithm (Levenberg, 1944; Marquardt, 1963); the LM algorithm represents a hybrid optimization approach between a pure gradient-descent method and the Gauss-Newton method, which is more powerful but less robust. We refer the readers to Thévenaz *et al.* (1998) and to Zavorin and Le Moigne (2005) for more details on this search algorithm. The original TRU algorithm is implemented using a coarse-to-fine spline pyramid (see Subsection 11.2.2); we will denote this implementation as *TRU-SplC*. In these experiments, we combined the LM search with the two Simoncelli pyramids, based on low-pass filters as well as band-pass filters (subbands *L* and *B*, respectively, in Fig. 11.2); these two versions will be denoted as *TRU-SimL* and *TRU-SimB*, respectively.

There are certain advantages and disadvantages of using a Simoncelli pyramid as opposed to the spline pyramid. On the one hand, the band-pass pyramid may

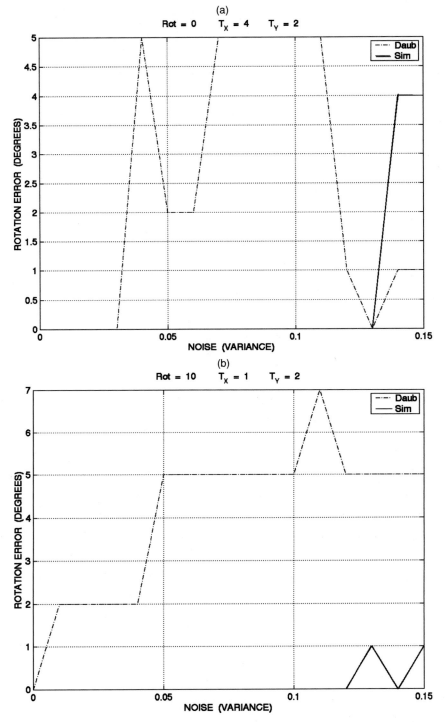

Figure 11.9. Rotation error as a function of noise: (a) $\theta = 0°$, $t_x = 4$, $t_y = 2$, (b) $\theta = 10°$, $t_x = 1$, $t_y = 2$, and (c) $\theta = 10°$, $t_x = 20$, $t_y = 30$. (Source: Le Moigne and Zavorin, 2000, © SPIE, reprinted with permission.)

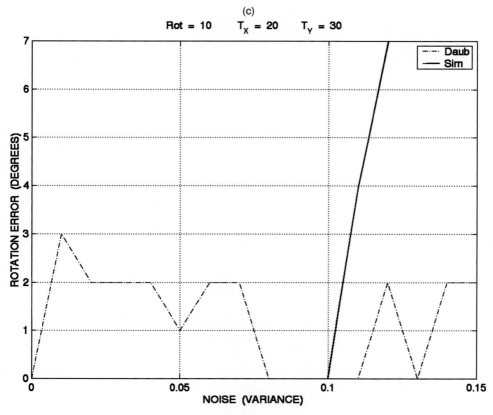

Figure 11.9. (*cont.*)

provide more accurate results because it combines noise-removal properties of a low-pass filter with high-frequency extraction of important edge- and contour-like features. In addition, both Simoncelli pyramids provide an option of using something other than the original image at the finest resolution level for the final adjustment of the result. On the other hand, they are more expensive to compute than the Spline pyramid.

For our experiments, three synthetic datasets were utilized:

(1) *Warping & noise:* The first dataset was created by combining geometric warping with noise, where shifts in x and y as well as rotation vary simultaneously between 0 and 20 (pixels for shifts and degrees for rotations), and noise varies between -15 dB and 20 dB.
(2) *Warping & PSF*: The second dataset was created by combining geometric warping with radiometric variations using the PSF described in Subsection 11.4.2. Again, shifts and

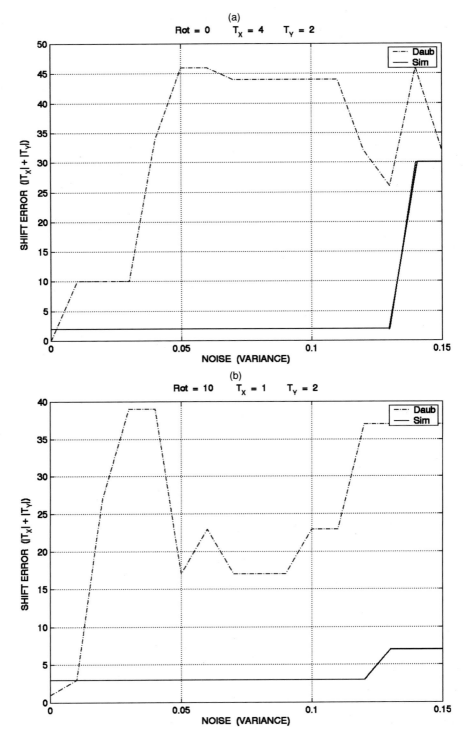

Figure 11.10. Shift error as a function of noise: (a) $\theta = 0°$, $t_x = 4$, $t_y = 2$, (b) $\theta = 10°$, $t_x = 1$, $t_y = 2$, and (c) $\theta = 10°$, $t_x = 20$, $t_y = 30$. (Source: Le Moigne and Zavorin, 2000, © SPIE, reprinted with permission.)

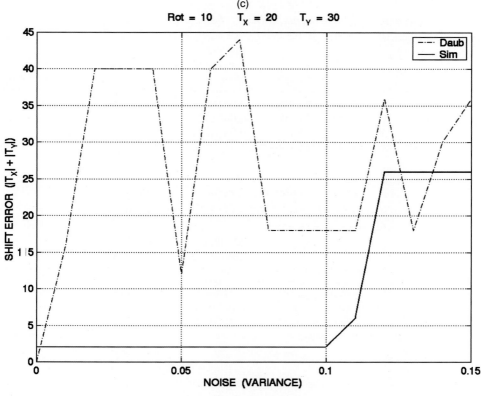

Figure 11.10. (*cont.*)

rotation values vary between 0 and 20, before the warped image is convolved with the PSF.

(3) *Warping & PSF & noise*: The third dataset was created by combining geometric warping with radiometric variations followed by adding noise. Again, shifts and rotation values are varied between 0 and 20, and noise is varied between $-15\,\text{dB}$ and $20\,\text{dB}$.

Results for the three datasets were presented in detail in Zavorin and Le Moigne (2005), and are summarized here in Tables 11.2–11.4. Overall, the experiments show that for the first test dataset (warping & noise), TRU-SimL has a much larger region of convergence than TRU-SplC and TRU-SimB (51% compared to 22.9% and 21.5%, respectively), indicating that even when starting farther from the solution, convergence can be expected for TRU-SimL. At the same time, as indicated in Table 11.2, when the algorithms did converge to a subpixel level, they all produced comparable final RMS errors. For the second dataset (warping & PSF), the regions of convergence for TRU-SplC and TRU-SimL are considerably larger than the one of TRU-SimB (28.9% and 29.8%, respectively, compared to 7.4%), so they may

Table 11.2 *Average error for converged region of test dataset (warping & noise)* *(source: Zavorin and Le Moigne, 2005, © IEEE, reprinted with permission)*

	Number of converged	Median converged error	Mean converged error	Standard deviation converged error
TRU-SplC	1657/7236 ≈ 22.9%	0.0219008	0.086284	0.148911
TRU-SimB	1552/7236 ≈ 21.5%	0.0411122	0.141976	0.212933
TRU-SimL	3693/7236 ≈ 51%	0.0214522	0.056263	0.109165

Table 11.3 *Average error for converged region of test dataset (warping & PSF)* *(source: Zavorin and Le Moigne, 2005, © IEEE, reprinted with permission)*

	Number of converged	Median converged error	Mean converged error	Standard deviation converged error
TRU-SplC	2831/9801 ≈ 28.9%	0.285565	0.300596	0.082751
TRU-SimB	725/9801 ≈ 7.4%	0.052036	0.065967	0.032609
TRU-SimL	2918/9801 ≈ 29.8%	0.320529	0.331067	0.091969

Table 11.4 *Average error for converged region of test dataset (warping & PSF & noise) (source: Zavorin and Le Moigne, 2005, © IEEE, reprinted with permission)*

	Number of converged	Median converged error	Mean converged error	Standard deviation converged error
TRU-SplC	1424/7236 ≈ 19.7%	0.216392	0.278916	0.168494
TRU-SimB	1415/7236 ≈ 19.6%	0.164233	0.252788	0.213441
TRU-SimL	4038/7236 ≈ 55.8%	0.243106	0.289142	0.142683

be more appropriate when misregistration between two images is large. However, when we inspect the average final errors produced by the three algorithms when they converge (Table 11.3), we observe that TRU-SimB generates significantly more accurate results. It appears from these tests, as well as others presented in Zavorin *et al.* (2002), Le Moigne *et al.* (2002b), and Zavorin *et al.* (2003), that TRU-SimB is less sensitive than TRU-SplC or TRU-SimL to the type of radiometric variations introduced by a simple PSF. For the third dataset (warping & PSF & noise), the results are similar to those of the first dataset in that TRU-SimL is better than

Table 11.5 *Summary of TRU test results (source: Zavorin and Le Moigne, 2005,*
© IEEE, reprinted with permission)

Test imagery	Largest radius of convergence	Best accuracy when converged	Best consistency
Synthetic, same radiometry, noisy/noiseless	TRU-SimL	About same	About same
Synthetic, different radiometry, noiseless	TRU-SplC/TRU-SimL	TRU-SimB	About same
Real, multisensor/multiterrain	TRU-SimB	Not applicable	TRU-SimB

TRU-SimB or TRU-SplC (55.8% compared to 19.7% and 19.6%, respectively). When convergence occurs, the results of the three algorithms are actually more comparable to each other than those of the first dataset. Of course, as expected, the overall performance of all algorithms deteriorates when radiometric variations are introduced.

The three algorithms were also tested on two real datasets and the full experiments are described in Zavorin and Le Moigne (2005). In conclusion, these experiments show that a gradient-based algorithm combined with the Simoncelli steerable low-pass pyramid (TRU-SimL) provides the best results in terms of radius of convergence, while in terms of accuracy and consistency, the best results are obtained using TRU-SimB, based on the Simoncelli band-pass pyramid. Table 11.5 summarizes these findings. This suggests that in order to obtain the best results, a hybrid method may be used, where TRU-SimL would be applied at the early stage of the optimization (e.g., at coarse pyramid levels) followed by TRU-SimB for fine tuning.

11.5 Conclusion

In this chapter, we showed how wavelet transforms provide features that can be used for registration purposes. We first introduced orthogonal wavelets in a separable multiresolution scheme and showed that, although this type of decomposition was not translation- or rotation-invariant, it could be used to obtain subpixel registration accuracy. Nevertheless, experiments show that overcomplete steerable filters are more robust than orthogonal wavelets and that when compared to spline wavelets in an L_2-based gradient descent scheme, the low-pass steerable subbands have a better region of convergence in the presence of noise. On the other hand, Simoncelli low-pass and spline pyramids perform similarly with respect to various geometric or radiometric transformations. Finally, we showed that when the algorithms converge, i.e., when the algorithms get close to the final solution, the band-pass steerable subbands generate significantly more accurate results. See Table 11.5 for a summary of these conclusions.

The studies presented in this chapter focus on the basic features that can be generated from wavelets, and the experiments were performed using exhaustive search and a gradient-descent method; other chapters show how wavelets are integrated in the registration process, either by grouping the basic features into meaningful geometric shapes or with different similarity metrics or different search strategies; see Chapters 8, 9, 10, and 14.

References

Brown, L. (1992). A survey of image registration techniques. *ACM Computing Surveys*, **24**(4), 325–376.

Chan, A. K., Chui, C., Le Moigne, J., Lee, H. J., Liu, J. C., and El-Ghazawi, T. A. (1995). The performance impact of data placement for wavelet decomposition of two-dimensional image data on SIMD machines. In *Proceedings of the Fifth Symposium on the Frontiers of Massively Parallel Computation*, McLean, VA, pp. 246–251.

Chui, C. K. (1992). *An Introduction to Wavelets*. San Diego, CA: Academic Press.

Coifman, R. R. and Donoho, D. L. (1995). Translation-invariant de-noising. In A. Antoniadis, and G. Oppenheim, eds., *Wavelets and Statistics. Lecture Notes in Statistics*, Vol. 103. New York: Springer Verlag, pp. 125–150.

Daubechies, I. (1988). Orthonormal bases of compactly supported wavelets. *Communications of Pure and Applied Mathematics*, **41**, 909–996.

Daubechies, I. (1992). *Ten Lectures on Wavelets*. Philadelphia, PA: Society for Industrial and Applied Mathematics.

Djamdji, J.-P., Bijaoui, A., and Maniere, R. (1993). Geometrical registration of images: The multiresolution approach. *Photogrammetric Engineering & Remote Sensing*, **59**(5), 645–653.

El-Ghazawi, T. A. and Le Moigne, J. (2005). Performance of the wavelet decomposition on massively parallel architectures. *International Journal of Computers and their Applications*, **27**(2), 72–81.

El-Ghazawi, T. A., Chalermwat, P., and Le Moigne, J. (1997). Wavelet-based image registration on parallel computers. In *Proceedings of the ACM/IEEE Conference on Supercomputing*, San Jose, CA Unpaginated CR-ROM. 9pp.

Heil, C. and Walnut, D. F., eds. (2006). *Fundamental Papers in Wavelet Theory*. Princeton, NJ: Princeton University Press.

Karasaridis, A. and Simoncelli, E. (1996). A filter design technique for steerable pyramid image transforms. In *Proceedings of the IEEE International Conference on Acoustics, Speech and Signal Processing*, Atlanta, GA, pp. 2387–2390.

Le Moigne, J. (1994). Parallel registration of multisensor remotely sensed imagery using wavelet coefficients. In *Proceedings of the SPIE Conference on Wavelet Applications*, Orlando, FL, Vol. 2242, pp. 432–443.

Le Moigne, J. and Zavorin, I. (1999). An application of rotation- and translation-invariant overcomplete wavelets to the registration of remotely sensed imagery. In

Proceedings of the SPIE on Wavelet Applications VI, Orlando, FL, Vol. 3723, pp. 130–140.

Le Moigne, J. and Zavorin, I. (2000). Use of wavelet for image registration. In *Proceedings of the SPIE Wavelet Applications VII*, Orlando, FL, Vol. 4056, pp. 99–108.

Le Moigne, J., Campbell, W. J., and Cromp, R. F. (2002a). An automated parallel image registration technique based on the correlation of wavelet features. *IEEE Transactions on Geoscience and Remote Sensing*, **40**(8), 1849–1864.

Le Moigne, J., Cole-Rhodes, A., Eastman, R., Johnson, K., Morisette, J. T., Netanyahu, N. S., Stone, H. S., and Zavorin, I. (2001). Multi-sensor image registration of Earth remotely sensed imagery. In *Proceedings of the SPIE Conference on Image and Signal Processing for Remote Sensing VII*, Toulouse, France, Vol. 4541, pp. 1–10.

Le Moigne, J., Cole-Rhodes, A., Eastman, R., Johnson, K., Morisette, J., Netanyahu, N., Stone, H., and Zavorin, I. (2002b). Multi-sensor image registration for on-the-ground or on-board science data processing. In *Proceedings of the NASA Science Data Workshop*, Greenbelt, MD.

Levenberg, K. (1944). A method for the solution for certain non-linear problems in least squares. *The Quarterly of Applied Mathematics*, **2**, 164–168.

Liu, Z., Ho, Y. K., Tsukada, K., Hanasaki, K., Dai, Y., and Li, L. (2002). Using multiple orientational filters of steerable pyramid for image registration. *Information Fusion*, **3**(3), 203–214.

Magarey, J. and Kingsbury, N. (1997). Motion estimation using a complex-valued wavelet transform. *IEEE Transactions on Signal Processing*, **46**(4), 1069–1084.

Mallat, S. G. (1989). A theory for multiresolution signal decomposition: The wavelet representation. *IEEE Transactions on Pattern Analysis and Machine Intelligence*, **11**(7), 674–693.

Marquardt, D. W. (1963). An algorithm for least-squares estimation of nonlinear parameters. *Journal of the Society for Industrial and Applied Mathematics*, **11**, 431–441.

Schelkens, P., Skodras, A., and Ebrahimi, T. eds. (2009). *The JPEG 2000 Suite*. Wiley-IS&T Series in Imaging Science and Technology. Chichester, UK: John Wiley & Sons.

Schoenberg, I. J. (1969). Cardinal interpolation and spline functions. *Journal of Approximation Theory*, **2**, 167–206.

Shannon, C. E. (1949). Communication in the presence of noise. *Proceedings of the Institute of Radio Engineers*, **37**(1), 10–21; also in *Proceedings of the IEEE*, **86**(2), 1998, 447–457.

Simoncelli, E. P. and Freeman, W. T. (1995). The steerable pyramid: A flexible architecture for multi-scale derivative computation, In *Proceedings of the Second Annual IEEE International Conference on Image Processing*, Washington, DC, Vol. 3, pp. 444–447.

Simoncelli, E. P., Freeman, W. T., Adelson, E. H., and Heeger, D. J. (1992). Shiftable multiscale transforms. *IEEE Transactions on Information Theory*, **38**(2), 587–607.

Stone, H. S., Le Moigne, J., and McGuire, M. (1999). The translation sensitivity of wavelet-based registration. *IEEE Transactions on Pattern Analysis and Machine Intelligence*, **21**(10), 1074–1081.

Strang, G. (1989). Wavelets and dilation equations: A brief introduction. *SIAM Review*, **31**(4), 614–627.

Thévenaz, P., Ruttimann, U. E., and Unser, M. (1998). A pyramid approach to subpixel registration based on intensity. *IEEE Transactions on Image Processing*, **7**(1), 27–41.

Unser, M., Aldroubi, A., and Eden, M. (1993a). The L_2-polynomial spline pyramid. *IEEE Transactions on Pattern Analysis and Machine Intelligence*, **15**(4), 364–379.

Unser, M. A., Aldroubi, A., and Gerfen, C. R. (1993b). A multiresolution image registration procedure using spline pyramids. In *Proceedings of the SPIE Conference on Mathematical Imaging: Wavelet Applications in Signal and Image Processing*, San Diego, CA, Vol. 2034, pp. 160–170.

Zavorin, I. and Le Moigne, J. (2005). Use of multiresolution wavelets feature pyramids. *IEEE Transactions on Image Processing*, **14**(6), 770–782.

Zavorin, I., Stone, H. S., and Le Moigne, J. (2002). Iterative pyramid-based approach to subpixel registration of multisensor satellite imagery. In *Proceedings of the SPIE Conference on Earth Observing Systems VII*, Seattle, WA, Vol. 4814, pp. 435–446.

Zavorin, I., Stone, H. S., and Le Moigne, J. (2003). Evaluating performance of automatic techniques for subpixel registration of remotely sensed imagery. In *Proceedings of the SPIE Conference on Image Processing: Algorithms and Systems II*, Santa Clara, CA, Vol. 5014, pp. 306–316.

12

Gradient descent approaches to image registration

ARLENE A. COLE-RHODES AND ROGER D. EASTMAN

Abstract

This chapter covers a general class of image registration algorithms that apply numerical optimization to similarity measures relating to cumulative functions of image intensities. An example of these algorithms is an algorithm minimizing the least-squares difference in image intensities due to an iterative gradient-descent approach. Algorithms in this class, which work well in 2D and 3D, can be applied simultaneously to multiple bands in an image pair and images with significant radiometric differences to accurately recover subpixel transformations. The algorithms discussed differ in the specific similarity measure, the numerical method used for optimization, and the actual computation used. The similarity measure can vary from a measure that uses a radiometric function to account for nonlinear image intensity differences in the least-squares equations, to one that is based on mutual information, which accounts for image intensity differences not accounted for by a standard functional model. The numerical methods considered are basic recursive descent, a method based on Levenberg-Marquardt's technique, and Spall's algorithm. This chapter relates to the above registration algorithms and classifies them by their various elements. It also analyzes the image classes for which variants of these algorithms apply best.

12.1 Introduction

We consider in this chapter a class of image registration algorithms that apply numerical techniques for optimizing some similarity measures that relate only to the image intensities (or a function of the image intensities) of an image pair. Such registration methods are referred to as *area-based* (Zitová and Flusser, 2003), and they deal directly with windowed portions of the images, without attempting first to detect salient features. Three of the most widely used similarity measures are *cross-correlation* (CC), the *sum of squared differences* (SSD), which is the essentially least-squares error norm, and *mutual information* (MI). Each of these measures is calculated by directly matching the raw image intensities of all the pixels in the image pair. Both CC and SSD are sensitive to intensity changes introduced by noise, varying illumination (lighting conditions), and/or different sensor types. Yet, they can be modified by using a radiometric function to account for nonlinear differences in the image intensities. Roche *et al.* (1998) proposed the correlation ratio as a metric for the registration of multimodal images, where the intensity dependence can be represented by some linear or nonlinear function, and compared its performance to CC and MI. Mutual information provides an alternative similarity metric when the intensity differences cannot be accounted for by using a functional model, such as in the case of registering multisensor data, or under conditions for which the image intensities vary with the geometric transformation. MI, which originates from *information theory*, is a measure of the statistical dependency between two datasets. It has been shown to be well suited for registering multimodal data in both remote sensing and medical applications, and has been used widely. Although SSD and CC have a wider basin of convergence than MI, the latter has a sharper peak than either of the former measures. Thus, MI yields a more accurate result for optimization, provided that the starting point is sufficiently close to the solution. A major disadvantage of these area-based similarity measures is their high computational cost, which involves all the pixels of the image.

For a given similarity measure, the registration problem is a multidimensional optimization problem with respect to the parameters specified in the geometric transformation. For cross-correlation and mutual information the optimization amounts to maximization, while the SSD measure needs to be minimized. An *exhaustive search* is the only method guaranteed to find the global optimum of the similarity measure. However, it is computationally demanding, as the computational cost increases with the number of degrees of freedom in the parameter space. For an image template with a total of N pixels, an exhaustive search over a parameter space of dimension d would require $O(N^d)$ evaluations of the similarity metric, if using integer displacements. In contrast, an iterative gradient descent algorithm requires a single function evaluation at each iteration, and its computational complexity depends on the metric type and the associated gradient computation. Thus, it is more efficient to obtain a solution to the registration problem by using an iterative process, such as gradient descent (or ascent) for the minimization (or maximization) of the specified similarity measure. Area-based methods lend themselves to fully-automated registration, as far as handling complex transformations in 2D and 3D, and achieving subpixel accuracy. However, the computational cost of the search can grow even more prohibitive as the complexity of the transformation increases. In addition to exhaustive search and gradient descent methods, other standard well-documented search strategies have been applied to image registration. These include, for example, Brent's method and other one-dimensional searches, Powell's method, the simplex method, genetic methods, and simulated annealing. Press *et al.* (2007) provide general descriptions of these algorithms. Hierarchical approaches, which use a coarse-to-fine multiresolution search strategy, may also be included to speed up the registration process and avoid local minima.

12.2 Area-based similarity measures

As noted, the most commonly used similarity measures which require optimization are cross-correlation (CC), the sum of squared differences (SSD), and mutual information (MI).

12.2.1 Correlation-like metrics

Cross-correlation is the classic example of an area-based measure. The *normalized cross-correlation* measure is computed by using the raw intensity values of the reference and sensed images, with the aim of finding its maximum. The correlation ratio has also been applied to multimodal images by specifying some function to represent the intensity dependence between the pair of images (Zitová and Flusser, 2003).

The correlation coefficient is defined as

$$C(A, B) = \frac{\sum\limits_{ij} (a_{ij} - \text{mean}(A))(b_{ij} - \text{mean}(B))}{\sqrt{\sum\limits_{ij} (a_{ij} - \text{mean}(A))^2 \sum\limits_{ij} (b_{ij} - \text{mean}(B))^2}}, \tag{12.1}$$

where the sums are taken over the rows and columns of the two images, and a_{ij}, b_{ij} are the pixel values of images A and B, respectively, at row i and column j. This statistic measures the correlation on an absolute scale in the range $[-1, 1]$. In general, the measure provides a degree of the similarity between the two images, where a value of 1 corresponds to identical images. For an image template containing N pixels, the cost of a single computation for correlation is $O(N^2)$. For additional details on correlation-based methods, see Chapter 4.

12.2.2 Sum of squared differences (SSD)

Consider two images, A and B, which differ by a geometric transform $Q_\mathbf{p}$ specified by a set of parameters $\mathbf{p} = [t_x \ t_y \ \theta]^T$. The latter corresponds to translation and rotation in 2D, i.e., $d = 3$. The SSD measure minimizes the L_2-norm of the error between corresponding pixel values of image A and the transformed image $Q_{\mathbf{p}}(B)$. Thus in terms of pixel indices, this measure can be expressed as follows:

$$SSD(A, B) = \sum_{ij}([E(\mathbf{p})]_{ij})^2 = \sum_{ij}(A_{ij} - [Q_{\mathbf{p}}(B)]_{ij})^2, \qquad (12.2)$$

where the sums are taken over the rows and columns of the overlap between image A and the transformed image $Q_{\mathbf{p}}(B)$. The SSD similarity measure is fast and easy to implement with a wide basin of convergence. Its gradient, too, is easily derived and well understood. Lucas and Kanade (1981) made one of the earliest practical attempts to efficiently align a template to a reference image using the SSD measure. Their method is often referred to as the optical flow algorithm, and it was first proposed for the computation of stereo image disparities. It has become one of the most widely used techniques in computer vision for tracking an image patch. The method involves an iterative solution based on using both the Jacobian and an approximation to the Hessian of the SSD function. Baker and Matthews (2004) discuss the different implementations of this gradient descent algorithm based on SSD, and provide details on the computational complexity of each. Irani and Peleg (1991) used SSD for the registration of image sequences with the aim of computing a combined super-resolution image. Thévenaz *et al.* (1998) used this metric for the registration of medical images, specifically calibrated MRI and PET images with negligible radiometric differences. Eastman and Le Moigne (2001) applied this metric for the registration of remote sensing images acquired under different conditions.

12.2.3 Mutual information metric

Mutual information was independently proposed by Viola and Wells (1995), Collignon *et al.* (1995), and Studholme *et al.* (1995) as a similarity measure for image registration. It is based on the entropy measure from *information theory*, and has been widely adopted, especially for the registration of multimodal medical images. It is defined by

$$I(A, B) = \sum_a \sum_b p_{A,B}(a, b) \log \frac{p_{A,B}(a, b)}{p_A(a)p_B(b)}, \qquad (12.3)$$

where $p_{A,B}(a, b)$ is the joint PDF of the random variables A and B (which correspond to the image intensities), and $p_A(a)$ and $p_B(b)$ are the marginal PDFs of A and B, respectively. These PDFs can be computed from the joint histogram of the images. Additional details on the use of the mutual information metric for registration can be found in Chapter 6. MI is computationally simple to implement;

in fact, it is only slightly more expensive than SSD. Its registration performance, however, is superior since it is robust to noise and illumination changes in the images. For an image template containing N pixels, the cost of computing the joint histogram is $O(N)$ (i.e., similar to the cost of SSD), and the cost of computing the mutual information from the joint histogram is $O(K)$, where K is the number of histogram bins (typically $K \leq 256$). MI has been optimized using many of the different optimization schemes mentioned in Section 12.1 with varying degrees of success. These schemes are described in detail in the survey paper by Pluim *et al.* (2003). MI, however, is a nonlinear function which does not lend itself easily to an analytic computation of the derivative, which makes the problem of optimization using gradient descent more challenging. An analytic derivative of MI was derived, though, in Maes *et al.* (1999) using partial volume interpolation, and in Thévenaz and Unser (2000) using B-spline Parzen windowing. Dowson and Bowden (2006) published general analytic derivatives for MI based on these and other common sampling methods for populating the histogram used in computing this similarity measure. They compared the registration performance of the MI family of metrics to the SSD and CC measures. Kern and Pattichis (2007) proposed a mathematical model for the MI surface in order to predict the capture range of the MI optimization (i.e., the range of parameter values for which convergence to the peak is guaranteed) when using gradient-type search algorithms, and related this to the spectral characteristics of the images to be registered.

12.3 Gradient descent search strategies

In this section we focus on the minimization of some similarity function $f(.)$. (Note that maximization can be equivalently achieved by minimizing the negative of this function.) Assuming that the similarity function is convex in the neighborhood of the optimum (i.e., locally parabolic), we discuss the use of various gradient descent optimization methods. Most nonlinear minimization and parameter estimation algorithms consist of two steps: (1) Minimization of the function $f(.)$ using a linear or quadratic approximation around the current estimate of the parameters, and (2) update of the parameter estimates.

12.3.1 Gradient descent approaches

The optimization methods classified as Newton-type (gradient) descent techniques include the Levenberg-Marquardt optimization method, Newton-Raphson's iterative procedure, and other steepest descent and stochastic search methods. The parameter update is given by equations of the form $\mathbf{p}_{k+1} = \mathbf{p}_k + \Delta \mathbf{p}_k$. Specifically, the following iterative equations,

$$\mathbf{p}_{k+1} = \mathbf{p}_k - \lambda_k \mathbf{g}_k, \tag{12.4}$$

$$\mathbf{p}_{k+1} = \mathbf{p}_k - \lambda_k \mathbf{H}_k^{-1} \mathbf{g}_k, \tag{12.5}$$

$$\mathbf{p}_{k+1} = \mathbf{p}_k - (\mathbf{H}_k + \lambda_k diag[\mathbf{H}_k])^{-1} \mathbf{g}_k, \tag{12.6}$$

correspond to the methods of steepest descent, Newton-Raphson, and Levenberg-Marquardt, respectively, where $\mathbf{p} = [t_x \ t_y \ \theta]^T$ and λ_k denote, respectively, the parameter vector (which consists of a translation and rotation in 2D) and the step size at iteration k. The gradient (or Jacobian) $\mathbf{g}_k$ and the Hessian matrix $\mathbf{H}_k$ are derived with respect to the similarity function $f(.)$ and evaluated at $\mathbf{p}_k$. In particular, they are given by $\mathbf{g}_k = \frac{\partial f}{\partial \mathbf{p}}\big|_{\mathbf{p}=\mathbf{p}_k} = \left(\frac{\partial f}{\partial p_1} \frac{\partial f}{\partial p_2} \cdots \frac{\partial f}{\partial p_d} \right)\big|_{\mathbf{p}=\mathbf{p}_k}$ and $[\mathbf{H}_k]_{ij} = \frac{\partial^2 f}{\partial p_i \partial p_j}\big|_{\mathbf{p}=\mathbf{p}_k}$, where $\mathbf{p} = (p_1, p_2, \ldots, p_d)$ specifies the d parameters of the similarity function defined by the search space, $d = 3$ in this case. Note that the diagonal matrix of the Hessian at the kth iteration is given by $diag[\mathbf{H}_k] = diag\{ \frac{\partial^2 f}{\partial p_i^2}\big|_{\mathbf{p}=\mathbf{p}_k}, i = 1, \ldots, d \}$, i.e., it consists only of the diagonal elements of $\mathbf{H}_k$.

Newton-type methods are contrasted with optimization methods that minimize the similarity measure along a directed line using Brent's algorithm. Steepest descent methods compute a search direction that guarantees a local decrease in the value of the optimization function. Using the gradient or an estimate of the gradient is a common way to determine this direction. Such methods usually assure convergence to a local minimum of the optimization function. To obtain a global minimum, multiresolution approaches (Zitová and Flusser, 2003) can be implemented to help avoid undesired local minimum with any of the above methods. In general, Newton's method performs well close to the optimum, while steepest descent methods, that ignore the local curvature information contained in the Hessian matrix, will perform better further away from the minimum. The Levenberg-Marquardt method combines steepest descent with the Newton-Raphson method by varying the step-size value λ_k from large to small. Newton's method is referred to as a quasi-Newton method if an approximation to the Hessian $\mathbf{H}_k$ is used in the update of Eq. (12.5). (The Lucas-Kanade algorithm falls into this category.) A steepest descent algorithm that uses a noisy approximation of the Jacobian $\mathbf{g}_k$ is referred to as a stochastic gradient method; the *simultaneous perturbation stochastic approximation* (SPSA) algorithm falls into this category. See Spall (2003) and Chapter 6 for further details.

Baker and Matthews (2004) reviewed the Lucas-Kanade type algorithms and classified such gradient descent algorithms as either *additive* or *compositional*, depending, respectively, on whether an additive increment or an incremental warp for the parameters was estimated. They showed that both these approaches are equivalent to the first order, and also considered the type of approximation used for gradient descent.

12.3.2 Multiresolution approaches

A hierarchical search is often added to the basic algorithm, with the optimization being implemented in a multiresolution manner. The pyramid framework from low-to-high (or coarse-to-fine) resolution, allows for the handling of more complex transformations in the registration process. This approach has been found to decrease the sensitivity of the scheme to local minima in the similarity function, and improve also the algorithm's running time. A rough estimate of registration parameters is found fairly quickly using the smaller downsampled images. This estimate is subsequently refined by using images of increasing resolution as one moves up the pyramid. Registration of the higher-resolution images will be faster if reasonable estimates are obtained from the lower-resolution images of the pyramid. A multiresolution pyramid of images can be obtained by downsampling the images to a number of lower-resolution levels. Alternatively, different wavelet decompositions can be applied to obtain a multiresolution pyramid. Examples of implementations of multiresolution registration schemes based on gradient descent methods can be found in Thévenaz *et al.* (1998), Maes *et al.* (1999), Thévenaz and Unser (2000), Eastman and LeMoigne (2001), and Cole-Rhodes *et al.* (2003).

12.4 Experiments

The sum of squared differences (SSD) metric in Eq. (12.2) minimizes the L_2-norm of the difference between the intensity values of a pair of images which also differ by a geometric transform $Q_{\mathbf{p}}$. The gradient descent algorithms proposed by Lucas and Kanade (1981), Irani and Peleg (1991), and Thévenaz *et al.* (1998) were reviewed and incorporated into a common framework by Eastman and Le Moigne (2001). The three algorithms were compared in terms of the type of transformations allowed, type of image warping, computation of the image derivatives, updating of the registration parameters, implementation of the image pyramid (if using multiresolution), and the termination condition used. Table 12.1 below, from Eastman and Le Moigne (2001), summarizes the similarities and differences between the three algorithms.

Eastman and Le Moigne (2001) further applied a Lucas-Kanade type algorithm to the registration of multimodal remote sensing images which may also differ radiometrically. Their paper discusses approaches for dealing with these radiometric differences in the image intensities, and shows how to minimize the L_2-norm of the difference between the image intensity values using the SSD metric. The parameter vector is updated using the Newton-Raphson iterative procedure of Eq. (12.5), where the additive update, evaluated at the current parameter estimate

Table 12.1 *Comparison of three gradient descent algorithms (source: Eastman and Le Moigne, 2001, © FUSION 2001, reprinted with permission)*

Algorithm element	Peleg *et al.*	Thévenaz *et al.*	Lucas and Kanade
Allowed transforms	2D translation and rotation	3D translation, affine map, contrast	2D affine map and translation
Computation of image pyramid	Gaussian	L_2-spline pyramid (bicubic B-spline)	Not described aside from "smoothing"
Computation of image derivatives	Not described	From bicubic spline approximation	First-order image differences
Solution to normal equations	Not described	Levenberg-Marquardt with SVD	Matrix inverse
Image warping	Bilinear	Bicubic	Bilinear
Weighting (for one pixel's contribution to summations)	Not used	Binary masking based on measurement error	Inverse of difference of image derivatives
Termination condition	Change in parameters goes close to zero	Error term is small; relative gain in term is small; relative change in parameters is small	Not described

$\mathbf{p}_k$, is given by

$$\Delta\mathbf{p} = -\mathbf{H}^{-1}\mathbf{g}.$$

The gradient vector for the SSD measure is given by $\mathbf{g} = \sum_{ij} \left\{ \left[\frac{\partial B}{\partial \mathbf{p}}\right]_{ij} [E(\mathbf{p})]_{ij} \right\}$, where $[E(\mathbf{p})]_{ij} = (A_{ij} - [Q_{\mathbf{p}}(B)]_{ij})$, and the gradient of the sensed image B is given by $\left[\frac{\partial B}{\partial \mathbf{p}}\right]^T = \left(\frac{\partial B}{\partial p_1} \frac{\partial B}{\partial p_2} \cdots \frac{\partial B}{\partial p_d}\right)$. The Hessian $\mathbf{H}$ is specified by the first order Gauss-Newton approximation, $\mathbf{H} = \sum \left\{ \left[\frac{\partial B}{\partial \mathbf{p}}\right]\left[\frac{\partial B}{\partial \mathbf{p}}\right]^T \right\}$, which is summed over the rows and columns of the image B. Eastman and Le Moigne (2001) discussed the case of translation and rotation in 2D, where the image derivative is given by $\left[\frac{\partial B}{\partial \mathbf{p}}\right]^T = (B_x \ B_y \ B_\theta)$, and B_θ is specified by its linear approximation $B_\theta = xB_y - yB_x$.

A registration example for a pair of multisensor images according to the framework of Table 12.1 is provided using this steepest descent algorithm for translations only. Image warping was done using bicubic interpolation, and the image pyramid was updated using a Gaussian kernel. The procedure terminated once the solution estimate changed by less than 0.1%. The multisensor images used for illustrating this registration method are:

(1) 6784×7552 Landsat-7/ETM+ with an IFOV of 31.25 meters, and
(2) 2048×2048 IKONOS with an IFOV of 3.91 meters.

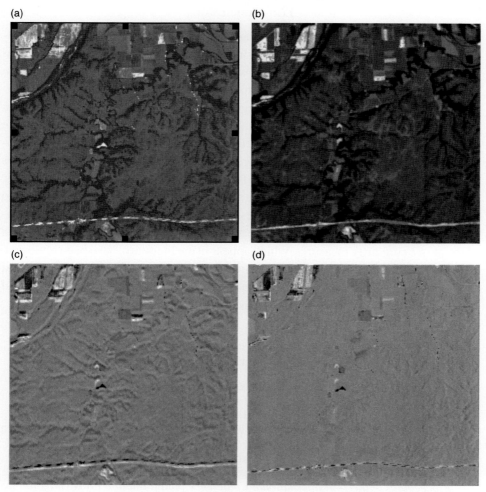

Figure 12.1. Downscaling IKONOS to ETM (spatial resolution ratio of 3.91: 31.25) with resulting image size 256 × 256: (a) Reference image, (b) target image, (c), (d) difference before and after registration, respectively. (Source: Eastman and Le Moigne, 2001, © *FUSION 2001*, reprinted with permission.)

Figures 12.1(a) and 12.1(b) show, respectively, the IKONOS image used as the reference image, and the Landsat/ETM image used as the sensed image. Figures 12.1(c) and 12.1(d) show, respectively, the image difference before and after the registration. For this IKONOS–ETM image pair, the successful registration is illustrated by the highway across the bottom of Fig. 12.1(c). This segment was brought into a fairly good alignment, as can be seen in Fig. 12.1(d), despite down-sampling artifacts in the difference along the highway.

12.5 Conclusions

We reviewed in this chapter some similarity measures and gradient descent methods for the registration of remote sensing images. The image registration techniques reviewed form a general class, which is based on the optimization of an intensity-based similarity measure. Three of the most commonly used intensity-based similarity measures for image registration are cross-correlation (CC), sum of squared differences (SSD), and mutual information (MI). These measures were compared and contrasted, in terms of their computational efficiency and applicability for monomodal and multimodal images. The registration of multimodal images acquired by different sensors, under different conditions, and at different times, has proved particularly challenging due to radiometric differences, in addition to geometric differences, which may be present between the images. Successful application of Newton-type gradient descent algorithms for the numerical optimization of the above similarity measures, which appear in the literature, have been reviewed. Although gradient descent approaches are more efficient in the search for a global optimum than exhaustive search techniques, their possible convergence to a local (rather than global) optimum requires special consideration.

References

Baker, S. and Matthews, I. (2004). Lucas-Kanade 20 years on: A unifying framework. *International Journal of Computer Vision*, **56**(3), 221–255.

Cole-Rhodes, A., Johnson, K., Le Moigne, J., and Zavorin, I. (2003). Multiresolution registration of remote sensing imagery by optimization of mutual information using a stochastic gradient. *IEEE Transactions on Image Processing*, **12**(12), 1495–1511.

Collignon, A., Maes, F., Delaere, D., Vandermeulen, D., Suetens, P., and Marchal, G. (1995). Automated multimodality image registration based on information theory. In *Proceedings of the 14th International Conference on Information Processing in Medical Imaging*, Ile de Berder, France; *Computational Imaging and Vision*, Y. Bizais, C. Barillot, and R. Di Paola, eds. Dordrecht, The Netherlands: Kluwer Academic Publishers, pp. 263–274.

Dowson, N. and Bowden, R. (2006). A unifying framework for mutual information methods for use in non-linear optimisation. In *Proceedings of the Ninth European Conference on Computer Vision*, Graz, Austria, Part I; *Lecture Notes in Computer Science*, A. Leonardis, H. Bischof, and A. Pinz, eds. Berlin: Springer, Vol. 3951, pp. 365–378.

Eastman, R. and Le Moigne, J. (2001). Gradient descent techniques for multitemporal and multi-sensor image registration of remotely sensed imagery. In *Proceedings of the Fourth International Conference on Information Fusion*, Montreal, Canada.

Irani, M. and Peleg, S. (1991). Improving resolution by image registration. *CVGIP: Graphical Models and Image Processing*, **53**(3), 231–239.

Kern, J. and Pattichis, M. (2007). Robust multispectral image registration using mutual-information models. *IEEE Transactions on Geoscience and Remote Sensing*, **45**(5), 1494–1505.

Lucas, B. D. and Kanade, T. (1981). An iterative image registration technique with an application to stereo vision. In *Proceedings of the DARPA Image Understanding Workshop*, Vancouver, British Columbia, Canada, pp. 121–130.

Maes, F., Vandermeulen, D., and Suetens, P. (1999). Comparative evaluation of multiresolution optimization strategies for multimodality image registration by maximization of mutual information. *Medical Image Analysis*, **3**(4), 272–286.

Pluim, J. P., Maintz, J. B., and Viergever, M. A. (2003). Mutual information-based registration of medical images: A survey. *IEEE Transactions on Medical Imaging*, **22**(8), 986–1004.

Press, W., Teukolsky, S., Vetterling, W., and Flannery, B. (2007). *Numerical Recipes*: *The Art of Scientific Computing*, 3rd edn. Cambridge: Cambridge University Press.

Roche, A., Malandain, G., Pennec, X., and Ayache, N. (1998). The correlation ratio as a new similarity measure for multimodal image registration. In *Proceedings of the First International Conference on Medical Image Computing and Computer-Assisted Intervention*, Cambridge, MA; *Lecture Notes in Computer Science*, W. M. Wells III, A. C. F. Colchester, and S. L. Delp, eds. Berlin: Springer, Vol. 1496, pp. 1115–1124.

Spall, J. C. (2003). *Introduction to Stochastic Search and Optimization: Estimation, Simulation, and Control*. Hoboken, NJ: Wiley.

Studholme, C., Hill, D. L., and Hawkes, D. J. (1995). Automated 3D registration of truncated MR and CT images of the head. In *Proceedings of the British Machine Vision Conference*, Birmingham, UK, pp. 27–36.

Thévenaz, P. and Unser, M. (2000). Optimization of mutual information for multiresolution image registration. *IEEE Transactions on Image Processing*, **9**(12), 2083–2099.

Thévenaz, P., Ruttimann, U. E., and Unser, M. (1998). A pyramid approach to subpixel registration based on intensity. *IEEE Transactions on Image Processing*, **7**(1), 27–41.

Viola, P. and Wells III, W. M. (1995). Alignment by maximization of mutual information. In *Proceedings of the Fifth International Conference on Computer Vision*, Cambridge, MA, pp. 16–23.

Zitová, B. and Flusser, J. (2003). Image registration methods: A survey. *Image and Vision Computing*, **21**, 977–1000.

13

Bounding the performance of image registration

MIN XU AND PRAMOD K. VARSHNEY

Abstract

Performance bounds can be used as a performance benchmark for any image registration approach. These bounds provide insights into the accuracy limits that a registration algorithm can achieve from a statistical point of view, that is, they indicate the best achievable performance of image registration algorithms. In this chapter, we present the Cramér-Rao lower bounds (CRLBs) for a wide variety of transformation models, including translation, rotation, rigid-body, and affine transformations. Illustrative examples are presented to examine the performance of the registration algorithms with respect to the corresponding bounds.

13.1 Introduction

Image registration is a crucial step in all image analysis tasks in which the final information is obtained from the combination of various data sources, as in image fusion, change detection, multichannel image restoration, and object recognition. See, for example, Brown (1992) and Zitová and Flusser (2003). The accuracy of image registration affects the performance of image fusion or change detection in applications involving multiple imaging sensors. For example, the effect of registration errors on the accuracy of change detection has been investigated by Townshend *et al.* (1992), Dai and Khorram (1998), and Sundaresan *et al.* (2007). An accurate and robust image registration algorithm is, therefore, highly desirable.

The purpose of image registration is to find the transformation parameters, so that the two given images that represent the same scene are aligned. There are many factors that might affect the performance of registration algorithms. Since the two images may originate from different sensors, they represent, therefore, different features of the same scene, which increases the difficulty of registration. Further, illumination effects and sensor noise also hamper image registration. An effective image registration algorithm should take into account all of these factors and be robust in the presence of these effects.

Although many approaches have been proposed for image registration (e.g., Brown, 1992; Zitová and Flusser, 2003; Varshney *et al.*, 2005), very little attention has been paid to evaluating their performance (Yetik and Nehorai, 2006). Performance analysis for image registration is usually performed visually. However, visual inspection is not a satisfactory evaluation method, and hence one needs to explore other evaluation techniques, such as the use of performance bounds for the estimation of transformation parameters.

Analytical evaluation of the performance of registration algorithms is often difficult because the transformation parameters appear in the image data in a nonlinear manner. Investigation of the performance bounds, especially tight bounds, will help us evaluate the efficiency of image registration techniques (Yetik and Nehorai, 2006). Also, it will help determine parameter regions where good or poor registration is expected (Yetik and Nehorai, 2006), as well as interpret the effect of translation and rotation uncertainties on the registration error.

Image registration is essentially a parameter estimation problem. A tight performance bound for image registration gives the minimum achievable *mean squared error* (MSE) for the estimation of the transformation parameters for a specific pair of images. In other words, the bound tells us what the best achievable performance for any image registration algorithm is for the specific image pair, that is, no image registration algorithm can produce a registration error smaller than that bound. It can be used as a performance benchmark for any image registration approach. The distance or gap between the actual registration error of an image registration approach and a tight bound indicates the efficiency of that registration approach and the room for improvement. Moreover, without implementing any image registration approach, we are able to get some useful insights about the image registration ability for the particular pair of images from the bounds. Thus, it is an important problem to obtain tight and robust performance bounds for image registration.

The goal of this chapter is to provide an overview of the existing methods for obtaining performance bounds for image registration. In the next section, a statistical approach to image modeling, that serves as the basis for deriving the performance bounds, is presented. In Section 13.3 we present the Cramér-Rao bounds for a wide variety of transformation models, including translation only, rigid-body transformation, skew transformation, and affine transformation models. In addition, we give a brief introduction to other bounds, for example, the Bhattacharyya bound (Bhattacharyya, 1943) and the Ziv-Zakai bound (Ziv and Zakai, 1969; Zakai *et al.*, 1975) for rigid-body transformation. Some other bounds derived for specific registration techniques, such as feature-based techniques, are also provided. Illustrative examples are presented to examine the performance of the registration algorithms with respect to the corresponding bounds. A summary and suggestions for future work are given in Section 13.4.

13.2 Image modeling

Denoting (x, y) as the spatial location, the observed intensity of the pixel $I(x, y)$ in an image can be modeled as a Gaussian random variable with the mean equal to the noise-free intensity at (x, y) and the variance equal to that of the noise (Robinson and Milanfar, 2004; Yetik and Nehorai, 2006). This Gaussian model has been used under numerous scenarios and in a variety of imaging systems.

Image registration is defined as a spatial mapping between two images so that they become aligned. Let $I_1(x, y)$ and $I_2(x, y)$ denote the two images to be registered. As noted, each image can be considered as a noise-free image plus noise. It is assumed that the misalignment transformation is applied to the noise-free image. Here, we only consider the registration of images of the same modality. Thus, their corresponding noise-free images are assumed to only have geometric distortion and the corresponding pixels in the two noise-free images are assumed to have the same intensity values. For the registration of images of different modalities, the intensity of a pixel in one image is a nonlinear function of that of the corresponding pixel in the other image. This makes the derivation of the bound more complicated.

Assuming $u(x, y)$ and $v(x, y)$ to be the transformed coordinates, the reference image and floating image can be modeled, respectively, as

$$
\begin{aligned}
I_1(x, y) &= f(x, y) + n(x, y), \\
I_2(x, y) &= f(u(x, y), v(x, y)) + n(x, y).
\end{aligned}
\tag{13.1}
$$

We assume here that the noise $n(x, y)$ is independent and identically distributed (i.i.d.) Gaussian noise with zero mean and variance N. Thus, each pixel intensity is considered as a Gaussian random variable with the mean equal to the noise-free pixel intensity and variance equal to N. Based on this, the joint probability density function (PDF) of the reference and floating images can be written as

$$
P(I_1, I_2) = \frac{1}{(2\pi N)^{\frac{M}{2}}} e^{-\frac{1}{2N} \sum_{\mathbf{r}} [I_1(\mathbf{r}) - f(\mathbf{r})]^2 + [I_2(\mathbf{r}) - f(u(\mathbf{r}), v(\mathbf{r}))]^2},
\tag{13.2}
$$

where M is the total number of pixels of the image and the vector $\mathbf{r}$ denotes (x, y). This yields

$$
\ln P(I_1, I_2) = -\frac{M}{2} \ln(2\pi N) - \frac{1}{2N} \sum_{\mathbf{r}} \left[I_1(\mathbf{r}) - f(\mathbf{r}) \right]^2
$$
$$
+ \left[I_2(\mathbf{r}) - f(u(\mathbf{r}), v(\mathbf{r})) \right]^2.
\tag{13.3}
$$

The above model is the basic statistical model we use to derive the performance bounds on image registration in the following sections. Note that this model is idealized, in the sense that it does not account for misregistration effects encountered in practice, such as aliasing. In order to be consistent with the model, in our illustrative example, we use the bicubic interpolation method to create correct synthetic images that are alias free.

13.3 Performance bounds on image registration

13.3.1 Cramér-Rao lower bound for intensity-based registration

The CRLB is typically used to investigate the fundamental limits on the accuracy of parameter estimation (Kay, 1993). It is a lower bound on the error variance of the best estimator for the parameter using the given system.

Suppose there is a parameter vector $\phi = \{\phi_1, \ldots, \phi_k\}$ to be estimated. It is assumed that the PDF of the data $P(s; \phi)$ satisfies the "regularity" condition (Kay, 1993), i.e.,

$$E\left[\frac{\partial \ln P(s; \phi)}{\partial \phi}\right] = 0 \quad \text{for all } \phi, \tag{13.4}$$

where the expectation is taken with respect to $P(s; \phi)$. Then, the unbiased version of the CRLB is defined as the inverse of the *Fisher information matrix* (FIM), which can be expressed (Kay, 1993) as

$$E\left(\phi_i - \hat{\phi}_i\right)^2 \geq \text{CRLB}\left(\hat{\phi}_i\right) = FI^{-1}(i, i) \quad \text{for} \quad i = 1, 2, \ldots, K, \tag{13.5}$$

where ϕ_i is the ith parameter, and FI is the FIM of size $K \times K$. The latter is defined as

$$FI_{ij} = -E\left[\frac{\partial^2 \ln P(s; \phi)}{\partial \phi_i \partial \phi_j}\right] \quad \text{for} \quad i, j = 1, 2, \ldots, K, \tag{13.6}$$

where the derivative is evaluated at the true value of ϕ and the expectation is taken with respect to $P(s; \phi)$. The Fisher information matrix provides a measure of the information content of the observed data relative to particular parameters.

In an image registration problem, suppose that there are K transformation parameters to be estimated. The two images are the observed data. The partial derivative of the logarithm of the PDF with respect to the ϕ_i is obtained as

$$\frac{\partial \ln P(I_1, I_2)}{\partial \phi_i} = \frac{1}{N} \sum_{\mathbf{r}} \left[I_2(\mathbf{r}) - f(u(\mathbf{r}), v(\mathbf{r}))\right] \cdot \frac{\partial f}{\partial \phi_i}. \tag{13.7}$$

Because $E\left[I_2(\mathbf{r}) - f(u(\mathbf{r}), v(\mathbf{r}))\right] = 0$, we obtain $E\left[\frac{\partial \ln P(s; \phi)}{\partial \phi_i}\right] = 0$ and, therefore, the "regularity" condition is satisfied. Further, the second derivatives are given by

$$\frac{\partial^2 \ln P\,(I_1, I_2)}{\partial \phi_i \partial \phi_j} = \frac{1}{N} \sum_{\mathbf{r}} \left\{ [I_2\,(\mathbf{r}) - f\,(u\,(\mathbf{r}),\, v\,(\mathbf{r}))] \cdot \frac{\partial^2 f}{\partial \phi_i \partial \phi_j} - \frac{\partial f}{\partial \phi_i} \frac{\partial f}{\partial \phi_j} \right\}.$$

(13.8)

Thus, as $E\,[I_2\,(\mathbf{r})] = f\,(u\,(\mathbf{r}),\, v\,(\mathbf{r}))$, the expectation of Eq. (13.8) becomes

$$E\left[\frac{\partial^2 \ln P\,(I_1, I_2)}{\partial \phi_i \partial \phi_j}\right] = \frac{1}{N} \sum_{\mathbf{r}} -\frac{\partial f}{\partial \phi_i} \frac{\partial f}{\partial \phi_j}$$

$$= -\frac{1}{N} \sum_{\mathbf{r}} \left[f_u\,(\mathbf{r}) \frac{\partial u}{\partial \phi_i} + f_v\,(\mathbf{r}) \frac{\partial v}{\partial \phi_i} \right]$$

$$\times \left[f_u\,(\mathbf{r}) \frac{\partial u}{\partial \phi_j} + f_v\,(\mathbf{r}) \frac{\partial v}{\partial \phi_j} \right], \quad (13.9)$$

where $f_u(\mathbf{r})$ and $f_v(\mathbf{r})$ denote, respectively, the gradient components of the noise-free image f along the horizontal and vertical axes. The gradient of the image can be obtained by using derivative filters, such as $[1 - 1]$ and $[1 - 1]^T$ for the horizontal axis and vertical axis, respectively. In practice, we only have noisy images I_1 and I_2 available. Thus, preprocessing is required to remove the noise from the image I_2. After the noise-free image is obtained, its gradient along both axes can be calculated.

Based on Eqs. (13.6), (13.8), the (i, j)th term of the FIM is obtained as

$$FI_{ij} = \frac{1}{N} \sum_{\mathbf{r}} \left[f_u\,(\mathbf{r}) \frac{\partial u}{\partial \phi_i} + f_v\,(\mathbf{r}) \frac{\partial v}{\partial \phi_i} \right] \left[f_u\,(\mathbf{r}) \frac{\partial u}{\partial \phi_j} + f_v\,(\mathbf{r}) \frac{\partial v}{\partial \phi_j} \right]$$

$$\text{for} \quad i, j = 1, 2, \ldots, K.$$

(13.10)

Next, we present the CRLB for several special cases of image misalignments.

13.3.1.1 Translation-only distortion

Translation misalignment is the simplest case in image registration. Assuming that the floating image has a translation of x_0 and y_0 in the x and y directions, respectively, the mapping function can be written as

$$u = x - x_0,$$
$$v = y - y_0.$$

(13.11)

Hence, the intensities of the two images can be expressed as

$$I_1\,(x, y) = f\,(x, y) + n\,(x, y),$$
$$I_2\,(x, y) = f\,(x - x_0,\, y - y_0) + n\,(x, y).$$

(13.12)

In the case of translation, the transformation parameters to be estimated are given by $\phi = \{x_0, y_0\}$.

Based on Eq. (13.2), the expectation of the derivative of the PDF with respect to the transformation parameters is equal to zero in this case, which indicates that the regularity condition is satisfied. Thus, the CRLB for the image registration problem in this case can be obtained by Eq. (13.5).

Substituting Eq. (13.11) into Eq. (13.10), the FIM of the translations in the x and y directions can be expressed by

$$FI = \frac{1}{N} \begin{bmatrix} \sum_{\mathbf{r}} f_u^2(\mathbf{r}) & \sum_{\mathbf{r}} f_u(\mathbf{r}) f_v(\mathbf{r}) \\ \sum_{\mathbf{r}} f_v(\mathbf{r}) f_u(\mathbf{r}) & \sum_{\mathbf{r}} f_v^2(\mathbf{r}) \end{bmatrix}. \tag{13.13}$$

See Robinson and Milanfar (2004) and Yetik and Nehorai (2006). Thus, the CRLB for the estimates $\hat{x}_0$ and $\hat{y}_0$ for the translation only case can be written as

$$\text{CRLB}(\hat{x}_0) = FI^{-1}(1, 1),$$
$$\text{CRLB}(\hat{y}_0) = FI^{-1}(2, 2). \tag{13.14}$$

The above derived CRLBs serve as performance limits for image registration algorithms. For illustration purposes, we conducted Monte Carlo simulation experiments, where the actual mean squared errors were computed for a set of image registration algorithms and compared against the bounds. Figure 13.1 shows a Landsat MSI band 1 image of size 100×100, which was used for the simulations. Specifically, we generated a pair of images with random horizontal and vertical translations (along the x- and y-axes, respectively). These random translations were uniformly distributed in the intervals $[0°, 10°]$ and $[0°, 10°]$, respectively, using bicubic interpolation. White Gaussian noise was then added to the image pair prior to registration. Methods based on the *mutual information* (MI) (Chen, 2002; Chen and Varshney, 2003) and *mean squared differences* (MSD) (Varshney *et al.*, 2005) similarity measures were used to estimate the translation errors. We used exhaustive search in the search space $[0°, 10°] \times [0°, 10°]$ to find the optimal parameters which maximized MI and minimized MSD, respectively. The process was repeated 100 times at each noise power value. We then computed the *root mean square error* (RMSE) of the translation estimates and compared them with the square root of the CRLB. The results are depicted in Fig. 13.2. Specifically, the bounds shown provide limits as to the best achievable performance levels. In other words, no image registration algorithm can achieve – for this image pair – an MSE that is smaller than the bound shown at each noise level. The MSD-based method yielded smaller registration errors than the MI-based method. The gap between the registration errors and the above bounds implies that there may exist better registration

Figure 13.1. Original Landsat image.

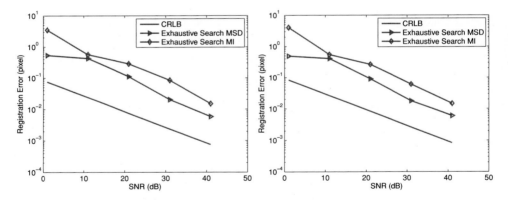

Figure 13.2. Registration errors and CRLB in the translation-only case.

algorithms for this image pair. Also, there may exist tighter bounds based on more accurate image models.

Since image derivatives are involved in the calculation of the CRLB expression, the bounds of interest are affected by image content. Larger image derivatives produce lower bounds, which is very reasonable because they correspond to higher frequencies in the image. An image consisting of higher frequencies contains more features, and is thus registered more accurately. In other words, the more an image is influenced by the unknown transformation parameters, the better we should be able to estimate the parameters or register the image.

13.3.1.2 Rigid-body transformation distortion

In this subsection, we consider the more general case of misalignment that is specified in terms of rigid-body transformation. Misalignment error defined as a rigid-body transformation is expressed by the following mapping:

$$
\begin{aligned}
u &= \cos\theta_0 x + \sin\theta_0 y + x_0, \\
v &= -\sin\theta_0 x + \cos\theta_0 y + y_0.
\end{aligned}
\tag{13.15}
$$

The parameters to be estimated for registration are the rotation θ_0, and the translations x_0 and y_0. We assume that the scale does not need to be estimated here. Otherwise, it would be a straightforward extension.

Under the assumption that the regularity condition holds, the FIM of the rigid-body transformation can be expressed as

$$
FI = \frac{1}{N} \sum_{\mathbf{r}} \left\{ \left[\left(f_u(\mathbf{r}) \frac{du}{d\theta_0} + f_v(\mathbf{r}) \frac{dv}{d\theta_0} \right) \quad f_u(\mathbf{r}) f_v(\mathbf{r}) \right] \right.
$$
$$
\left. \bullet \left[\left(f_u(\mathbf{r}) \frac{du}{d\theta_0} + f_v(\mathbf{r}) \frac{dv}{d\theta_0} \right) \quad f_u(\mathbf{r}) f_v(\mathbf{r}) \right]^T \right\},
\tag{13.16}
$$

where $\frac{du}{d\theta_0} = -\sin\theta_0 x + \cos\theta_0 y$, $\frac{dv}{d\theta_0} = -\cos\theta_0 x - \sin\theta_0 y$, and $\bullet$ denotes the vector dot product (Yetik and Nehorai, 2006). Thus, the CRLB of $\phi = \{\theta_0, x_0, y_0\}$ for a given θ_0 is given by

$$
\text{CRLB}\left(\hat{\phi}_i | \theta_0\right) = FI^{-1}(i, i), \quad i = 1, 2, 3,
\tag{13.17}
$$

where ϕ_i is the ith element of the vector ϕ.

According to the above result, the CRLB is a function of the rotation angle θ_0. Assuming the rotation angle to be uniformly distributed in $[0, \Delta]$, the expected FIM for rigid-body transformation misalignment is given by

$$
E_{\theta_0}\left(FI_{ij}\right) = \frac{1}{\Delta} \int_0^\Delta FI_{ij} d\theta_0
\tag{13.18}
$$

where FI_{ij} denotes the (i, j)th element of the FIM given in (13.16). From this, the averaged or Bayesian CRLB can be obtained from Van Trees (1968) as

$$
\text{CRLB}_{\text{ave}}\left(\hat{\phi}_i\right) = \left(E_{\theta_0}(FI)\right)^{-1}(i, i), \quad i = 1, 2, 3.
\tag{13.19}
$$

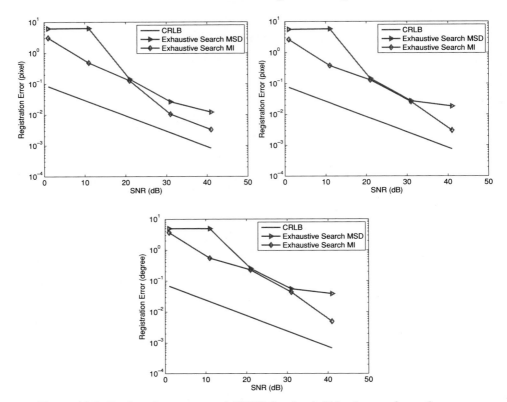

Figure 13.3. Registration errors and CRLB for the rigid-body transformation case.

In a manner similar to that of the translation-only distortion case, we generated a pair of images using the same image in Fig. 13.1 with random horizontal and vertical translations and random rotation uniformly distributed in the intervals $[0°, 10°]$, $[0°, 10°]$, and $[0°, 10°]$, respectively. The MI- and MSD-based methods were used to estimate the translation and rotation errors. For both translation and rotation errors, the averaged CRLB was obtained using Eq. (13.19) under the assumption that θ_0 is uniformly distributed in $[0°, 10°]$. The square root of the CRLB and the RMSE of the two registration methods are shown in Fig. 13.3. According to these results, the MI-based method produces smaller registration errors than the MSD method. However, there is still a gap between the errors obtained by the MI-based method and the averaged CRLB, which indicates that a tighter performance bound is desired, that is, there may exist a better registration algorithm for this particular image, which can achieve higher registration accuracy than the MI-based method.

13.3.1.3 Other types of distortion

Skew distortion is a nonrigid transformation with the mapping function

$$
\begin{aligned}
u &= x + w_x y, \\
v &= w_y x + y.
\end{aligned}
\tag{13.20}
$$

It can be shown that the FIM corresponding to the skew distortion $\phi = \{w_x, w_y\}$ is given by (Yetik and Nehorai, 2006)

$$
FI = \sum_{\mathbf{r}} \begin{bmatrix} y^2 f_u^2(\mathbf{r}) & xy f_u(\mathbf{r}) f_v(\mathbf{r}) \\ xy f_u(\mathbf{r}) f_v(\mathbf{r}) & x^2 f_v^2(\mathbf{r}) \end{bmatrix}.
\tag{13.21}
$$

Recall that the CRLB is the inverse of the FIM. From Eq. (13.21) we can see that while the CRLB of skew distortion is a function of the derivatives of the image, it is independent of the skew parameters w_x and w_y.

Affine transformation distortions are the most general form of geometric deformation mapping parallel lines to parallel lines (Yetik and Nehorai, 2006). In general, they are expressed by

$$
\begin{aligned}
u &= d_1 x + d_2 y + d_3, \\
v &= d_4 x + d_5 y + d_6.
\end{aligned}
\tag{13.22}
$$

The parameters to be estimated are $\phi = \{d_1, d_2, \ldots, d_6\}$. The FIM for this case is

$$
FI = \sum_{\mathbf{r}} \begin{bmatrix} f_u^2(\mathbf{r}) K_3 & f_u(\mathbf{r}) f_v(\mathbf{r}) K_3 \\ f_u(\mathbf{r}) f_v(\mathbf{r}) K_3 & f_v^2(\mathbf{r}) K_3 \end{bmatrix},
\tag{13.23}
$$

where

$$
K_3 = \begin{bmatrix} x^2 & xy & x \\ xy & y^2 & y \\ x & y & 1 \end{bmatrix}.
\tag{13.24}
$$

See Yetik and Nehorai (2006). Also, the CRLB of the affine transformation is the inverse of the FIM in Eq. (13.23).

13.3.2 Other performance bounds for intensity-based registration

CRLB is used typically as a benchmark for MSE performance in estimation problems. It is relatively simple to calculate. However, it is known to be an optimistic bound that may not be tight enough to provide meaningful insight into the achievable estimator performance. There are other lower bounds that may serve as better indicators, for certain SNR ranges, for performance estimation (Bell *et al.*, 1997).

For this reason, the Bhattacharyya bound (BB) and the Ziv-Zakai bound (ZZB) were investigated in Xu and Varshney (2006) for the estimation of the transformation parameters in image registration.

The Bhattacharyya bound is a generalized lower bound on the MSE error of an estimator. While the CRLB is given by the inverse of the FIM, the BB is defined as the inverse of a higher-order FIM (Weinstein and Weiss, 1988; Lu and Krogmeier, 2002). The first-order BB is equal to the CRLB. Increasing the order yields a tighter bound, but requires more computational effort.

In contrast to the CRLB and the BB, which consider the parameters to be deterministic, the Ziv-Zakai bound, proposed originally in Ziv and Zakai (Ziv and Zakai 1969; Zakai *et al.*, 1975), is a Bayesian bound, which assumes that the parameter to be estimated is a random variable with known a-priori distribution. It provides a bound on the global MSE averaged over the a-priori probability density function. Details regarding the performance bounds for image registration based on the BB and the ZZB can be found in Xu and Varshney (2006) and Xu *et al.* (2009).

13.3.3 CRLB for control point-based image registration

Control point-based image registration methods have been most widely used. They typically involve extraction of geometric features, also known as ground control points (GCPs), such as intersections and landmarks clearly visible in both images. Image registration is performed then by matching of the control points in the two images and estimating the transformation parameters.

Let $\mathbf{f} = [\mathbf{f}_1, \mathbf{f}_2, \ldots, \mathbf{f}_l]^T$ and $\mathbf{g} = [\mathbf{g}_1, \mathbf{g}_2, \ldots, \mathbf{g}_l]^T$ represent the coordinates of the l points corresponding to the l features in the two images I_1 and I_2, respectively. The $\mathbf{f}$ and $\mathbf{g}$ are modeled as the noise-free set of real-world coordinates $\mathbf{p} = [\mathbf{p}_1, \mathbf{p}_2, \ldots, \mathbf{p}_l]^T$ plus noise. Thus, the statistical model is given by (Yetik and Nehorai, 2006)

$$\begin{aligned} \mathbf{f}_i &= C\mathbf{p}_i + \mathbf{n}_i, \\ \mathbf{g}_i &= D\mathbf{p}_i + \mathbf{n}_i, \end{aligned} \tag{13.25}$$

where

$$\mathbf{f}_i = \begin{bmatrix} f_{ix} & f_{iy} \end{bmatrix}^T, \tag{13.26}$$

$$\mathbf{g}_i = \begin{bmatrix} g_{ix} & g_{iy} \end{bmatrix}^T, \tag{13.27}$$

and

$$\mathbf{p}_i = \begin{bmatrix} p_{ix} & p_{iy} & 1 \end{bmatrix}^T, \tag{13.28}$$

and where

$$C = \begin{bmatrix} 1 & 0 & 0 \\ 0 & 1 & 0 \end{bmatrix}. \tag{13.29}$$

The matrix D denotes the transformation matrix that maps the coordinates. For a rigid-body transformation, it is expressed as

$$D = \begin{bmatrix} \cos\theta_0 & \sin\theta_0 & x_0 \\ -\sin\theta_0 & \cos\theta_0 & y_0 \end{bmatrix}. \tag{13.30}$$

The variable $\mathbf{n}_i$ represents an i.i.d. Gaussian noise with zero mean and variance $\mathbf{N} = \begin{bmatrix} N & 0 \\ 0 & N \end{bmatrix}$. Because the noise-free coordinate set $\mathbf{p}$ is assumed deterministic, the coordinates of the feature points in the two images $\mathbf{f}$ and $\mathbf{g}$ become i.i.d. Gaussian random variables with mean $\mathbf{p}$ and variance $\mathbf{N}$, and the joint PDF of $\mathbf{f}$ and $\mathbf{g}$ can be written as

$$P(\mathbf{f}, \mathbf{g}) = \frac{1}{(2\pi N)^{\frac{l}{2}}} e^{-\frac{1}{2}\sum_{i=1}^{l}(\mathbf{f}_i - C\mathbf{p}_i)\mathbf{N}^{-1}(\mathbf{f}_i - C\mathbf{p}_i)^T + (\mathbf{g}_i - D\mathbf{p}_i)\mathbf{N}^{-1}(\mathbf{g}_i - D\mathbf{p}_i)^T}. \tag{13.31}$$

Thus, substituting Eq. (13.31) into Eq. (13.6), the FIM obtained for a rigid-body transformation distortion is (Yetik and Nehorai, 2006)

$$FI_{ij} = \sum_{i=1}^{l} \left(\frac{\partial D}{\partial \phi_i}\mathbf{p}_i\right)^T \mathbf{N}^{-1} \left(\frac{\partial D}{\partial \phi_j}\mathbf{p}_i\right). \tag{13.32}$$

The CRLB can be obtained then for the parameters ϕ from the inverse of the FIM.

The CRLB for translations, in the special case of translation only distortion using control point-based image registration, reduces to (Yetik and Nehorai, 2006)

$$\mathrm{CRLB} = \frac{1}{l}\begin{bmatrix} N & 0 \\ 0 & N \end{bmatrix}. \tag{13.33}$$

The CRLB for the translations and rotation of a rigid-body transformation distortion is given by (Yetik and Nehorai, 2006)

$$\mathrm{CRLB} = N \cdot \begin{bmatrix} \sum_{i=1}^{l} p_{ix}^2 + p_{iy}^2 & \sum_{i=1}^{l} E_{1i} & \sum_{i=1}^{l} E_{2i} \\ \sum_{i=1}^{l} E_{1i} & l & 0 \\ \sum_{i=1}^{l} E_{2i} & 0 & l \end{bmatrix}^{-1}, \tag{13.34}$$

where $E_{1i} = -\sin\theta_0 p_{ix} + \cos\theta_0 p_{iy}$ and $E_{2i} = -\sin\theta_0 p_{iy} - \cos\theta_0 p_{ix}$. The CRLB results for the skew distortion and affine transformation cases are available in Yetik and Nehorai (2006).

13.4 Summary

In this chapter, performance bounds, specifically the Cramér-Rao lower bound (CRLB), have been evaluated to predict the performance limits of image registration algorithms for a given pair of images. In particular, it gives the minimum achievable MSE of the transformation parameters, such as translations and rotation, for image registration algorithms with respect to a specific image scene. Thus, for a given image pair and a known transformation type, one can obtain the minimum achievable registration error in terms of the MSE of the transformation parameters. Details on how to derive the CRLB for several types of distortion have been provided.

In practice, the bound can be used to evaluate how well the image registration approaches are likely to perform for a particular image pair. By comparing the bound with the actual MSE of a registration algorithm, a small gap between these errors implies the superiority of the registration algorithm in question, while a large gap indicates unsatisfactory performance. In addition, one may have a set of reference images available for registration of an acquired image. The bounds can help us select a reference image which can provide the best registration. The image which produces the smallest bound should be selected, as it may produce the smallest registration error. Thus, the bounds provide a useful tool for the performance analysis, as well as performance enhancement of image registration.

Note that the image model considered here is an ideal statistical model. In practice, however, this model has some limitation. We assumed that each pixel in the image has i.i.d. Gaussian distribution as in Robinson and Milanfar (2004) and Yetik and Nehorai (2006). However, as we know, the pixels in an image are spatially correlated. In particular, in high-resolution images, the neighboring image pixels are highly correlated. Thus, it is important to incorporate the spatial correlation into an image model. In future work, one can incorporate spatial correlation into the image model. The development of performance bounds based on this model is desired. Moreover, when images are sampled below the Nyquist rate, aliasing noise will be introduced (Stone *et al.*, 1999). In particular, images of scenes with sharp edges typically suffer from alias noise, and the effect of alias noise is to shift the perceived edges from their actual position to a different position. This shifting is not modeled by Gaussian noise. We do not consider the aliasing effect in our model. It would be interesting to study this issue based on the bounds. In addition, performance bounds for images of different modalities need to be derived in the future.

References

Bell, K. L., Steinberg, Y., Ephraim, Y., and Van Trees, H. L. (1997). Extended Ziv-Zakai lower bound for vector parameter estimation. *IEEE Transactions on Information Theory*, **43**(2), 624–637.

Bhattacharyya, A. K. (1943). On a measure of divergence between two statistical populations defined by their probability distributions. *Bulletin of the Culcutta Mathematical Society*, **35**, 99–110.

Brown, L. G. (1992). A survey of image registration techniques. *ACM Computing Surveys*, **24**(4), 325–376.

Chen, H. (2002). Mutual Information-Based Image Registration with Applications. Ph.D. dissertation, Department of Electrical Engineering and Computer Science, Syracuse University, New York.

Chen, H. and Varshney, P. K. (2003). Mutual information-based CT-MR brain image registration using generalized partial volume joint histogram estimation. *IEEE Transactions on Medical Imaging*, **22**(9), 1111–1119.

Dai, X. and Khorram, S. (1998). The effects of image misregistration on the accuracy of remotely sensed change detection. *IEEE Transactions on Geoscience and Remote Sensing*, **36**(5), 1566–1577.

Kay, S. M. (1993). *Fundamentals of Statistical Signal Processing: Estimation Theory*. Upper Saddle River, NJ: Prentice Hall, Inc.

Lu, F. and Krogmeier, J. V. (2002). Modified Bhattacharyya bounds and their application to timing estimation. In *Proceedings of the IEEE Wireless Communications and Networking Conference*, Orlando, FL, Vol. 1, pp. 244–248.

Robinson, D. and Milanfar, P. (2004). Fundamental performance limits in image registration. *IEEE Transactions on Image Processing*, **13**(9), 1185–1199.

Stone, H. S., Orchard, M., and Change, E.-C. (1999). Subpixel registration of images. In *Proceedings of the Thirty-Third Asilomar Conference on Signals, Systems, and Computers*, Monterey, CA, Vol. 2, pp. 1446–1452.

Sundaresan, A., Varshney, P. K., and Arora, M. K. (2007). Robustness of change detection algorithms in the presence of registration errors. *Journal of Photogrammetric Engineering and Remote Sensing*, **73**(4), 375–383.

Townshend, J. R. G., Justice, C. O., and Gurney, C. (1992). The impact of misregistration on change detection. *IEEE Transactions on Geoscience and Remote Sensing*, **30**, 1054–1060.

Van Trees, H. L. (1968). *Detection, Estimation, and Modulation Theory, Part I*. New York: Wiley.

Varshney, P. K., Kumar, B., Xu, M., Drozd, A., and Kasperovich, I. (2005). Image registration: A tutorial. In *Proceedings of the NATO Advanced Science Institutes Series*, Albena, Bulgaria.

Weinstein, E. and Weiss, A. J. (1988). A general class of lower bounds in parameter estimation. *IEEE Transactions on Information Theory*, **34**(2), 338–342.

Xu, M. and Varshney, P. K. (2006). Tighter performance bounds on image registration. In *Proceedings of the IEEE International Conference on Acoustics, Speech, and Signal Processing*, Toulouse, France, Vol. II, pp. 777–780.

Xu, M., Chen, H., and Varshney, P. K. (2009). Ziv-Zakai bounds on image registration. *IEEE Transactions on Signal Processing*, **57**(5), 1745–1755.

Yetik, I. S. and Nehorai, A. (2006). Performance bound on image registration. *IEEE Transactions on Signal Processing*, **54**(5), 1737–1749.

Zakai, M., Chazan, D., and Ziv, J. (1975). Improved lower bounds on signal parameter estimation. *IEEE Transactions on Information Theory*, **21**(1), 90–93.

Zitová, B. and Flusser, J. (2003). Image registration methods: A survey. *Image and Vision Computing*, **21**(11), 977–1000.

Ziv, J. and Zakai, M. (1969). Some lower bounds on signal parameter estimation. *IEEE Transactions on Information Theory*, **15**(3), 386–391.

Part IV

Applications and Operational Systems

14

Multitemporal and multisensor image registration

JACQUELINE LE MOIGNE, ARLENE A. COLE-RHODES,
ROGER D. EASTMAN, NATHAN S. NETANYAHU, HAROLD S. STONE,
ILYA ZAVORIN, AND JEFFREY T. MORISETTE

Abstract

Registration of multiple source imagery is one of the most important issues when dealing with Earth science remote sensing data where information from multiple sensors exhibiting various resolutions must be integrated. Issues ranging from different sensor geometries, different spectral responses, to various illumination conditions, various seasons and various amounts of noise, need to be dealt with when designing a new image registration algorithm. This chapter represents a first attempt at characterizing a framework that addresses these issues, in which possible choices for the three components of any registration algorithm are validated and combined to provide different registration algorithms. A few of these algorithms were tested on three different types of datasets – synthetic, multitemporal and multispectral. This chapter presents the results of these experiments and introduces a prototype registration toolbox.

14.1 Introduction

In Chapter 1, we showed how the analysis of Earth science data for applications, such as the study of global environmental changes, involves the comparison, fusion, and integration of multiple types of remotely sensed data at various temporal, spectral, and spatial resolutions. For such applications, the first required step is fast and automatic image registration which can provide precision correction of satellite imagery, band-to-band calibration, and data reduction for ease of transmission. Furthermore, future decision support systems, intelligent sensors and adaptive constellations will rely on real- or near-real-time interpretation of Earth observation data, performed both onboard and at ground-based stations. The more expert the system and far-reaching the application, the more important will it be to obtain timely and accurately registered data.

293

In Chapter 3, we surveyed many different registration techniques developed for different applications, for example, military, medical, as well as remote sensing, from aircraft and from spacecraft. Despite the wide variety of algorithms available for image registration, no commercial software seems to fully respond to the needs of Earth and space data registration. Many of the methods presented in Chapter 3 are or may be applicable to remote sensing problems but, with such a wide choice, it is necessary to develop a framework to evaluate their performance on well-chosen remote sensing data. Our objective is to carry out systematic studies to support image registration users in selecting appropriate techniques for a remote sensing application based on accuracy and suitability for that application. We carry this out by surveying, designing, and developing different components of the registration process to enable the evaluation of their performance on well-chosen multiple source data, to provide quantitative intercomparison, and to eventually build an operational image registration toolbox. Of course, the main evaluation of automatic image registration algorithms is performed with regards to their accuracy, but it is also useful to relate the accuracy to "initial conditions," that is, the distance between the initial navigation geolocation and the correct result. As was described in Chapter 1, depending on the quality of the navigation model and of the ephemeris data, such initial geolocation may be accurate from within one pixel to tens of pixels. Other ways to evaluate image registration algorithms are in terms of their range of application, geometric and radiometric, and their robustness or reliability. In this chapter, we present the first steps of such an evaluation using representative registration components to build a few image registration algorithms. First, a potential evaluation framework is described. Then, choices for the different components of image registration are reviewed, and the algorithms combining these components are described along with their evaluation on several test datasets. The first two datasets are synthetic datasets incorporating geometric warping, noise, and radiometric distortion. The algorithms are tested utilizing multitemporal data from the Landsat instrument, and multisensor data acquired over several Earth Observing System (EOS) Land Validation Core Sites. These last datasets include data from the IKONOS, Landsat-7, Moderate Resolution Imaging Spectroradiometer (MODIS), and Sea-viewing Wide Field-of-view Sensor (SeaWiFS) sensors featuring multiple spatial and spectral resolutions.

14.2 A framework for the evaluation of image registration of remote sensing data

The NASA Goddard Image Registration Group was started in 1999 with the goal of developing and assessing image registration methodologies that will enable accurate multisource integration. In our work, we assume that the data have already been corrected according to a navigation model and that they are at a level equivalent to the EOS Level 1B (see Chapter 1 for a definition of EOS data levels). Assuming that the results of the systematic correction are accurate within a few or a few tens of pixels, our precision-correction algorithms utilize selected image features or control points to refine this geolocation accuracy within one pixel or a subpixel.

Our studies have been following the first two steps (or components) that define registration algorithms as described in Brown (1992). These are summarized in Chapter 3:

(1) Extraction of features to be used in the matching process.
(2) Feature matching strategy and metrics.
(3) Resampling or indexing of the data.

For alignment, we consider transformations that vary from translation only in x and y to rotation, scale, and translation (RST).

The long-term goal of our research is to build a modular image registration framework based on these first two components. The concept guiding the development of this framework is that various components of the registration process can be combined in several ways in order to reach optimum registration on a given type of data and under given circumstances. Thereby, the purpose of this framework is twofold:

(1) It represents a testing framework for:
  • Assessment of various combinations of components as a function of the applications.
  • Assessment of a new registration component compared to other known ones.
(2) Eventually, it could be the basis of a registration tool where a user could "schedule" a combination of components as a function of the application at hand, the available computational resources, and the required registration accuracy.

Many choices are available for each of the three components defined above. Our experiments deal with components 1 and 2, first focusing on various types of features utilizing only correlation-based methods and then looking at these features with different similarity metrics and strategies.

14.2.1 Feature extraction

14.2.1.1 Correlation-based experiments

Using cross-correlation as a similarity metric, features such as gray levels, edges, and Daubechies wavelet coefficients were compared using monosensor data (Le Moigne *et al.*, 1998). *Gray level* features were matched using either a basic spatial correlation or a phase correlation. When using *edge features*, the registration was performed in an iterative manner, first estimating independently the parameters of a rigid transformation on the center region of the two images, and then iteratively refining these parameters using larger and larger portions of the images (Le Moigne *et al.*, 1997). *Wavelet features* were also extracted and registered after decomposing both images with a discrete orthonormal basis of wavelets (Daubechies' least asymmetric filters; see Daubechies, 1992) in a multiresolution fashion. Low-pass features, which provide a compressed version of the original data and some texture information, and high-pass features, which provide detailed edge-like information, were both considered as potential registration features (see Le Moigne *et al.*, 2002a, and Chapter 11). This work was focused on correlation-based methods combined with an exhaustive search. One of the main drawbacks of this method is the prohibitive amount of computation required when the number of transformation parameters increases (e.g., affine transformation vs. shift-only), or when the size of the data increases (full-size scenes vs. small portions; multiband processing vs. monoband). To answer some of these concerns, we investigated different types of similarity metrics and different types of feature matching strategies (Subsection 14.2.2).

For this first evaluation, we used three datasets: Two synthetic datasets for which the true transformation parameters were known, and one dataset for which no ground truth was available but manual registration was computed. Accuracy and computation times were used as evaluation criteria. Results showed that, as expected, edges or edge-like features like wavelets are more robust to noise, local intensity variations or time-of-the day conditions than original gray level values. On the other hand, when only looking for translation in cloud-free data, phase correlation provides a fast and accurate answer. Comparing edges and wavelets, orthogonal wavelet-based registration is usually faster, although not always as accurate as a full-resolution edge-based registration. This lack of consistent accuracy of orthogonal wavelets is mainly due to the lack of translation invariance, and is presented in more detail in the second set of experiments.

14.2.1.2 Wavelet-based experiments

Chapter 11 describes the set of experiments that we performed using wavelets or wavelet-like features. These experiments verified that separable orthogonal wavelet transforms are not translation- and rotation-invariant. By lack of translation (resp. rotation) invariance, we mean that the wavelet transform does not commute with the translation (resp. rotation) operator. The two studies described in Chapter 11 showed that:

(1) Low-pass subbands of orthogonal wavelets are relatively insensitive to translation, provided that the features of interest have an extent at least twice the size of the

wavelet filters, while high-pass subbands are more sensitive to translation, although peak correlations are high enough to be useful (Stone *et al.*, 1999).

(2) Simoncelli's steerable filters perform better than Daubechies' filters. Rotation errors obtained with steerable filters are minimum, independent of rotation size or noise amount. Noise studies also reinforced the results that steerable filters show a better robustness to larger amounts of noise than do orthogonal filters (Zavorin and Le Moigne, 2005).

14.2.2 Feature matching

We then considered various similarity metrics and various matching strategies that can be utilized for feature matching of remote sensing data. As an alternative to correlation, mutual information (MI) is another similarity metric that was first introduced in Maes *et al.* (1997) and was used very successfully for medical image registration. Mutual information, or relative entropy, is a basic concept from information theory which measures the statistical dependence between two random variables; or, equivalently, it measures the amount of information that one variable contains about another. Experiments described in Cole-Rhodes *et al.* (2003) show that mutual information may be better suited for subpixel registration as it produces consistently sharper optimum peaks than correlation, thereby yielding higher accuracy.

Mutual information is particularly efficient when used in conjunction with an optimization method, for example, a steepest gradient-based type method like the one described in Irani and Peleg (1991) or a Levenberg-Marquardt optimization like the one utilized for medical image registration and described in Thévenaz *et al.* (1998). Different optimization methods are described in Chapter 12 and in Eastman and Le Moigne (2001). Gray levels, edge magnitudes or low-frequency wavelet information could be used as input to these optimization methods. In Cole-Rhodes *et al.* (2003), mutual information was combined with a stochastic gradient search and the results showed that mutual information was generally found to optimize with one-third the number of iterations required by correlation.

We also studied the use of a statistically robust feature matching method based on the use of nearest-neighbor matching and a generalized Hausdorff distance metric (Mount *et al.*, 1999; Netanyahu *et al.*, 2004). This method (also described in Chapter 8) is based on the principle of point mapping with feedback. Specifically, given corresponding sets of control points in the reference and input images within a prespecified transformation (e.g., rigid, affine), this method derives a computationally efficient algorithm to match these point patterns. The algorithms described use the partial Hausdorff distance and compute the matching transformation either by using a geometric branch-and-bound search of transformation space or by

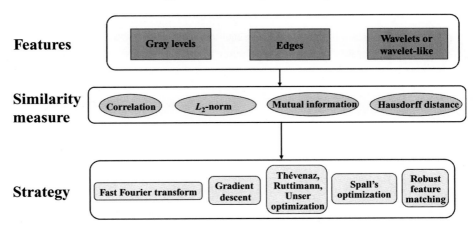

Figure 14.1. Modular approach to image registration combining various choices for feature extraction, similarity metrics, and matching strategy. (Source: Le Moigne *et al.*, 2003, © IEEE, reprinted with permission.)

considering point alignments. This method has been applied successfully to multitemporal Landsat data using Simoncelli's overcomplete wavelet features; results are described in Subsection 14.3.2.2 (see also Netanyahu *et al.*, 2004, for details).

14.2.3 Testing framework

The investigations described in Subsections 14.2.1 and 14.4.2 led to a first version of a testing framework illustrated in Fig. 14.1, in which a registration algorithm is defined as the combination of a set of features, a similarity measure, and a matching strategy. In this framework, *features* can be either gray levels, low-pass features from Simoncelli steerable filters decomposition or from a spline decomposition, or Simoncelli band-pass features; *similarity metrics* can be either cross-correlation, the L_2-norm, mutual information or a Hausdorff distance; *matching strategies* are based either on a fast Fourier correlation, one of the three optimization methods (steepest gradient descent, a Levenberg-Marquardt technique, and a stochastic gradient algorithm), or a robust feature matching approach.

By combining these different components, five algorithms were developed and tested. They are compared in Section 14.3:

- **Method 1**: Gray levels matched by fast Fourier correlation (Stone *et al.*, 2001). We will label it *FFC*.
- **Method 2**: Spline or Simoncelli pyramid features matched by optimization and an L_2-norm using the algorithm developed by Thévenaz *et al.* (1998) and Zavorin and Le Moigne (2005). We will label it *TRU*.

- **Method 3**: Spline or Simoncelli pyramid features matched by optimization and a mutual information criterion using the algorithm developed by Thévenaz *et al.* (1998) and Zavorin and Le Moigne (2005). We will label it *TRUMI*.
- **Method 4**: Spline or Simoncelli pyramid features matched by optimization of the mutual information criterion using the Spall algorithm (Cole-Rhodes *et al.*, 2003). We will label it *SPSA*.
- **Method 5**: Simoncelli wavelet features using a robust feature matching algorithm and a generalized Hausdorff distance (Netanyahu *et al.*, 2004). We will label it *RFM*.

For some of the Methods (1 and 5), registration is computed on individual subimages and then integrated by computing a global transformation. For Methods 2–4, registration is computed on the entire images but iteratively, using pyramid decompositions. Another method, called *GGD*, was based on gray-level matching using a gradient-descent algorithm with an L_2-norm. It was utilized as a reference in early experiments, but since it can be considered as a special case of Method 2, TRU, it will not be systematically evaluated in most of the experiments described in Section 14.3. More details on GGD can be found in Eastman and Le Moigne (2001).

14.3 Comparative studies

In this section, we describe systematic studies that were performed to compare the five algorithms defined in Subsection 14.2.3. Assessing an image registration algorithm for subpixel accuracy and for robustness to noise and to initial conditions, using remote sensing data, presents some difficulty since often ground truth is not available. Interleaving two images for visual assessment can detect gross mismatches and global misalignment but is difficult to extend to quantitative subpixel evaluation. In a few cases, a limited number of control points are known with highly accurate and absolute Ground Positioning System (GPS) information and are used to compute an approximation of the algorithm's accuracy. Manual registration can be used to calculate the unknown transformation but it is uncertain if it is accurate enough to test subpixel accuracy on small regions. Another approach is to generate synthetic image pairs by matching one image against a transformed and resampled version of itself with or without added noise. To avoid some resampling issues, this can be done by using high-resolution imagery and downsampling to a lower resolution using an appropriate point spread function to generate both images in a pair. While useful, this approach is limited in realistically modeling noise, temporal scene changes, or cross-sensor issues. Yet another approach is to use circular registration results on natural imagery when three or more overlapping images are available. In this case, the transformations should compose to yield the identity – for three images registered pairwise by T_1, T_2, and T_3, the composition $T_1 \circ T_2 \circ T_3$ should be close to the identity transformation (Le Moigne *et al.*, 2002b).

For our experiments, we utilize three types of test data – synthetic data, multitemporal data, and multisensor data. This section describes the test datasets, followed by the corresponding experiments and the results obtained when testing the five algorithms described in Subsection 14.2.3.

14.3.1 Test data

14.3.1.1 Synthetic datasets

Our goal is to evaluate the strength of various algorithms when applied to many types of satellite data, so test data sets should include imagery from different platforms, with different spatial and spectral resolutions, taken at various dates. The disadvantage of such data is that in the majority of cases, ground truth, if available at all, is approximate at best. Therefore, in our experiments, we first use synthetic images created by a controlled process, designed to emulate real data (Zavorin and Le Moigne, 2005). Three types of transformations were applied in various combinations to a given "source" image to produce synthetic test data, namely, (1) geometric warping, (2) radiometric variations, and (3) addition of noise:

- *Geometric warping* was introduced by simply applying an RST transformation with predetermined amounts of shift, rotation and/or scale to the source image. The resulting warped image is radiometrically identical to the source. The scale was fixed at a value close to 0.95 while "bundling together" the different shifts and rotations. This was done by varying an auxiliary parameter α and assigning its value to t_x, t_y and θ. The use of this parameter decreases the amount of required computations, thus making the experiments faster and easier to interpret, while still keeping the essence of significantly varying shifts and rotations.
- *Radiometric variations* were introduced to mimic how an instrument would actually process a scene. To do this an image representing the "real" scene is convolved with a *point spread function* (PSF) (Lyon *et al.*, 1997). The PSF may or may not correspond to a specific sensor, but it is very important that it does not introduce any geometric warping to the image. In this chapter, we use a simple PSF that was constructed by convolving with itself a 512-by-512 image that was "black" except for the 5-by-5 "white" center. A similar approach for synthetic image generation was used in Stone *et al.* (2001) and Foroosh *et al.* (2002), where Gaussian point spread functions were applied. This general approach can potentially be used to synthesize various multisensor satellite data.
- *Gaussian noise* was added to emulate imperfections of optical systems and of models used in preprocessing of satellite data. The amount of additional noise is usually specified in terms of signal-to-noise ratio (SNR). The SNR of n dB is defined as:

$$n = 10 \cdot \log_{10} \frac{Var\,(\text{image})}{Var\,(\text{noise})}.$$

By using these transformations, two synthetic datasets were created:

(1) *Warping & noise* (or "SameRadNoisy"): The first dataset was created by combining geometric warping with noise. The auxiliary parameter α, defining the shifts and rotation, was varied between 0 and 1 (corresponding to shifts between 0 and 1 pixels, and rotations between $0°$ and $1°$), with a step of 0.025, while noise varies between -15 dB and 20 dB with a step of 1 dB. The parameter α was chosen relatively small, compared to the experiments reported in Zavorin and Le Moigne (2005), to ensure convergence in most cases. A total of 1476 ref-input pairs was generated for this dataset.

(2) *Warping & PSF* (or "DiffRadNoiseless"): The second dataset was created by combining geometric warping with radiometric variations using the PSF described above. Again shifts and rotation were varied, using the auxiliary parameter α, before the warped image was convolved with the PSF. A total of 3321 ref-input pairs was generated for this case.

Figures 14.2(a)–(c) show the original image and examples of the synthetic images created from it.

14.3.1.2 Multitemporal datasets

The multitemporal datasets have been acquired over two different areas: (1) The Washington DC/Baltimore area (Landsat WRS-2 Path 15, Row 33) and (2) Central Virginia (VA) (Path 15, Row 34). A multitemporal dataset of Landsat-5/Thematic Mapper (TM) and Landsat-7/Enhanced Thematic Mapper (ETM+) images was assembled for each region (see Table 14.1 and Netanyahu *et al.*, 2004). For each region, one ETM+ scene was picked as the "reference" scene; the systematic navigational information provided with the reference scenes was considered to be "the truth." A number of reference chips (eight 256×256 pixel subregions for Washington DC/Baltimore and six 256×256 pixel subregions for Central Virginia) were extracted from these reference scenes. For an operational system, it is reasonable to assume that a database would include between 5 and 10 well-distributed "reference chips" per Landsat scene; they are usually defined as small subimages representing well-contrasted visual landmarks, such as bridges, city grids, islands, or high-curvature points in coastlines, and correspond to cloud-free, different seasons and/or different reflectance conditions of each landmark area. With regards to our multitemporal datasets, we have only one reference chip for each landmark area, so only one season and one radiometry are available for reference; therefore these datasets present the following challenges. The Washington DC/Baltimore data involves multiple sensors (ETM+ and TM), and although the band definitions of these sensors are identical, their spectral responses are different; thus those scenes present a challenge due to spectral differences. On the

(a)   (b)

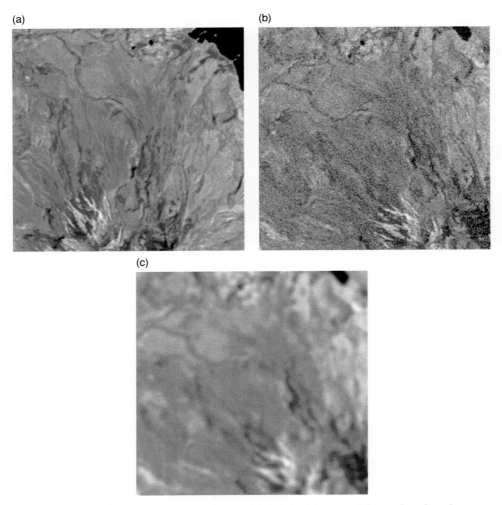

(c)

Figure 14.2. Synthetic image samples: (a) Original image, (b) warping & noise, and (c) warping & PSF. (Source: Le Moigne *et al.*, 2004, © IEEE, reprinted with permission.)

other hand, the Central Virginia dataset spans multiple seasons, and thus presents a challenge in matching features that have very different appearances due to seasonal effects.

All scenes, reference and input, were projected using a WGS-84 model (National Imagery and Mapping Agency, 2000). Using the Universal Transverse Mercator (UTM) coordinates of the four corners of each chip and the UTM coordinates of the four corners of each input scene, a corresponding window was extracted for each chip and each input scene. Figures 14.3 to 14.6 show four of the reference

Table 14.1 *Multitemporal Landsat datasets (source: Netanyahu* et al., *2004,*
© *IEEE, reprinted with permission)*

Location	Acquisition date	Platform/sensor
Washington, DC (P15 R33)	*July 28, 1999 (990728) (reference)*	*Landsat-7 ETM+*
	August 27, 1984 (840827)	Landsat-5 TM
	May 16, 1987 (870516)	Landsat-5 TM
	August 12, 1990 (900812)	Landsat-5 TM
	July 11, 1996 (960711)	Landsat-5 TM
Central Virginia (P15 R34)	*October 7, 1999 (991007) (reference)*	*Landsat-7 ETM+*
	August 4, 1999 (990804)	Landsat-7 ETM+
	November 8, 1999 (991108)	Landsat-7 ETM+
	February 28, 2000 (000228)	Landsat-7 ETM+
	August 22, 2000 (000822)	Landsat-7 ETM+

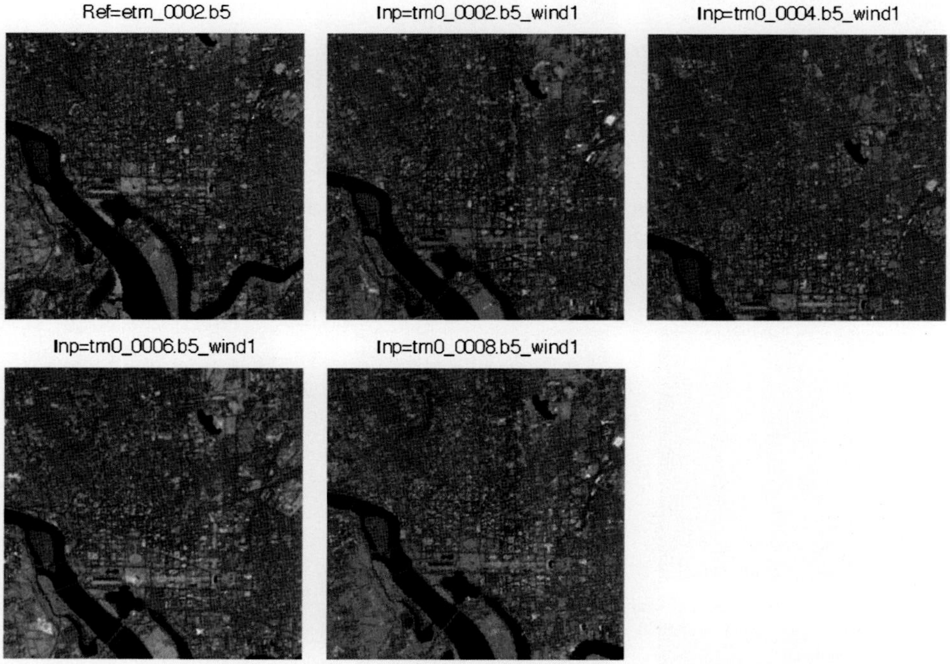

Figure 14.3. Washington DC/Baltimore area: Landsat multitemporal dataset. A
reference chip and four input subwindows.

chips and, for each, four corresponding windows from the input scenes, for both
the Washington DC/Baltimore and the Central Virginia areas.

For the experiments presented below, each input Landsat-5/ETM or Landsat-
7/ETM+ window is registered to its corresponding chip. In our work, we also
assume that the transformation between incoming Landsat scenes and reference

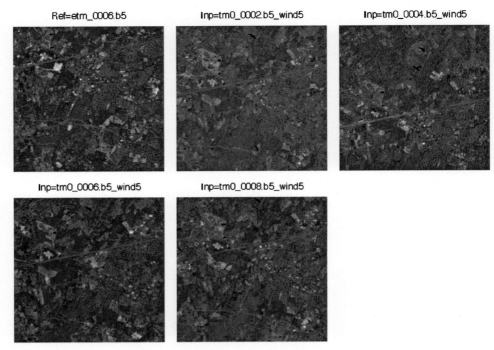

Figure 14.4. Washington DC/Baltimore area: Landsat multitemporal dataset. Another reference chip and four input subwindows.

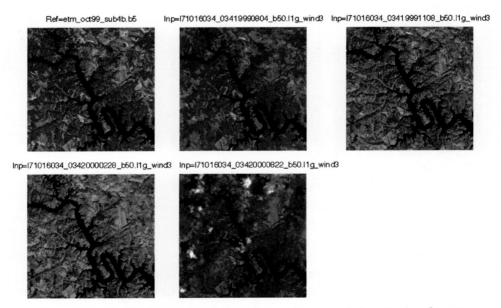

Figure 14.5. Central Virginia area: Landsat multitemporal dataset. A reference chip and four input subwindows.

Table 14.2 *Manual registration of all multitemporal datasets (source: Netanyahu et al., 2004, © IEEE, reprinted with permission)*

Reference scenes	Input scenes	Manual ground truth		
		θ	t_x	t_y
DC – 990728	840827	0.026	5.15	46.26
	870516	0.034	8.58	45.99
	900812	0.029	15.86	33.51
	960711	0.031	8.11	103.18
VA – 991007	990804	0.002	0.04	3.86
	991108	0.002	1.20	13.53
	000228	0.008	1.26	2.44
	000822	0.011	0.35	9.78

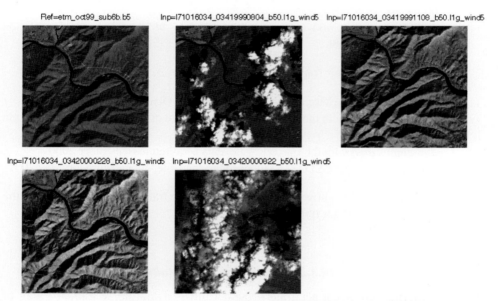

Figure 14.6. Central Virginia area: Landsat multitemporal dataset. Another reference chip and four input subwindows.

chips is limited to the composition of a rotation and a translation. Then, for each pair of scenes, a global registration can be computed with a generalized robust estimator that combines all previous local registrations (see Netanyahu *et al.*, 2004, for more details). Manual registration is available for this dataset to compute algorithm accuracies; see Table 14.2. According to the manual ground truth, the DC datasets present much larger transformations, with rotations of about

0.03 radians and shifts varying between 33 and 103 pixels in the vertical direction. On the other hand, the transformations of the VA datasets have rotations ranging from 0.002 radians to 0.011 radians, with the largest translation shift of about 13 pixels.

14.3.1.3 Multisensor datasets

The multisensor datasets used for this study were acquired by four different sensors over four of the MODIS Validation Core Sites (Morisette *et al.*, 2002). The four sensors and their respective bands and spatial resolutions are:

(1) IKONOS bands 3 (red; 632–698 nm) and 4 (near-infrared (NIR); 757–853 nm), at a spatial resolution of 4 meters per pixel.
(2) Landsat-7/ETM+ bands 3 (red; 630–690 nm) and 4 (NIR; 750–900 nm), at a spatial resolution of 30 meters per pixel.
(3) MODIS bands 1 (red; 620–670 nm) and 2 (NIR; 841–876 nm), at a spatial resolution of 500 meters per pixel.
(4) SeaWiFS bands 6 (red; 660–680 nm) and 8 (NIR; 845–885 nm), at a spatial resolution of 1000 meters per pixel.

The four sites represent four different types of terrain in the United States:

(1) *A coastal area* with the Virginia site, data acquired in October 2001.
(2) *An agricultural area* with the Konza Prairie in the state of Kansas, data acquired from July to August 2001.
(3) *A mountainous area* with the Cascades site, data acquired in September 2000.
(4) *An urban area* with the USDA, Greenbelt, Maryland site, data acquired in May 2001.

Figures 14.7 to 14.9 show some examples of extracted subimages from the IKONOS, Landsat, and MODIS sensors.

14.3.2 Experiments and results

For all experiments, when accurate ground truth is available, a standard way of assessing registration accuracy is by using the root mean square (RMS) error. Details of how to compute RMS are given in Zavorin and Le Moigne (2005), but briefly, if a ground truth transformation is given by $(t_{x_1}, t_{y_1}, \theta_1)$ and a computed transformation is given by $(t_{x_2}, t_{y_2}, \theta_2)$, then the RMS error is given by the following:

$$E = \frac{(N_x^2 + N_y^2)}{3} \cdot 2\cos\theta_\varepsilon + \left(t_{x_\varepsilon}^2 + t_{y_\varepsilon}^2\right)$$
$$+ (N_x t_{x_\varepsilon} + N_y t_{y_\varepsilon})\cos\theta_\varepsilon - (N_x t_{y_\varepsilon} - N_y t_{x_\varepsilon})\cos\theta_\varepsilon, \quad (14.1)$$

ETM+

IKONOS

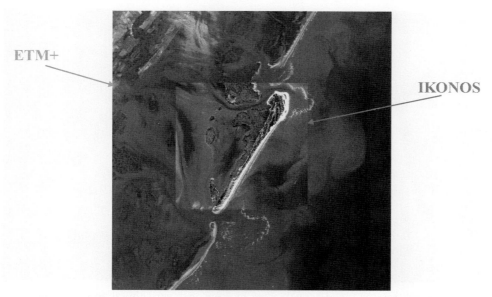

Figure 14.7. ETM+ and IKONOS data of the Virginia coastal area. See color plates section. (Source: Le Moigne *et al.*, 2004, © IEEE, reprinted with permission.)

ETM+

IKONOS

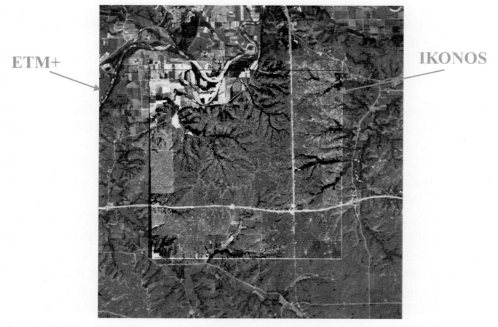

Figure 14.8. ETM+ and IKONOS data of the Cascades mountainous area. See color plates section. (Source: Le Moigne *et al.*, 2004, © IEEE, reprinted with permission.)

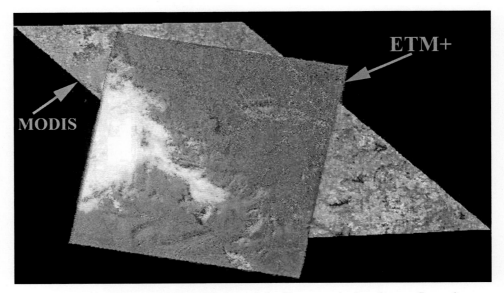

Figure 14.9. ETM+ and MODIS data of the Konza agricultural area. See color plates section.

where $(t_{x_\varepsilon}, t_{y_\varepsilon}, \theta_\varepsilon)$ represents the "error" transformation between $(t_{x_1}, t_{y_1}, \theta_1)$ and $(t_{x_2}, t_{y_2}, \theta_2)$, N_x represents the size of the images in the x direction, and N_y represents the size of the images in the y direction.

14.3.2.1 Synthetic data experiments

Using the two synthetic datasets described in Subsection 14.3.1.1, we tested four of the algorithms defined in Subsection 14.2.3, namely FFC, TRU, TRUMI, and SPSA. For each of the three last algorithms, we utilized three wavelet-like types of features – spline (*SplC*), Simoncelli low-pass (*SimL*) and Simoncelli band-pass (*SimB*). For all three variations of all algorithms, the initial guess was chosen as $(t_x, t_y, \theta) = (0, 0, 0)$, and four pyramid levels were utilized.

Accuracy is measured using Eq. (14.1), so that the RMS error is obtained as a function of either shifts, rotation, and noise (for the first group of datasets), or as a function of shifts, rotation, and radiometric difference (for the second group). Each time an algorithm is run on a pair of images, the resulting RMS error is computed and compared to several thresholds, {0.025, 0.05, 0.075, 0.1, 0.2, 0.25, 0.5, 0.75, 0.75, 1}. Table 14.3 shows the results obtained in this experiment, where for each data type, algorithm, and threshold value, an integer value between 0 and 100 represents the percentage of cases for which the RMS error was below the corresponding threshold value, out of all cases tested. Essentially, the bigger the

Table 14.3 *Summary of experimental results for synthetic data; percentage of cases for which RMS is below threshold*

Thresh	TRU SplC	TRU SimB	TRU SimL	TRUMI SplC	TRUMI SimB	TRUMI SimL	SPSA SplC	SPSA SimB	SPSA SimL	FFC Gray
SameRadNoisy										
0.025	*27*	31	**35**	1	0	2	2	3	12	10
0.05	*35*	**47**	41	25	26	29	11	16	28	33
0.075	39	**53**	*43*	33	42	41	18	26	37	44
0.1	42	**59**	44	41	52	45	23	34	41	*51*
0.2	51	**75**	50	61	70	*62*	45	52	54	58
0.25	53	**80**	53	*67*	74	65	50	58	58	60
0.5	63	**91**	63	*81*	85	75	63	72	75	67
0.75	68	**95**	68	*83*	90	80	70	80	83	70
1	74	**97**	72	83	93	83	74	83	*88*	71
DiffRadNoiseless										
0.025	**1**	2	0	**1**	1	1	1	**3**	1	1
0.05	1	**28**	0	5	5	*9*	1	19	6	3
0.075	1	**47**	0	17	14	*19*	5	39	18	9
0.1	1	**63**	0	27	23	29	19	54	*44*	20
0.2	1	86	1	59	51	60	*85*	80	**98**	41
0.25	20	86	17	72	62	76	89	86	**100**	43
0.5	83	86	85	86	74	97	90	92	**100**	50
0.75	88	86	91	86	74	97	90	92	**100**	56
1	90	86	*93*	86	74	97	90	92	**100**	60

Legend: **Best** (bold); second best (underlined); *third best* (italic).

number the better is the algorithm. The table also shows, for each threshold, the best algorithm (in bold), the second best (underlined), and the third best one (in italic).

In summary, the results show that for the "SameRadNoisy" dataset, TRU with SimB features performs consistently the best for nearly all thresholds, with an accuracy of 0.25 pixels 80% of the time, and TRUMI using SimB features being second best most of the time. For the "DiffRadNoiseless" dataset, TRU with SimB is best for smaller thresholds, which means that when it converges, it is more accurate, but SPSA with SimL converges more often for higher thresholds and we can say that it reaches accuracies of 0.2 pixels with a 98% probability. Overall, these results show that:

(1) Simoncelli-based methods outperform those with the spline pyramid.
(2) TRUMI (based on mutual information) does not really perform better than TRU (based on an L_2-norm).

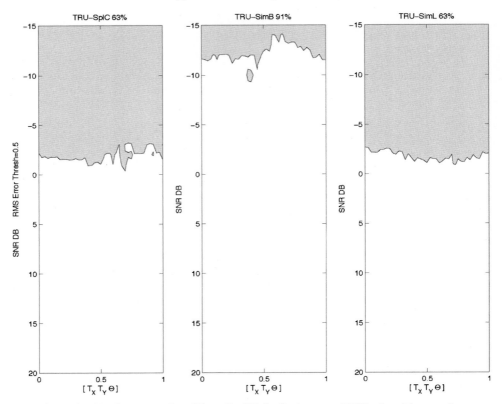

Figure 14.10. Contour plot "SameRadNoisy" dataset; TRU algorithm and a threshold of 0.5.

(3) SimL performs better than SimB for images of different radiometry, but overall all other algorithms seem to perform more poorly for different radiometries than for noisy conditions.

Figures 14.10 to 14.13 show the contour plots corresponding to the results obtained by the four algorithms for the "SameRadNoisy" dataset and a threshold of 0.5, where the white areas depict the regions of convergence of the algorithms with an error less than the threshold. As expected from the results shown in Table 14.3, the plots show that TRU used with SimB (Fig. 14.10) has the largest convergence region, followed by TRUMI with any features (Fig. 14.11). Similarly, Figs. 14.14–14.17 (pages 314 to 317) show the contour plots for the "DiffRadNoiseless" dataset and a threshold of 0.2. Again, as expected from Table 14.3, the plots show that for this dataset, SPSA has the largest region of convergence for all types of features, followed by TRU-SimB.

These experiments do not include a study of the sensitivity of the different algorithms to the initial conditions, but previous results reported in Le Moigne

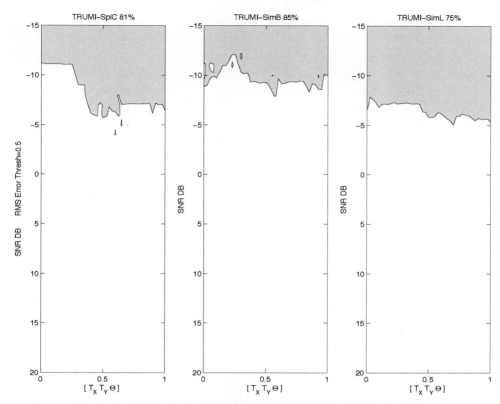

Figure 14.11. Contour plot "SameRadNoisy" dataset; TRUMI algorithm and a threshold of 0.5.

et al. (2004) showed that SimB was more sensitive to the initial guess than SplC or SimL and that SPSA was more robust to initial conditions than TRU or TRUMI.

14.3.2.2 Multitemporal experiments

FFC and optimization-based methods Similarly to the experiments performed on the synthetic datasets, for both DC and VA datasets, we compare the four algorithms, FFC, TRU, TRUMI, and SPSA, using the three types of wavelets, SplC, SimB, and SimL for the three latter ones. In this case, not only do we compare the accuracy of the different algorithms but we also assess the sensitivity of the optimization-based methods to the initial conditions, by setting the initial guess of the ground truth to the values given in Table 14.2; that is, if $(t_{x_0}, t_{y_0}, \theta_0)$ is the ground truth between Scenes 1 and 2, the registration is started with the initial guess $(d \cdot t_{x_0}, d \cdot t_{y_0}, d \cdot \theta_0)$ with d taking the successive values $\{0.0, 0.1, 0.2, 0.3, \ldots, 0.9, 1.0\}$.

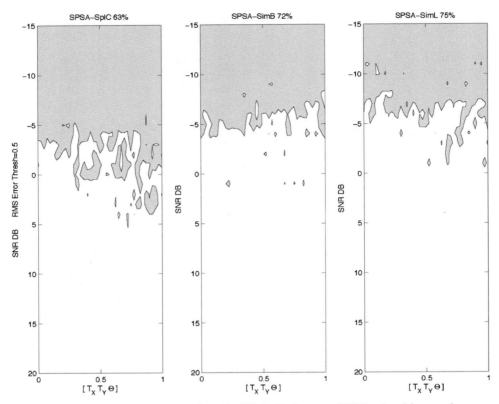

Figure 14.12. Contour plot "SameRadNoisy" dataset; SPSA algorithm and a threshold of 0.5.

Results show that for the VA dataset, all four algorithms perform well. There is very little difference between their accuracy regardless of the initial guess or of the wavelet type used. Tables 14.4a–14.4d (pages 318 to 321) show these results for $d = 0.0$ and the four algorithms.

For the DC dataset, unlike the results obtained for VA, we observe that TRU, TRUMI, and SPSA exhibit significant sensitivity to the initial guess, while FFC is essentially insensitive and produces overall the best results. Table 14.5 (page 322) shows the numbers of correct runs, out of the total of 32 (4 scenes by 8 chip-window pairs) for each algorithm, each pyramid type, and each value of d between 0.0 and 1.0. The results show the following:

- Except for FFC, which is the most insensitive to the initial guess, the algorithms improve performance more or less monotonically as the initial guess gets closer to the ground truth.

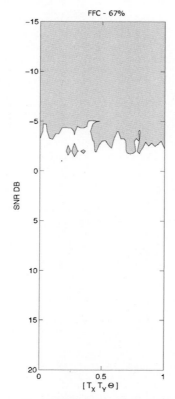

Figure 14.13. Contour plot "SameRadNoisy" dataset; FFC algorithm and a threshold of 0.5.

- TRU, TRUMI, and SPSA are comparable in terms of initial guess sensitivity, with SPSA performing slightly better than the other two.
- Among the three pyramids, SimL seems to perform best in terms of initial guess sensitivity.

These very different results between the DC dataset and the VA dataset might be explained by the characteristics described for the DC area, different sensors and different seasons, as well as by the fact that the DC images tend to have higher frequencies than the VA images. When the images contain high frequencies there is more probability for the optimization algorithms to fall into a local optimum, especially when the initial guess is not very close to the correct solution. Also, FFC is not a local algorithm, it finds the best correlation for each chip, wherever that correlation lies in the image; the algorithm diminishes the effect of false features by correlating a large number of chips, and removing the outliers. The difference

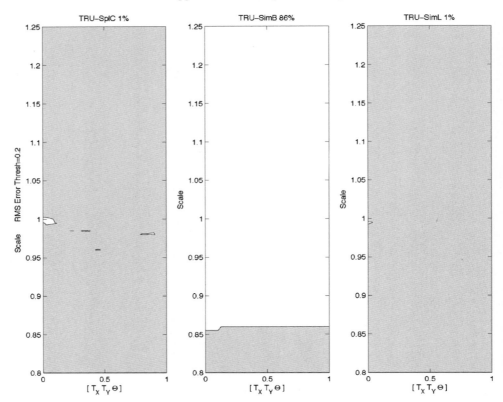

Figure 14.14. Contour plot "DiffRadNoiseless" dataset; TRU algorithm and a threshold of 0.2.

in results between the VA and DC datasets can also be explained by the amount of shifts between reference and input scenes in the DC area being much larger than the ones found in the Virginia area (probably due to the fact that the input DC scenes were acquired by Landsat-5 and the reference by Landsat-7, which has much better navigation capabilities).

Robust feature matching Although the RFM algorithm was not tested simultaneously with the previous algorithms on the multitemporal datasets, results previously obtained in Netanyahu *et al.* (2004), which are summarized in Tables 14.6a and 14.6b (page 322), show the following:

- The rotation angle obtained in each case is very small (on the order of a few hundredths of a degree at most). Thus the affine transformation computed can be viewed as essentially "translation only," which is in accordance with Landsat's specifications.

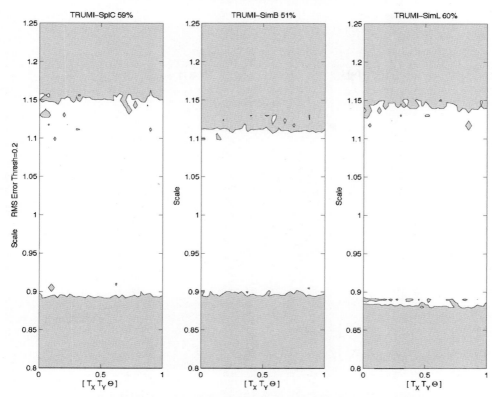

Figure 14.15. Contour plot "DiffRadNoiseless" dataset; TRUMI algorithm and a threshold of 0.2.

- There is a very good agreement between the transformation parameters obtained by RFM and the ground truth shown in Table 14.2. The average errors for the shifts in *x* and *y* were 0.21 and 0.59, respectively, for the DC scenes, and 0.26 and 0.49, respectively, for the Virginia scenes.

RFM was not studied for its sensitivity to initial conditions and this will need to be investigated and compared to the other methods in the future. More details on this method can be found in Netanyahu *et al.* (2004) and in Chapter 8.

14.3.2.3 Multisensor experiments

Algorithm comparison For these experiments, multisensor registrations were performed in "cascade": IKONOS to ETM+, ETM+ to MODIS, and MODIS to SeaWiFS. Wavelet decomposition was utilized, not only to compute registration features, but also to bring various spatial resolution data to similar resolutions, by

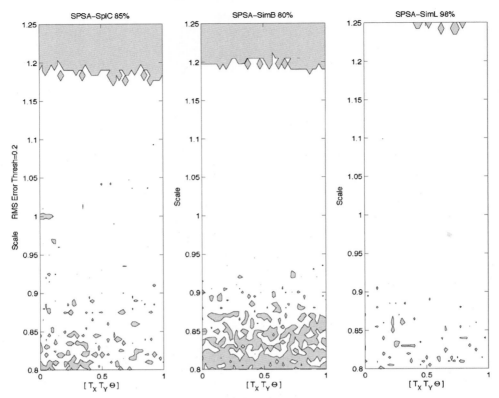

Figure 14.16. Contour plot "DiffRadNoiseless" dataset; SPSA algorithm and a threshold of 0.2.

performing recursive decimation by 2. For example, after three levels of wavelet decomposition, the IKONOS spatial resolution was brought to 32 meters that, compared to the Landsat spatial resolution, corresponds to a scaling of about 1.07. This is the scaling expected when registering IKONOS to Landsat data.

For all scenes, we extracted subimages from the original images so that their dimensions in x and y were multiples of 2^L, where L is the maximum number of wavelet decomposition levels used in the registration process.

We first performed a comparison of the algorithms with the Konza agricultural dataset, using exhaustive search: in order to simplify this comparison, we resampled the IKONOS and ETM+ data to the respective spatial resolutions of 3.91 and 31.25 meters, using the commercial software, PCI®. This slight alteration in the resolution of the data enables us to obtain compatible spatial resolutions by performing recursive decimation by 2 of the wavelet transform, and therefore to only search for translations and rotations. Overall, we considered eight different

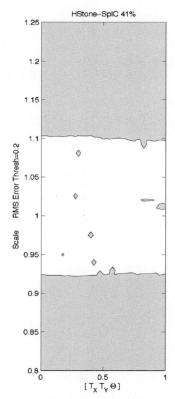

Figure 14.17. Contour plot "DiffRadNoiseless" dataset; FFC algorithm and a threshold of 0.2.

subimages, two for each band of the four sensors. We then performed manual registration for two pairs of data (the two bands of IKONOS and ETM data) and found the transformation $t_x = 2$, $t_y = 0$, and $\theta = 0°$. Then, five methods were applied: (a) FFC, (b) GGD, defined in Subsection 14.2.3, (c) exhaustive search using Simoncelli band-pass and correlation, (d) exhaustive search using Simoncelli band-pass and mutual information, and (e) RFM, defined in Subsection 14.2.3. Results are shown in Table 14.7 (page 323). All methods confirm the coregistration of the two ETM bands, red and NIR; Methods (b), (c), and (d) all found the correct transformation for the registration of IKONOS to ETM, while Methods (a) and (e) had a 1- or 2-pixel misregistration; all algorithms discovered a misregistration between the MODIS and the SeaWiFS datasets, which was then confirmed manually to be -8 pixels in the x direction. Overall, all results obtained by the five algorithms were similar within $0.5°$ in rotation and 1 pixel in translation. More details can be found in Le Moigne *et al.* (2001).

Table 14.4a *Results of multitemporal experiments for the Central Virginia area using TRU and an initial guess of (0, 0, 0)*

TRU algorithm	SplC				SimB				SimL			
	t_x	t_y	θ	RMS error	t_x	t_y	θ	RMS error	t_x	t_y	θ	RMS error
990804												
Chip1	0.312	−3.650	0.026	0.280	0.365	−3.711	0.008	0.328	0.257	−3.565	0.072	0.282
Chip2	0.089	−3.897	0.029	0.229	0.110	−3.924	0.007	0.255	0.078	−3.917	0.030	0.247
Chip3	2.249	−1.214	−1.248	3.941	0.313	−3.870	−0.024	0.334	2.582	−1.898	−1.437	3.962
Chip4	0.035	−3.928	0.021	0.251	0.031	−3.902	0.016	0.224	0.067	−3.958	0.019	0.282
Chip5	−0.082	−4.399	−0.431	1.060	0.270	−3.791	−0.014	0.255	−0.354	−6.767	−1.961	4.691
Chip6	0.309	−3.875	−0.231	0.519	0.111	−3.919	−0.055	0.266	0.408	−4.192	−0.067	0.639
MEDIAN	**0.199**	**−3.886**	**−0.105**	**0.400**	**0.190**	**−3.886**	**−0.004**	**0.260**	**0.167**	**−3.937**	**−0.024**	**0.461**
991108												
Chip1	1.137	−13.535	0.012	0.064	1.130	−13.526	0.030	0.081	1.097	−13.526	0.012	0.102
Chip2	1.416	−13.669	0.007	0.258	1.407	−13.678	0.003	0.255	1.429	−13.696	−0.001	0.282
Chip3	1.381	−13.599	0.015	0.198	1.363	−13.600	0.026	0.187	1.373	−13.587	0.017	0.187
Chip4	1.322	−13.404	−0.005	0.175	1.298	−13.405	0.001	0.158	1.307	−13.367	−0.012	0.194
Chip5	1.270	−13.704	−0.002	0.187	1.254	−13.702	0.000	0.180	1.222	−13.727	−0.007	0.199
Chip6	1.229	−13.481	−0.018	0.064	1.208	−13.451	−0.009	0.081	1.234	−13.458	−0.026	0.091
MEDIAN	**1.296**	**−13.567**	**0.003**	**0.181**	**1.276**	**−13.563**	**0.002**	*0.169*	**1.270**	**−13.557**	**−0.004**	**0.191**
000228												
Chip1	−1.177	−2.528	−0.019	0.129	−1.145	−2.517	0.005	0.138	−1.204	−2.544	−0.030	0.134
Chip2	−1.136	−2.692	−0.002	0.281	−1.132	−2.682	−0.016	0.276	−1.106	−2.684	−0.007	0.289
Chip3	−1.096	−2.535	0.027	0.193	−1.117	−2.546	0.041	0.189	−1.031	−2.480	0.010	0.233
Chip4	−1.051	−2.287	0.019	0.260	−1.076	−2.277	0.014	0.246	−1.028	−2.260	0.014	0.294
Chip5	−1.112	−2.753	0.019	0.347	−1.120	−2.739	0.021	0.331	−1.105	−2.797	0.019	0.391
Chip6	−1.149	−2.365	0.016	0.135	−1.145	−2.337	0.012	0.155	−1.085	−2.503	0.048	0.202
MEDIAN	**−1.124**	**−2.531**	**0.017**	**0.227**	**−1.126**	**−2.531**	**0.013**	*0.218*	**−1.095**	**−2.524**	**0.012**	**0.261**
000822												
Chip1	−3.152	−11.838	1.588	4.749	0.433	−9.565	0.047	0.242	−19.539	−13.491	8.162	23.808
Chip2	0.164	−9.774	0.005	0.188	0.166	−9.798	0.009	0.185	0.111	−9.779	−0.007	0.244
Chip3	0.356	−9.642	−0.024	0.151	0.385	−9.674	−0.010	0.117	0.309	−9.648	−0.027	0.156
Chip4	0.432	−9.329	0.026	0.460	0.432	−9.362	0.024	0.427	0.441	−9.286	0.013	0.503
Chip5	0.358	−9.667	0.022	0.115	0.327	−9.705	0.024	0.081	0.334	−9.710	0.023	0.075
Chip6	1.512	−21.277	−1.139	11.727	−2.589	−17.576	−2.707	9.809	1.026	−21.184	−2.552	12.310
MEDIAN	**0.357**	**−9.720**	**0.013**	**0.324**	**0.356**	**−9.689**	**0.016**	*0.213*	**0.322**	**−9.744**	**0.003**	**0.373**

Legend: Best Median RMS error shown in italics.

318

Table 14.4b *Results of multitemporal experiments for the Central Virginia area using TRUMI and an initial guess of (0, 0, 0)*

TRUMI algorithm	SplC				SimB				SimL			
	t_x	t_y	θ	RMS error	t_x	t_y	θ	RMS error	t_x	t_y	θ	RMS error
990804												
Chip1	0.407	−3.670	−0.014	0.367	0.306	−3.721	−0.010	0.269	0.471	−3.582	−0.008	0.441
Chip2	0.176	−3.956	0.020	0.311	0.121	−3.959	0.016	0.293	0.132	−3.940	0.010	0.276
Chip3	0.262	−3.826	0.023	0.271	21.653	−232.311	0.249	229.652	0.222	−3.834	0.016	0.241
Chip4	0.063	−3.945	0.017	0.268	0.025	−3.909	0.017	0.232	0.043	−3.981	0.010	0.302
Chip5	0.246	−3.886	−0.066	0.310	0.336	−3.820	−0.013	0.327	0.292	−3.691	−0.106	0.307
Chip6	0.213	−4.004	0.016	0.369	5.273	−12.356	−5.197	13.609	0.235	−4.255	0.077	0.624
MEDIAN	**0.230**	**−3.915**	**0.016**	**0.311**	**0.321**	**−3.934**	**0.003**	**0.310**	**0.229**	**−3.887**	**0.010**	***0.304***
991108												
Wind1	1.249	−13.508	0.015	0.061	1.151	−13.509	0.017	0.057	1.133	−13.483	0.000	0.083
Wind2	1.403	−13.633	0.007	0.228	1.397	−13.694	0.012	0.258	1.419	−13.724	0.014	0.295
Wind3	1.393	−13.576	0.008	0.200	1.377	−13.592	0.019	0.194	1.399	−13.615	0.004	0.216
Wind4	1.355	−13.373	−0.005	0.219	1.300	−13.394	−0.002	0.168	1.328	−13.358	−0.011	0.214
Wind5	1.302	−13.655	0.007	0.162	1.248	−13.706	0.005	0.182	1.234	−13.735	−0.009	0.209
Wind6	1.083	1.016	−0.049	14.545	4.528	−1.524	−0.587	12.454	1.191	−13.468	−0.024	0.078
MEDIAN	**1.328**	**−13.542**	**0.007**	**0.210**	**1.338**	**−13.551**	**0.008**	***0.188***	**1.281**	**−13.549**	**−0.004**	**0.211**
000228												
Chip1	−1.155	−2.512	−0.009	0.130	−1.116	−2.510	−0.001	0.160	−1.162	−2.498	−0.030	0.131
Chip2	−1.179	−2.749	−0.002	0.319	−1.149	−2.725	−0.006	0.306	−1.165	−2.783	0.005	0.356
Chip3	−1.119	−2.660	0.013	0.262	−1.120	−2.598	0.039	0.219	−1.122	−2.665	0.008	0.263
Chip4	−1.031	−2.332	0.015	0.254	−1.064	−2.274	0.015	0.257	−1.002	−2.278	0.000	0.305
Chip5	−1.130	−2.792	0.014	0.376	−1.131	−2.766	0.024	0.352	−1.188	−2.875	0.009	0.441
Chip6	−1.142	−2.417	0.008	0.120	−1.168	−2.312	0.019	0.159	−1.181	−2.399	0.005	0.089
MEDIAN	**−1.136**	**−2.586**	**0.010**	**0.258**	**−1.125**	**−2.554**	**0.017**	***0.238***	**−1.164**	**−2.581**	**0.005**	**0.284**
000822												
Chip1	0.517	−9.525	0.039	0.312	0.436	−9.530	0.042	0.272	6.938	−48.263	10.038	43.298
Chip2	0.208	−9.815	0.005	0.148	0.169	−9.801	0.002	0.184	0.207	−9.830	0.004	0.153
Chip3	0.409	−9.692	−0.006	0.108	0.408	−9.695	−0.006	0.105	0.387	−9.684	−0.009	0.107
Chip4	0.467	−9.430	0.028	0.371	0.431	−9.407	0.024	0.383	0.469	−9.505	0.003	0.299
Chip5	0.350	−9.684	0.011	0.096	0.322	−9.712	0.023	0.076	0.330	−9.734	0.012	0.050
Chip6	−0.153	1.364	0.367	11.173	−48.823	−1.877	0.142	49.783	−10.486	7.762	−3.828	22.020
MEDIAN	**0.379**	**−9.604**	**0.020**	**0.230**	**0.365**	**−9.612**	**0.023**	**0.228**	**0.358**	**−9.709**	**0.003**	***0.226***

Legend: Best Median RMS error shown in italics.

Table 14.4c *Results of multitemporal experiments for the Central Virginia area using SPSA and an initial guess of (0, 0, 0)*

SPSA algorithm		SplC t_x	SplC t_y	SplC θ	SplC RMS error	SimB t_x	SimB t_y	SimB θ	SimB RMS error	SimL t_x	SimL t_y	SimL θ	SimL RMS error
990804													
Chip1	Wind1	0.390	−3.630	−0.002	0.354	0.284	−3.705	−0.017	0.246	0.440	−3.642	−0.010	0.401
Chip2	Wind2	0.178	−3.863	0.023	0.234	0.117	−3.940	0.015	0.273	0.106	−3.962	0.010	0.291
Chip3	Wind3	0.234	−3.883	0.022	0.285	0.218	−3.880	0.003	0.268	0.235	−3.868	0.019	0.275
Chip4	Wind4	0.032	−3.941	0.000	0.261	0.027	−3.898	0.019	0.221	0.018	−3.980	0.009	0.302
Chip5	Wind5	0.328	−3.805	−0.008	0.313	0.302	−3.781	−0.003	0.281	0.318	−3.819	−0.004	0.311
Chip6	Wind6	0.147	−4.026	0.012	0.364	0.132	−3.906	−0.007	0.245	0.244	−4.040	0.029	0.419
MEDIAN		**0.206**	**−3.873**	**0.006**	**0.299**	**0.175**	**−3.889**	**0.000**	*0.257*	**0.240**	**−3.915**	**0.010**	**0.306**
991108													
Chip1	Wind1	1.218	−13.519	0.034	0.064	1.148	−13.517	0.018	0.057	1.139	−13.479	0.016	0.081
Chip2	Wind2	1.388	−13.666	0.033	0.244	1.398	−13.687	−0.001	0.252	1.412	−13.722	0.013	0.289
Chip3	Wind3	1.380	−13.619	0.025	0.210	1.361	−13.613	0.019	0.187	1.390	−13.613	0.020	0.214
Chip4	Wind4	1.335	−13.424	−0.008	0.170	1.294	−13.402	−0.009	0.158	1.315	−13.373	−0.017	0.195
Chip5	Wind5	1.245	−13.735	0.004	0.210	1.244	−13.726	0.007	0.201	1.235	−13.765	−0.004	0.238
Chip6	Wind6	1.193	−13.519	−0.009	0.024	1.193	−13.472	−0.004	0.060	1.219	−13.504	−0.010	0.037
MEDIAN		**1.290**	**−13.569**	**0.015**	*0.190*	**1.269**	**−13.565**	**0.003**	*0.173*	**1.275**	**−13.559**	**0.004**	**0.204**
000228													
Chip1	Wind1	−1.156	−2.503	−0.020	0.130	−1.141	−2.506	0.003	0.136	−1.139	−2.507	−0.023	0.147
Chip2	Wind2	−1.203	−2.756	−0.006	0.322	−1.150	−2.733	−0.012	0.314	−1.141	−2.776	−0.006	0.357
Chip3	Wind3	−1.189	−2.616	0.013	0.190	−1.124	−2.612	0.044	0.230	−1.122	−2.675	0.016	0.273
Chip4	Wind4	−1.071	−2.303	0.012	0.234	−1.056	−2.292	0.020	0.253	−1.028	−2.290	0.005	0.277
Chip5	Wind5	−1.188	−2.800	0.040	0.372	−1.147	−2.773	0.027	0.354	−1.170	−2.855	0.021	0.426
Chip6	Wind6	−1.175	−2.396	0.015	0.097	−1.190	−2.321	0.019	0.140	−1.187	−2.377	0.008	0.096
MEDIAN		**−1.181**	**−2.560**	**0.013**	*0.212*	**−1.144**	**−2.559**	**0.020**	**0.241**	**−1.140**	**−2.591**	**0.007**	**0.275**
000822													
	Wind1	0.378	−9.499	0.003	0.283	0.357	−9.456	−0.015	0.328	0.482	−9.447	0.022	0.360
	Wind2	0.201	−9.842	0.010	0.161	0.185	−9.825	−0.005	0.176	0.185	−9.843	−0.001	0.180
	Wind3	0.418	−9.645	−0.005	0.153	0.373	−9.705	−0.001	0.080	0.392	−9.673	−0.002	0.117
	Wind4	0.408	−9.459	0.021	0.327	0.414	−9.369	0.022	0.417	0.428	−9.456	0.015	0.334
	Wind5	0.347	−9.741	0.016	0.041	0.318	−9.709	0.025	0.080	0.323	−9.729	0.019	0.059
	Wind6	−4.301	5.894	−1.409	16.603	−18.685	9.732	0.642	27.209	−9.858	11.557	0.500	23.636
MEDIAN		**0.362**	**−9.572**	**0.007**	*0.222*	**0.337**	**−9.581**	**0.010**	**0.252**	**0.358**	**−9.565**	**0.017**	**0.257**

Legend: Best Median RMS error shown in italics.

320

Table 14.4d *Results of multitemporal experiments for the Central Virginia area using FFC and an initial guess of (0, 0, 0)*

FFC Algorithm		t_x	t_y	θ	RMS error
990804					
Chip1	Wind1	0.218	−3.724	−0.053	0.204
Chip2	Wind2	0.038	−4.001	−0.008	0.322
Chip3	Wind3	0.056	−4.030	0.025	0.358
Chip4	Wind4	−0.001	−3.975	0.007	0.298
Chip5	Wind5	0.145	−4.022	−0.009	0.382
Chip6	Wind6	−0.022	−4.115	0.006	0.455
	MEDIAN	**0.047**	**−4.011**	**−0.001**	*0.340*
991108					
Chip1	Wind1	1.152	−13.537	0.007	0.114
Chip2	Wind2	1.297	−13.782	0.035	0.294
Chip3	Wind3	1.286	−13.675	0.070	0.247
Chip4	Wind4	1.138	−13.267	0.005	0.285
Chip5	Wind5	1.140	−13.913	0.027	0.398
Chip6	Wind6	0.975	−13.296	0.014	0.324
	MEDIAN	**1.146**	**−13.606**	**0.020**	*0.289*
000228					
Chip1	Wind1	−1.234	−2.348	−0.084	0.191
Chip2	Wind2	−1.078	−2.899	−0.052	0.504
Chip3	Wind3	−1.057	−2.600	0.030	0.272
Chip4	Wind4	−1.017	−2.083	0.025	0.434
Chip5	Wind5	−0.986	−2.891	−0.013	0.529
Chip6	Wind6	−1.268	−2.187	−0.001	0.263
	MEDIAN	**−1.068**	**−2.474**	**−0.007**	*0.353*
000822					
Chip1	Wind1	0.524	−9.513	−0.021	0.406
Chip2	Wind2	0.053	−9.677	−0.104	0.398
Chip3	Wind3	0.186	−9.848	−0.059	0.227
Chip4	Wind4	0.614	−9.594	−0.106	0.370
Chip5	Wind5	0.127	−9.887	0.021	0.262
Chip6	Wind6	−126.590	−45.147	76.912	273.851
	MEDIAN	**0.157**	**−9.763**	**−0.040**	*0.384*

Legend: Best Median RMS error shown in italics.

We further performed registrations of all ETM and IKONOS multisensor data using five of the algorithms defined in Subsection 14.2.3: (a) FFC, (b) TRU with SplC, (c) TRU with SimB, (d) TRU with SimL, and (e) SPSA with SimB (Le Moigne *et al.*, 2003). Overall, for each site, six different registrations are performed, corresponding to inter- and intra-sensor registrations, including cross-spectral (multimodal) matching. Results are shown in Tables 14.8 and 14.9 for two of the sites, Cascades-Mountainous and Virginia-Coast. For this study, no exact

Table 14.5 *Number of cases that converged (out of 32) for the DC dataset, running four algorithms with the initial guess varying between the origin (d = 0.0) and ground truth (d = 1.0)*

d	TRU			TRUMI			SPSA			FFC
	SplC	SimB	SimL	SplC	SimB	SimL	SplC	SimB	SimL	
0.0	7	5	12	10	2	14	5	3	7	30
0.1	8	4	14	8	4	12	5	4	11	30
0.2	8	6	16	8	7	15	7	5	15	30
0.3	8	8	16	11	6	19	11	12	17	30
0.4	10	14	21	10	9	17	16	16	20	30
0.5	15	19	25	15	12	21	17	17	24	30
0.6	16	23	27	15	16	25	22	26	27	30
0.7	22	26	28	20	26	29	24	27	28	30
0.8	24	31	31	27	28	30	31	29	32	30
0.9	30	32	31	29	32	31	32	32	32	30
1.0	31	32	31	32	32	31	32	32	32	30

Table 14.6a *Global transformation versus ground truth parameters for the four scenes in the DC/Baltimore area. The rotation angle is in degrees. (Source: Netanyahu et al., 2004, © IEEE, reprinted with permission)*

Scene	RFM registration			Manual ground truth			Absolute error		
	θ	t_x	t_y	θ	t_x	t_y	$|\Delta\theta|$	$|\Delta t_x|$	$|\Delta t_y|$
840827	0.031	4.72	−46.88	0.026	5.15	−46.26	0.005	0.43	0.62
870516	0.051	8.49	−45.62	0.034	8.58	−45.99	0.017	0.09	0.37
900812	0.019	17.97	−33.36	0.029	15.86	−33.51	0.010	0.11	0.15
960711	0.049	8.34	−101.97	0.031	8.11	−103.18	0.018	0.23	1.21

Table 14.6b *Global transformation versus ground truth parameters for the four scenes in the Virginia area. The rotation angle is in degrees. (Source: Netanyahu et al., 2004, © IEEE, reprinted with permission)*

Scene	RFM registration			Manual ground truth			Absolute error		
	θ	t_x	t_y	θ	t_x	t_y	$|\Delta\theta|$	$|\Delta t_x|$	$|\Delta t_y|$
990804	0.009	0.36	3.13	0.002	0.04	3.86	0.011	0.40	0.73
991108	0.000	1.00	13.00	0.002	1.20	13.53	0.002	0.20	0.53
000228	0.005	0.88	−2.32	0.008	1.26	2.44	0.003	0.38	0.12
000822	0.002	0.41	9.22	0.011	0.35	9.78	0.013	0.06	0.56

Table 14.7 Results of multisensor registration for the Konza agricultural area using the four different algorithms

Pair to register	FFC		GGD		SimB-correl		SimB-MI		RFM	
	θ	(t_x, t_y)	θ	(t_x, t_y)	θ	(t_x, t_y)	θ	(t_x, t_y)	θ	(t_x, t_y)
ETM-NIR/ETM-red	Rotation = 0, translation = (0, 0) computed by all methods, using seven subwindow pairs									
IKO-NIR/ETM-NIR	—	(2, 1)	0.0001	(1.99, −0.06)	0	(2, 0)	0	(2, 0)	0.00	(0.0, 0.0)
IKO-red/ETM-red	—	(2, 1)	−0.0015	(1.72, 0.28)	0	(2, 0)	0	(2, 0)	0.00	(0.0, 0.0)
ETM-NIR/MODIS-NIR	—	(−2, −4)	0.0033	(−1.78, −3.92)	0	(−2, −4)	0	(−2, −4)	0.00	(−3.0, 3.5)
ETM-red/MODIS-red	—	(−2, −4)	0.0016	(1.97, −3.90)	0	(−2, −4)	0	(−2, −4)	0.00	(−2.0, −3.5)
MODIS-NIR/SeaWiFS-NIR	—	(−9, 0)	0.0032	(−8.17, 0.27)	0	(−8, 0)	0	(−9, 0)	0.50	(−6.0, 2.0)
MODIS-red/SeaWiFS-red	—	(−9, 0)	0.0104	(−7.61, 0.57)	0	(−8, 0)	0	(−8, 0)	0.25	(−7.0, 1.0)

Table 14.8 *Results of five algorithms on the Cascades-Mountainous area (initial guess = (1.0, 0.0, 0.0, 0.0))*

CASCADES	FFC	TRU/SplC	TRU/SimB	TRU/SimL	SPSA/SimB	Median
1. IKO-red/IKO-NIR						
Scale	1.000	1.000	1.000	1.000	1.000	*1.000*
θ	0.000	0.001	0.001	0.001	0.018	*0.001*
t_x	0.014	−0.024	−0.036	−0.046	0.020	*−0.024*
t_y	0.014	−0.160	−0.183	−0.209	0.054	*−0.160*
2. IKO-red/ETM-red						
Scale	1.064	1.138*	1.064	1.179*	1.065	*1.065*
θ	0.092	1.567*	0.074	2.542*	0.130	*0.130*
t_x	8.674	10.918*	8.652	8.993*	8.777	*8.777*
t_y	10.162	15.750*	10.044	11.330*	10.039	*10.162*
3. IKO-red/ETM-NIR						
Scale	1.065	1.064*	1.065	1.000*	1.064	*1.064*
θ	0.088	0.091*	0.084	0.000*	0.114	*0.088*
t_x	8.694	8.542*	8.641	0.000*	8.898	*8.641*
t_y	10.217	10.153*	10.129	0.000*	10.224	*10.153*
4. IKO-NIR/ETM-red						
Scale	1.064	1.097*	1.150*	no convrg*	1.066	*1.081*
θ	0.039	−1.153*	2.108*	no convrg*	0.128	*0.083*
t_x	8.562	13.130*	3.150*	no convrg*	8.732	*8.647*
t_y	10.164	12.494*	9.572*	no convrg*	9.924	*10.044*
5. IKO-NIR/ETM-NIR						
Scale	1.065	1.065	1.065	1.065	1.065	*1.065*
θ	0.109	0.068	0.070	0.066	0.110	*0.070*
t_x	8.668	8.687	8.704	8.663	8.663	*8.668*
t_y	10.167	10.148	10.140	10.153	10.156	*10.153*
6. ETM-red/ETM-NIR						
Scale	1.000	1.000	1.000	1.000	1.000	*1.000*
θ	−0.001	0.000	0.000	0.000	0.093	*0.000*
t_x	0.079	0.000	0.000	0.000	0.734	*0.000*
t_y	−0.029	0.000	0.000	0.000	0.942	*0.000*

ground truth is available, but we expect the multimodal intra-sensor registrations to be scale $s = 1$, $t_x = 0$, $t_y = 0$, and $\theta = 0°$, with $s = 1.07$ for the IKONOS to Landsat registrations. All cases for which these default results are not obtained or which seem to fall far from the median result are flagged with an asterisk. The results of Tables 14.8 and 14.9 show that, as expected, the registrations based on gray levels are less reliable on cross-spectral data than those based on edge-like (band-pass)

Table 14.8 (*cont.*)

CASCADES	FFC	TRU/SplC	TRU/SimB	TRU/SimL	SPSA/SimB	Median
7. IKO-red to IKO-NIR to ETM-NIR to ETM-red						
t_x	8.761	8.663	8.668	8.617	9.417	
t_y	10.151	9.988	9.957	9.944	11.152	
Round-robin error $\|7-2\|$						
t_x	0.087	2.255	0.016	0.377	0.641	
t_y	0.010	5.762	0.087	1.387	1.112	
8. IKO-red to ETM-red to ETM-NIR						
t_x	8.754	10.918	8.652	8.993	9.511	
t_y	10.133	15.750	10.044	11.330	10.981	
Round-robin error $\|8-3\|$						
t_x	0.059	2.377	0.011	8.993	0.613	
t_y	0.085	5.597	0.085	11.330	0.757	
9. IKO-NIR to ETM-NIR to ETM-red						
t_x	8.747	8.687	8.704	8.663	9.397	
t_y	10.138	10.148	10.140	10.153	11.098	
Round-robin error $\|8-4\|$						
t_x	0.186	4.443	5.554	no convrg	0.665	
t_y	0.026	2.346	0.567	no convrg	1.174	

Legend: Least reliable results flagged with an asterisk.

features, but, when reliable, these results are more accurate. Also, most results are within 1/4 to 1/3 pixels of each other (by looking at the median values).

In the absence of ground truth, to assess the accuracy of the registrations in Tables 14.8 and 14.9, one can make use of a technique called *round-robin regis-tration*. The idea is to use three or more images of the same scene, and to form pairwise registrations of those images. For example, the pairwise registrations can be A to C, C to B, and A to B. The registrations of A to C and C to B give one calculation for the relative registration of A with B. That relative registration ideally should be identical to what is obtained when registering A directly to B. In reality, there is always some registration error in the round-robin registrations of A to B, B to C, and C to A. The value of round-robin registrations is that when the error estimate is low, e.g., a fraction of a pixel, there is great confidence that each of the pairwise registrations has low error. Conversely, if the error estimate is high, e.g., several pixels, then at least one and possibly more than one pairwise registration is off by several pixels. However, the analysis gives no indication as to which of the pairwise registrations has high error in the latter case.

Table 14.9 *Results of five algorithms on the Virginia-Coast area (initial guess =* *(1.0, 0.0, 0.0, 0.0))*

VA-COAST	FFC	TRU/SplC	TRU/SimB	TRU/SimL	SPSA/SimB	Median
1. IKO-red/IKO-NIR						
Scale	1.000	1.000	0.999	1.000	1.001	*1.000*
θ	−0.001	0.000	0.002	0.000	0.081	*0.000*
t_x	0.007	−0.148	0.052	−0.243	0.922	*0.007*
t_y	−0.054	−0.484	−0.560	−0.532	0.751	*−0.484*
2. IKO-red/ETM-red						
Scale	1.066	1.064	1.066	1.066	1.066	*1.066*
θ	0.001	0.030	0.019	0.045	0.104	*0.030*
t_x	12.858	13.357	12.944	13.100	13.024	*13.024*
t_y	13.172	12.957	13.200	13.222	14.138	*13.200*
3. IKO-red/ETM-NIR						
Scale	1.619	1.048*	1.075	1.049	1.066	*1.066*
θ	−0.121	−1.096*	−1.546	−1.041	0.010	*−1.041*
t_x	12.395	11.099*	8.465	11.099	12.216	*11.099*
t_y	12.218	9.276*	12.714	9.529	13.156	*12.218*
4. IKO-NIR/ETM-red						
Scale	1.061	1.055*	0.997*	1.097*	1.067	*1.061*
θ	−0.903	−1.095*	−0.665*	−1.342*	0.972	*−0.903*
t_x	10.329	27.921*	−2.465*	23.063*	16.090	*16.090*
t_y	11.549	6.665*	−3.043*	12.034*	16.097	*11.549*
5. IKO-NIR/ETM-NIR						
Scale	1.065	1.000*	1.066	1.064	1.066	*1.065*
θ	−0.109	−0.001*	0.011	0.024	0.006	*0.006*
t_x	12.591	−5.760*	12.861	13.123	12.856	*12.856*
t_y	12.898	9.914*	13.169	13.048	13.246	*13.048*
6. ETM-red/ETM-NIR						
Scale	1.000	0.098*	0.999	0.995	1.000	*0.999*
θ	0.002	−1.266*	0.003	−0.111	−0.002	*−0.002*
t_x	−0.067	−1.918*	−0.272	−0.374	0.851	*−0.272*
t_y	−0.014	−3.849*	0.358	−0.457	0.665	*−0.014*

Round-robin computations performed on the results in Tables 14.8 and 14.9 show that FFC and SPSA/SimB generally result in a smaller round-robin error than the other algorithms. Because round-robin error measures the cumulative error from several pairwise registrations, if only a single pairwise registration has significant error the round-robin error will be significant. For round-robin error to be small, each pairwise registration in the sequence of registrations should be small. For the Cascades mountainous region study summarized in Table 14.8, FFC

Table 14.9 (*cont.*)

VA-COAST	FFC	TRU/SplC	TRU/SimB	TRU/SimL	SPSA/SimB	Median
7. IKO-red to IKO-NIR to ETM-NIR to ETM-red						
t_x	12.531	−7.826	12.641	12.506	14.629	
t_y	12.831	5.581	12.967	12.059	14.662	
Round-robin error $|7 - 2|$						
t_x	0.326	21.183	0.302	0.594	1.604	
t_y	0.342	7.376	0.233	1.164	0.524	
8. IKO-red to ETM-red to ETM-NIR						
t_x	12.791	11.439	12.672	12.727	13.875	
t_y	13.159	9.108	13.558	12.766	14.803	
Round-robin error $|8 - 3|$						
t_x	0.395	0.340	4.206	1.628	1.659	
t_y	0.941	0.168	0.844	3.237	1.647	
9. IKO-NIR to ETM-NIR to ETM-red						
t_x	12.524	−7.678	12.589	12.749	13.706	
t_y	12.885	6.065	13.527	12.591	13.912	
Round-robin error $|8 - 4|$						
t_x	2.1949	35.5989	15.0546	10.3137	2.3837	
t_y	1.3357	0.6000	16.5696	0.5567	2.1856	

Legend: Least reliable results flagged with an asterisk.

and SPSA/SimB yielded very consistent results, with FFC producing results that were consistent within 0.18 pixels, and SPSA/SimB results being consistent within 1.11 pixels for all but one of the offsets. The results were much less robust for the Virginia Coast region summarized in Table 14.9, where FFC produced results that were within 2.2 pixels and SPSA/SimB results were within 2.4 pixels. The other algorithms generally produced results that revealed significant inconsistencies in the round-robin sense. Therefore, one or more of the pairwise registrations produced by the other algorithms was inaccurate, but no pairwise registration in the sequences for FFC and for SPSA/SimB were inaccurate by more than that indicated by the round-robin results.

The Virginia Coast region and the Cascades mountainous region produced very different results for the round-robin data. This is possibly due to the differences in the registration features available in the image sets. The mountainous region has many edges visible in each of the images in the set, and the edges provide excellent registration characteristics. Edges are less prevalent in the Virginia Coast dataset.

Since the TRU algorithm produced results less reliable than FFC and SPSA for all three types of features, further experiments were performed with TRU where the initial guess was given closer to the expected transformations. Tables 14.10

Table 14.10 *Results of TRU algorithm with three different features on the Cascades-Mountainous area (initial guess = (1.07, 0.0, 8.0, 10.0))*

CASCADES	TRU/SplC	TRU/SimB	TRU/SimL
1. IKO-red/IKO-NIR			
Scale	1.000	1.000	1.000
θ	0.001	0.001	0.001
t_x	−0.024	−0.036	−0.046
t_y	−0.160	−0.183	−0.209
2. IKO-red/ETM-red			
Scale	1.067	1.065	1.070
θ	0.015	0.065	0.074
t_x	8.384	8.626	9.225
t_y	10.225	10.083	10.423
3. IKO-red/ETM-NIR			
Scale	1.065	1.065	1.066
θ	0.054	0.078	0.044
t_x	8.292	8.470	8.207
t_y	10.315	10.133	10.235
4. IKO-NIR/ETM-red			
Scale	1.070	1.065	no convrg
θ	0.000	0.084	no convrg
t_x	8.000	8.641	no convrg
t_y	10.000	10.130	no convrg
5. IKO-NIR/ETM-NIR			
Scale	1.065	1.065	1.065
θ	0.068	0.070	0.066
t_x	8.687	8.704	8.662
t_y	10.148	10.140	10.153
6. ETM-red/ETM-NIR			
Scale	1.000	1.000	1.000
θ	0.000	0.000	0.000
t_x	0.000	0.000	0.000
t_y	0.000	0.000	0.000

and 14.11 show the results of these experiments, and for both areas, Cascades and Virginia, it can be seen that the results improved significantly for TRU used in combination with SimB, although the round-robin results are still better for Cascades than for Virginia. For the low-pass features, spline (SplC) and SimL features, the results improved much less significantly. These results are in agreement with the conclusions drawn from the synthetic data experiments.

Table 14.10 (*cont.*)

CASCADES	TRU/SplC	TRU/SimB	TRU/SimL
7. IKO-red to IKO-NIR to ETM-NIR to ETM-red			
t_x	8.663	8.668	8.616
t_y	9.988	9.957	9.944
Round-robin error $|7 - 2|$			
t_x	0.279	0.042	0.609
t_y	0.237	0.126	0.479
8. IKO-red to ETM-red to ETM-NIR			
t_x	8.384	8.626	9.225
t_y	10.225	10.083	10.423
Round-robin error $|8 - 3|$			
t_x	0.092	0.156	1.018
t_y	0.090	0.050	0.188
9. IKO-NIR to ETM-NIR to ETM-red			
t_x	8.687	8.704	8.662
t_y	10.148	10.140	10.153
Round-robin error $|8 - 4|$			
t_x	0.687	0.063	no convrg
t_y	0.148	0.010	no convrg

Overall, these experiments also show that using several algorithms in combination might be a solution to obtain accurate and robust multimodal registration, for example, by using as a final result the median values of all transformation parameters.

14.3.2.4 Subpixel accuracy assessment

This section discusses a technique for estimating registration accuracy in the absence of ground truth. The registration experiment uses a set of images from different spectra and different resolutions of the same Earth region. When registering images of different resolutions, the registration algorithm matches the coarse image to the fine image and to the nearest fine image pixel. Because the resolution of the fine image is a multiple of the resolution of the coarse image, the nearest pixel of the fine image corresponds to a fractional pixel (a *phase*) of the coarse image. To assess the accuracy of the registration, the registrations were compiled for a collection of image pairs such that there are two or sequences of pairwise registrations from which one can find the relative registration on an image A with respect to an image B, which permits the use of a round-robin analysis as discussed in the previous section. The results discussed in this section show a few instances

Table 14.11 *Results of TRU algorithm with 3 different features on the Virginia-Coast area (initial guess = (1.07, 0.0, 12.0, 12.0))*

VA-COAST	TRU/SplC	TRU/SimB	TRU/SimL
1. IKO-red/IKO-NIR			
Scale	1.000	0.999	1.000
θ	0.000	0.002	0.000
t_x	−0.148	0.052	−0.243
t_y	−0.484	−0.560	−0.532
2. IKO-red/ETM-red			
Scale	1.064	1.066	1.066
θ	0.049	0.019	0.039
t_x	13.179	12.944	13.126
t_y	13.050	13.200	13.176
3. IKO-red/ETM-NIR			
Scale	1.048	1.075	1.049
θ	−1.097	−1.546	−1.041
t_x	11.097	8.465	11.099
t_y	9.279	12.174	9.259
4. IKO-NIR/ETM-red			
Scale	1.100	1.075	1.117
θ	0.232	1.591	0.395
t_x	20.835	15.209	24.201
t_y	17.181	16.597	21.848
5. IKO-NIR/ETM-NIR			
Scale	1.215	1.066	1.064
θ	−0.396	0.011	0.015
t_x	17.695	12.861	13.127
t_y	24.171	13.169	13.126
6. ETM-red/ETM-NIR			
Scale	0.098	0.999	0.995
θ	−1.266	0.003	−0.111
t_x	−1.918	−0.272	−0.374
t_y	−3.849	0.358	−0.457

where the registration error is on the order of a tenth of a pixel, others where it is on the order of 1 or 2 pixels, and still others where the error is substantially higher.

In the experiment described in Le Moigne *et al.* (2002b), our objective was to register a coarse image to a fine image at the resolution of the fine image, and therefore to assess the subpixel registration capabilities of our algorithms. For this purpose, we utilized a multiphase filtering technique, in which all possible phases of the fine image are registered with respect to the coarse image. Each different

Table 14.11 (*cont.*)

VA-COAST	TRU/SplC	TRU/SimB	TRU/SimL
7. IKO-red to IKO-NIR to ETM-NIR to ETM-red			
t_x	15.629	12.641	12.510
t_y	19.838	12.967	12.137
Round-robin error $|7-2|$			
t_x	2.450	0.303	0.616
t_y	6.788	0.233	1.039
8. IKO-red to ETM-red to ETM-NIR			
t_x	11.261	12.672	12.752
t_y	9.201	13.558	12.719
Round-robin error $|8-3|$			
t_x	0.164	4.207	1.653
t_y	0.078	1.384	3.460
9. IKO-NIR to ETM-NIR to ETM-red			
t_x	15.777	12.589	12.753
t_y	20.322	13.527	12.669
Round-robin error $|8-4|$			
t_x	5.0580	2.6200	11.4475
t_y	3.1410	3.0700	9.1788

phase was filtered and downsampled to the coarse resolution. The phase that gives the best registration metric gives the registration to the resolution of the fine image. We registered this data using two different criteria, normalized correlation and mutual information.

In practice, we utilized two of the images prepared in the first part of the experiment described in Subsection 14.3.2.3, where the IKONOS and ETM+ data have been resampled to the respective spatial resolutions of 3.91 and 31.25 meters. IKONOS red and near-infrared (NIR) bands (of size 2048×2048) were shifted in the x and y directions by the amounts $\{0, \ldots, 7\}$, thus creating 64 images for each band, for a total of 128 images. We used the centered spline, SplC, filters (Unser *et al.*, 1993) to downsample with no offset bias. The 128 phase images were downsampled by 8 to a spatial resolution of 31.25 meters and dimensions of 256×256. At the coarse resolution, the integer pixel shifts now corresponded to subpixel shifts of $\{0, 1/8, \ldots, 7/8\}$. We constructed reference chips of size 128×128 from the ETM-red and ETM-NIR images by extraction at position (64, 64) of the initial images. Knowing from the results in Subsection 14.3.2.3 that the offset between the original downsampled IKONOS image and the ETM reference image is (2, 0), we expected to find the (x, y) offset of the IKONOS image to the ETM image at about (66, 64).

Table 14.12 *Correlation results. For each image pair, (X, Y) is the subpixel shift that gives the highest correlation, while for each of those offsets, (Peak X, Peak Y) represents the position of the correlation peak*

Pattern	Reference	Coarse resolution (X, Y) phase giving max correlation		Normalized correlation	Peak X giving max correlation	Peak Y	Relative offset X	Y
IKO-red	ETM-red	6/8	1/8	0.8521	67	65	66.2500	64.8750
IKO-red	ETM-NIR	0	3.5/8	0.2722	66	65	66.0000	64.5625
IKO-NIR	ETM-red	4/8	7/8	0.2191	67	66	66.5000	65.1250
IKO-NIR	ETM-NIR	6/8	1/8	0.8436	67	65	66.2500	64.8750

Table 14.13 *Mutual information results. For each image pair, (X, Y) is the subpixel shift that gives the highest MI, while for each of those offsets, (Peak X, Peak Y) represents the position of the MI Peak*

Pattern	Reference	Coarse resolution (X, Y) phase giving max mutual information		Mutual information	Peak X giving max mutual information	Peak Y	Relative offset X	Y
IKO-red	ETM-red	0	1/8	1.3017	66	65	66.0000	64.8750
IKO-red	ETM-NIR	6/8	3/8	0.3411	67	65	66.2500	64.6250
IKO-NIR	ETM-red	2/8	1/8	0.3826	66	65	65.7500	64.8750
IKO-NIR	ETM-NIR	7/8	2/8	1.1653	67	65	66.1250	64.7500

The complete experiment involved the registration of the 128×128 extracted red-band (resp. NIR-band) ETM chips to the 64 phased and downsampled 256×256 red-band (resp. NIR-band) IKONOS images. For each 128×128 ETM chip and for each of the 64 IKONOS phase images, we computed the maximum correlation (resp. mutual information) and the associated location at which this maximum occurred. Then, we found the phase that gave the maximum correlation (resp. mutual information) out of all phases, and we recorded the corresponding shift and the offset computed in this registration. These are shown in Tables 14.12 and 14.13 under "Coarse resolution (X, Y) phase" and under "(Peak X, Peak Y)." Notice that for all peaks, locations have been approximated to the nearest integers.

To find the relative offset of an IKONOS pattern to an ETM reference (in the last two columns of Tables 14.12 and 14.13), we subtracted the coarse resolution phase offset (X, Y) (in columns 3 and 4) from the corresponding peak offset (Peak X, Peak Y) (in columns 6 and 7). The two tables agree to within 0.25 pixels for all relative offsets except for the X offset from IKONOS-NIR to ETM-red. They disagree by about 0.75 pixels in that case. There are some small inconsistencies in the tables. If the two IKONOS images were registered to the nearest pixel before downsampling, the offsets that produce maximum correlation peaks should be identical when the downsampled patterns are registered to the same image. But when both IKONOS patterns are registered to ETM-red, the offsets that produce the highest correlations are different. The same phenomenon also occurs in the mutual information-based registrations. One explanation would be that IKONOS-red and IKONOS-NIR are misregistered by 1 or 2 pixels. Another possible explanation is that cross-spectral registration between IKONOS (downsampled by 8) with respect to ETM has an extra offset of 0.25 pixels when compared to within-spectrum registration. It is uncertain where this offset comes from. It is probably an artifact of the cross-spectral data wherein some edges in the image appear to be shifted because of the spectral responses, and these cause registration peaks to shift. More data are required to study this phenomenon. Overall, we can see that the average absolute difference between computed relative offsets and the expected (64, 66) is about 0.5 pixels for both correlation and mutual information metrics.

Another way to look at the data is to analyze the self-consistency of all four measurements. For this analysis, we computed the (x, y) offset of one of the images from the other three in two different ways. If the data are self-consistent, the answers should be the same. To do this, we established an x base point for IKONOS-red, and let this be $x = 0$. Then, we used the previous relative offsets shown in Tables 14.12 and 14.13 to determine (x, y) offsets for each of the other three images. Tables 14.14 and 14.15 show these results.

Note that both measures show a displacement of IKONOS-red from IKONOS-NIR of either 0.25 or 0.125 pixels, and the signs of the relative

Table 14.14 *Self-consistency study of the normalized correlation results*

Image name	Computed x	Computed y	Comes from registered pair
IKONOS-red	0	0	(Starting point)
IKONOS-NIR	−0.2500	−0.2500	IKO-red to ETM-red and ETM-red to IKO-NIR
IKONOS-NIR	−0.2500	−0.3125	IKO-red to ETM-NIR and ETM-NIR to IKO-NIR

Table 14.15 *Self-consistency study of the mutual information results*

Image name	Computed x	Computed y	Comes from registered pair
IKONOS-red	0	0	(Starting point)
IKONOS-NIR	0.2500	0.0000	IKO-red to ETM-red and ETM-red to IKO-NIR
IKONOS-NIR	0.1250	−0.1250	IKO-red to ETM-NIR and ETM-NIR to IKO-NIR

displacements differ for mutual information and normalized correlation registrations. For these two images, the two measures are self-consistent in their estimates of a relative displacement to within 1/8 of a coarse pixel.

14.4 Conclusions

The studies presented in this chapter investigated the use of various feature extraction and feature matching components for the purpose of remote sensing image data registration. Results were provided for a variety of test datasets, synthetic (including noise and radiometric variations), multitemporal, and multisensor. The performances of six different algorithms utilizing gray levels and wavelet-like features combined with correlation, mutual information, and partial Hausdorff distance as similarity metrics, and Fourier transform, optimization, and robust feature matching as search strategies were evaluated. Two of the metrics, correlation and mutual information, were further studied for subpixel registration.

Using synthetic data, we demonstrated that the algorithm based on a Levenberg-Marquardt optimization using the L_2-norm and band-pass wavelet-like features was the most accurate and the most robust to noise. Nevertheless, using Simoncelli's low-pass features with the same type of algorithm was less sensitive to the initial guess. Overall, an approach based on a stochastic gradient technique with a mutual information metric was more robust to initial conditions. If the transformations are very large and if the images contain many high-frequency features, the approach based on a global fast Fourier correlation of multiple chips seemed to work the best.

More generally, we can estimate two regions of interest in the space of registration parameters based on how the collection of registration algorithms we studied behaves for various parameter sets. We say that an algorithm converges for a set of registration parameters if it converges to a global optimum, that is, to the right answer when two images differ by that set of registration parameters. The first region of interest is the *region of convergence*, within which *all* the algorithms that we studied are likely to converge:

- If only shift, it is the region that ranges from −20 to 20 pixels.
- If only rotations, between −10° and 10°.
- If only scale, between 0.9 and 1.1.
- If rotation and shift, then it is when the shift is between −15 and 15 pixels and the rotation between −5° and 5°.
- If rotation, shift and scale, the region of convergence is defined by a shift between −10 and 10 pixels, a rotation between −5° and 5°, and a scale between 0.9 and 1.1.

The second region is the *region of divergence* within which *all* the algorithms will most likely diverge:

- If the shift is more than 30 pixels.
- If the rotation is more than 15°.
- If the scale is less than 0.8 or more than 1.2.
- If together, the rotation and shift are more than 20° and 10 pixels, respectively.
- If rotation, shift and scale, when the shift is larger than 15 pixels, the rotation more than 10°, and the scale less than 0.85 or more than 1.15.

The region between the two is the one where some algorithms converge and some do not. These regions were estimated fairly conservatively, based on a limited number of sample images. To get a more precise estimate, we would need to run more thorough testing with more images of various types.

Based on these first studies, we developed the first prototype of a web-based image registration toolbox (called TARA for "Toolbox for Automated Registration and Analysis") that is depicted in Figure 14.18. At present, this first prototype includes TRU, TRUMI and SPSA with the choice of spline, Simoncelli band-pass or Simoncelli low-pass features. The toolbox's interface is implemented in Java, and the algorithms are implemented in C or C++, but integrated in the toolbox as Java Native Interface (JNI) wrapped functions. The synthetic experiments enable us to define for each method an *applicability range* that will be provided as guidance to the users of the toolbox. At the same time, these methods continue to be tested

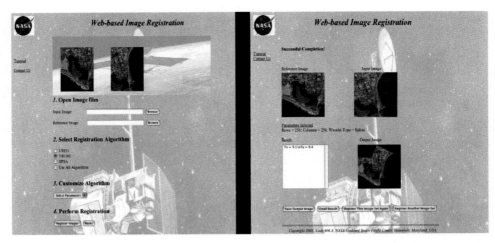

Figure 14.18. User interface of the TARA web-based image registration toolbox. See color plates section.

on other sensor data, for example, the ALI multispectral sensor and the Hyperion hyperspectral sensor, both carried on the EO-1 platform. Eventually, we hope that TARA will be used to assess other registration components and will be extended to include other methods, to compute more general transformations and to process other types of imagery, such as aerial images or other planetary data, for example, from the Moon or Mars.

Acknowledgements

The authors acknowledge the support of NASA under NRA-NAS2–37143, "Research in Intelligent Systems," and of the Goddard Internal Research and Development (IRAD) program, under which many of the experiments have been conducted. The authors also wish to thank Jeffrey Masek for the multitemporal data and the MODIS Validation Core Sites program for the multisensor data.

References

Brown, L. G. (1992). A survey of image registration techniques. *ACM Computing Surveys*, **24**(4), 325–376.

Cole-Rhodes, A. A., Johnson, K. L., Le Moigne, J., and Zavorin, I. (2003). Multiresolution registration of remote sensing imagery by optimization of mutual information using a stochastic gradient. *IEEE Transactions on Image Processing*, **12**(2), 1495–1511.

Daubechies, I. (1992). *Ten Lectures on Wavelets*. Philadelphia, PA: Society for Industrial and Applied Mathematics (SIAM).

Eastman, R. and Le Moigne, J. (2001). Gradient-descent techniques for multi-temporal and multi-sensor image registration of remotely sensed imagery. In *Proceedings of the 4th International Conference on Image Fusion*, Montreal, Canada.

Foroosh, H., Zerubia, J. B., and Berthod, M. (2002). Extension of phase correlation to subpixel registration. *IEEE Transactions on Image Processing*, **11**(3), 188–200.

Irani, M. and Peleg, S. (1991). Improving resolution by image registration, *Computer Vision, Graphics and Image Processing*, **53**(3), 231–239.

Le Moigne, J., Campbell, W. J., and Cromp, R. F. (2002a). An automated parallel image registration technique based on the correlation of wavelet features. *IEEE Transactions on Geoscience and Remote Sensing*, **40**(8), 1849–1864.

Le Moigne, J., Cole-Rhodes, A., Eastman, R., Johnson, K., Morisette, J., Netanyahu, N., Stone, H., and Zavorin, I. (2002b). Multisensor image registration for on-the-ground or on-board science data processing. In *Proceedings of the Science Data Workshop*, Greenbelt, MD.

Le Moigne, J., Cole-Rhodes, A., Eastman, R., Johnson, K., Morisette, J. T., Netanyahu, N. S., Stone, H. S., and Zavorin, I. (2001). Multisensor registration of Earth remotely sensed imagery. In *Proceedings of the SPIE Conference on Image and Signal Processing for Remote Sensing VII*, Toulouse, France, Vol. 4541, pp. 1–10.

Le Moigne, J., Cole-Rhodes, A., Eastman, R., Johnson, K., Morisette, J., Netanyahu, N. S., Stone, H., and Zavorin, I. (2003). Earth science imagery registration. In *Proceedings of the IEEE International Geoscience and Remote Sensing Symposium*, Toulouse, France, Vol. I, pp. 161–163.

Le Moigne, J., Cole-Rhodes, A., Eastman, R., Johnson, K., Morisette, J., Netanyahu, N. S., Stone, H., Zavorin, I., and Jain, P. (2004). A study of the sensitivity of automatic image registration algorithms to initial conditions. In *Proceedings of the IEEE International Geoscience and Remote Sensing Symposium*, Anchorage, AK, Vol. 2, pp. 1390–1393.

Le Moigne, J., El-Saleous, N., and Vermote, E. (1997). Iterative edge- and wavelet-based registration of AVHRR and GOES satellite imagery. In *Proceedings of the Image Registration Workshop*, NASA Goddard Space Flight Center, Greenbelt, MD, pp. 137–146.

Le Moigne, J., Xia, W., El-Ghazawi, T., Mareboyana, M., Netanyahu, N., Tilton, J. C., Campbell, W. J., and Cromp, R. F. (1998). First evaluation of automatic image registration methods. In *Proceedings of the IEEE International Geoscience and Remote Sensing Symposium*, Seattle, WA, Vol. 1, pp. 315–317.

Lyon, R. G., Dorband, J. E., and Hollis, J. M. (1997). Hubble space telescope faint object camera calculated point-spread functions. *Applied Optics*, **36**(8), 1752–1765.

Maes, F., Collignon, A., Vandermeulen, D., Marchal, G., and Suetens, P. (1997). Multimodality image registration by maximization of mutual information. *IEEE Transactions on Medical Imaging*, **16**(2), 187–198.

Morisette, J. T., Privette, J. L., and Justice, C. O. (2002). A framework for the validation of MODIS land products. *Remote Sensing of Environment*, **83**, 77–96.

Mount, D. M., Netanyahu, N. S., and Le Moigne, J. (1999). Efficient algorithms for robust feature matching. *Special Issue of Pattern Recognition on Image Registration*, **32**(1), 17–38.

National Imagery and Mapping Agency (2000). *Department of Defense World Geodetic System 1984: Its Definition and Relationships with Local Geodetic Systems*. Technical Report 8350.2, 3rd edn.

Netanyahu, N. S., Le Moigne, J., and Masek, J. G. (2004). Georegistration of Landsat data via robust matching of multiresolution features. *IEEE Transactions on Geoscience and Remote Sensing*, **42**(7), 1586–1600.

Stone, H. S., Le Moigne, J., and McGuire, M. (1999). The translation sensitivity of wavelet-based registration. *IEEE Transactions on Pattern Analysis and Machine Intelligence*, **21**(10), 1074–1081.

Stone, H. S., Orchard, M. T., Cheng, E.-C., and Martucci, S. A. (2001). A fast direct Fourier-based algorithm for subpixel registration of images. *IEEE Transactions on Geoscience and Remote Sensing*, **39**(10), 2235–2243.

Thévenaz, P., Ruttimann, U. E., and Unser, M. (1998). A pyramid approach to subpixel registration based on intensity. *IEEE Transactions on Image Processing*, **7**(1), 27–41.

Unser, M., Aldroubi, A., and Murry, E. (1993). The L_2 polynomial spline pyramid. *IEEE Transactions on Pattern Analysis and Machine Intelligence*, **15**(4), 364–379.

Zavorin, I. and Le Moigne, J. (2005). Use of multiresolution wavelets feature pyramids. *IEEE Transactions on Image Processing*, **14**(6), 770–782.

Figure 1.1. A reference image and its transformed image, extracted from an IKONOS scene acquired over Washington, DC. (IKONOS satellite imagery courtesy of GeoEye. Copyright 2009. All rights reserved.)

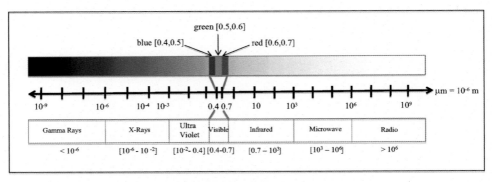

Figure 1.2. Electromagnetic spectrum.

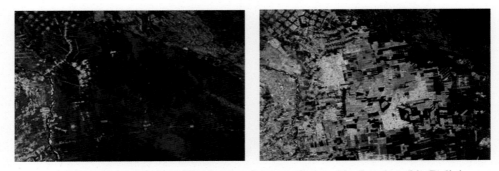

Figure 1.4. Human-induced land cover changes observed by Landsat-5 in Bolivia in 1984 and 1998. (Courtesy: Compton J. Tucker and the Landsat Project, NASA Goddard Space Flight Center.)

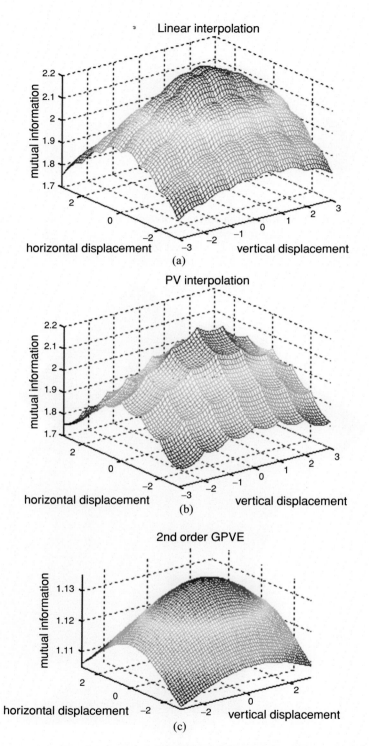

Figure 6.1. (a–c) Typical artifacts resulting from linear interpolation, partial volume interpolation and generalized partial volume estimation.

(a)   (b)

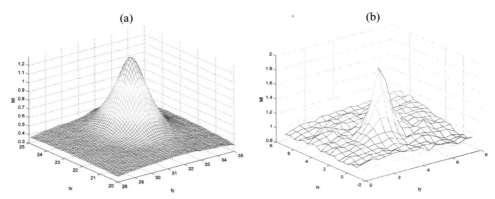

Figure 6.3. Mutual information surfaces obtained due to B-spline interpolation: (a) Subpixel MI surface for image pair of Fig. 6.2 (level 1 images), and (b) MI surface for coarse, level 4 images (derived from image pair of Fig. 6.2). (Source: Cole-Rhodes *et al.*, 2003, © IEEE, reprinted with permission.)

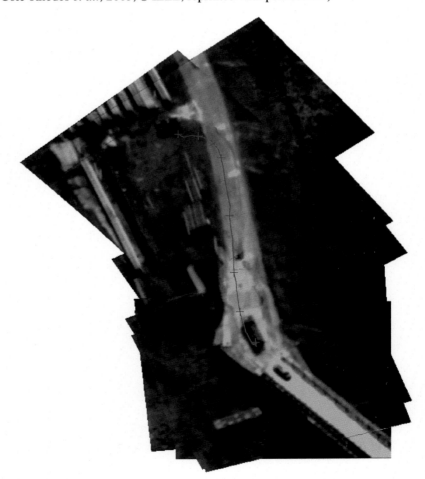

Figure 10.7. Mosaic created out of a long sequence with independent motions; trajectory of one independently moving feature point is shown.

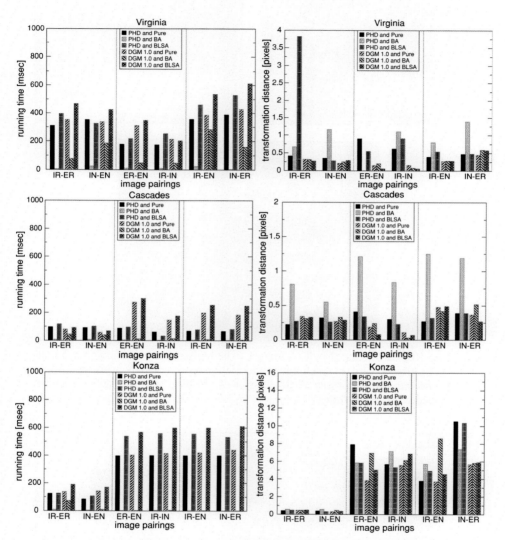

Figure 8.5. Results of Experiment 2 comparing the performance of the algorithms for various choices of sensor and spectral bands.

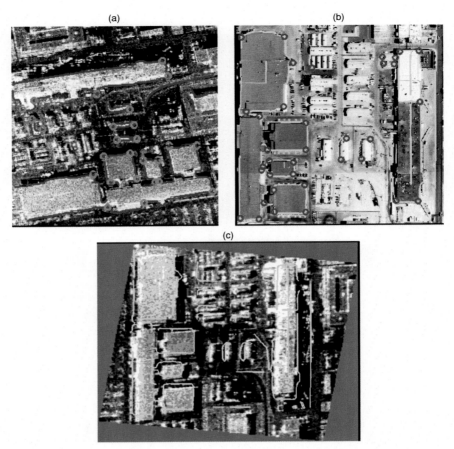

Figure 10.5. (a) Radar, (b) visual images to be registered (feature points are indicated by circles), and (c) resulting transformation represented by overlaying edges onto the radar image. (Source: Shekhar *et al.*, 1999, p. 49, © Elsevier Science, reprinted with permission.)

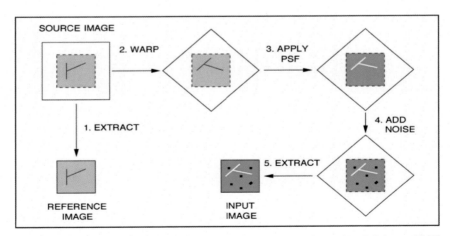

Figure 11.6. Synthetic image generation. (Source: Zavorin and Le Moigne, 2005, © IEEE, reprinted with permission.)

ETM+

IKONOS

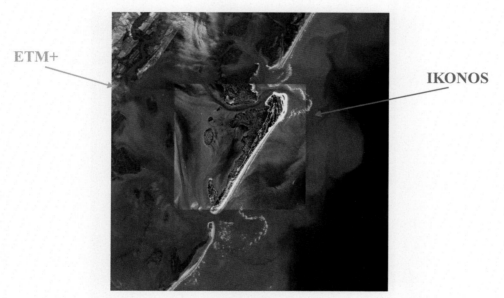

Figure 14.7. ETM+ and IKONOS data of the Virginia coastal area. (Source: Le Moigne *et al*., 2004, © IEEE, reprinted with permission.)

ETM+  IKONOS

Figure 14.8. ETM+ and IKONOS data of the Cascades mountainous area. (Source: Le Moigne *et al.*, 2004, © IEEE, reprinted with permission.)

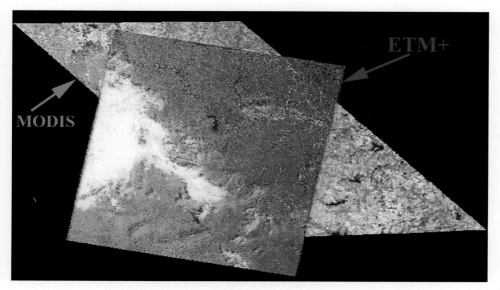

Figure 14.9. ETM+ and MODIS data of the Konza agricultural area.

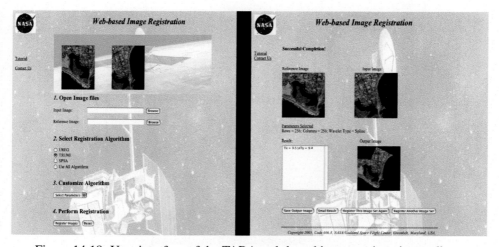

Figure 14.18. User interface of the TARA web-based image registration toolbox.

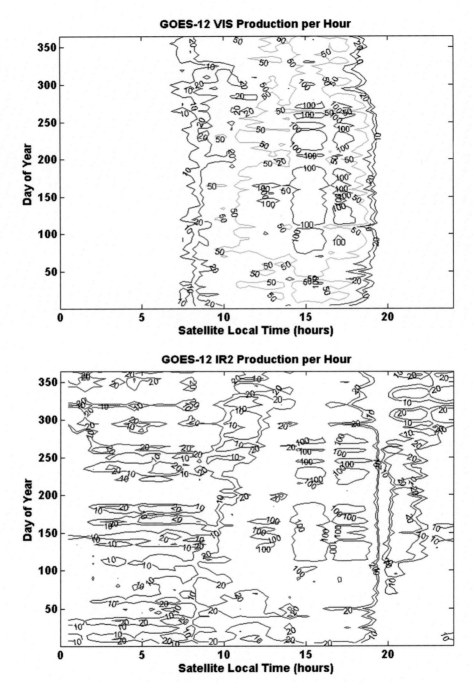

Figure 15.4. VIS and IR (channel 2) landmark production rates for a year of Replacement Product Monitor (RPM) operational experience spanning 2004 and 2005.

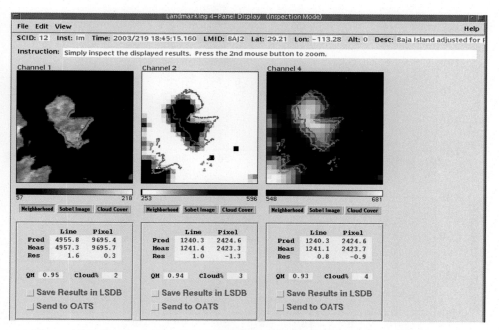

Figure 15.3. Screen capture from a Replacement Product Monitor (RPM) browse tool with one visible (channel 1) and two IR neighborhoods. The chips are shown in two locations, before and after matching with the land–water boundary edges.

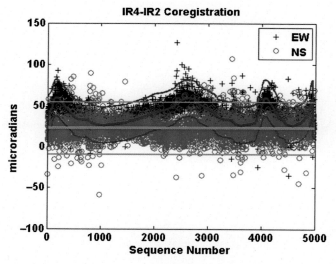

Figure 15.6. Measured channel-to-channel coregistration (differences in absolute errors at the same time and the same site in different bands), covering about three days of GOES-12 in-orbit experience.

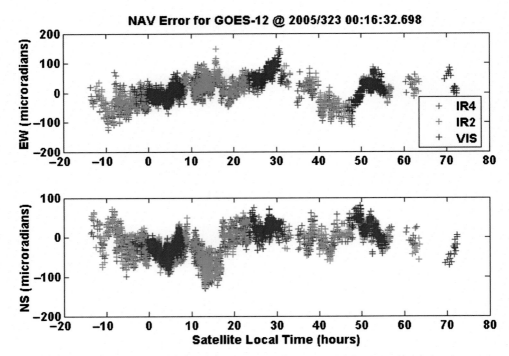

Figure 15.7. Absolute navigation error observed for GOES-12 during operations using platinum quality data from the RPM.

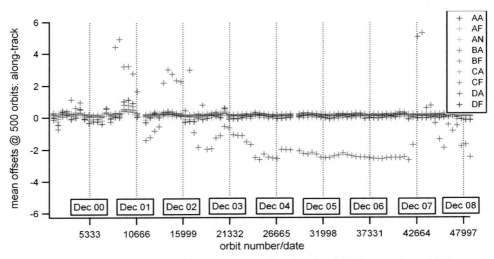

Figure 16.6. Global geometric performance for the nine MISR cameras; stability of pointing geometry shown with the mean values at 500 orbits plotted against orbit number.

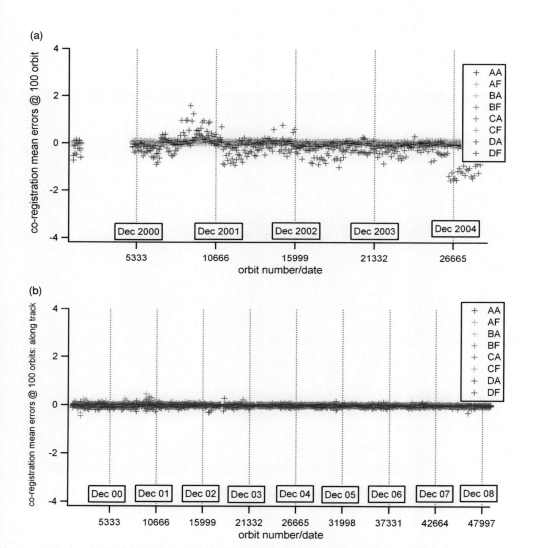

Figure 16.7. Global geometric performance for the nine MISR cameras; product accuracy given as mean coregistration error at 100 orbits plotted against orbit number: (a) Prior to update of standard production algorithm, and (b) final version product. (Source: Jovanovic *et al.*, 2007, © Elsevier science, reprinted with permission.)

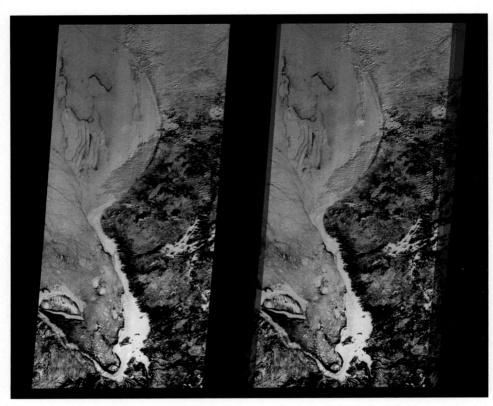

Figure 16.8. True color (left) and multiangle false color (right) images of the areas around Hudson and James Bays, Canada, generated from MISR data collected on February 24, 2000 during Terra orbit 995. These data were acquired shortly after the opening of the instrument's cover.

First Image   Second Image

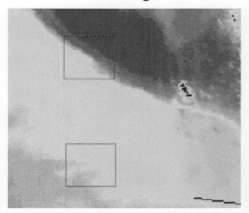 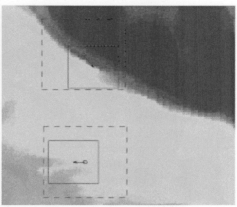

Figure 17.1. The MCC method illustrated for two grid locations. The solid boxes in the first image are the template subwindows; this is the data feature to search for within the dashed search windows in the second image. The second-image subwindows that provide the highest cross-correlations suggest the most likely displacements of the data features (solid boxes).

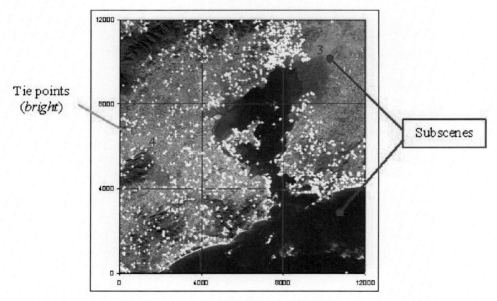

Figure 19.6 Uniformly distributed tie-points by subscenes in a 2.5-m resolution SPOT-5 image (displayed with pseudo natural colors). (Source: Baillarin *et al.*, 2005, reprinted with permission from the International Society for Photogrammetry and Remote sensing (ISPRS).)

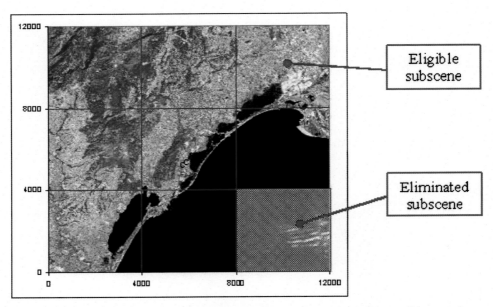

Figure 19.7 Eligible and eliminated subscenes in a 10-m resolution, multispectral SPOT-5 image of a coast landscape. (Source: Baillarin *et al.*, 2005, reprinted with permission from the International Society for Photogrammetry and Remote sensing (ISPRS).)

15

Georegistration of meteorological images

JAMES L. CARR

Abstract

Meteorological images are acquired routinely from Sun-synchronous and geostationary platforms to meet the needs of operational forecasting and climate researchers. In the future, polar regions may be covered from Molniya orbits as well. The use of geostationary and Molniya orbits permits rapid revisits to the same site for the purpose of observing the evolution of weather systems. The large orbital altitudes for these systems introduce special problems for controlling georegistration error. This chapter reviews the subject of georegistration for meteorological data, emphasizing approaches for measuring and controlling georegistration error.

15.1 Introduction to meteorological satellites

Meteorological satellite images are captured from Low Earth Orbit (LEO) or Geostationary Earth Orbit (GEO) using visible, infrared (IR), and microwave bands. The U.S. Polar Operational Environmental Satellite (POES) and the Defense Meteorological Satellite Program (DMSP) are two such LEO systems with operational histories reaching back into the 1960s. Future LEO meteorological remote sensing needs will be fulfilled by the U.S. National Polar-orbiting Operational Environmental Satellite System (NPOESS) and by the European MetOp satellites (a series of polar orbiting meteorological satellites operated by the European Organization for the Exploitation of Meteorological Satellites), using instruments from the USA and Europe. The U.S. Geostationary Operational Environmental Satellite (GOES) program and the European METEOSAT program are two such GEO programs with similarly long operational histories. Japan and India have also operated GEO weather satellites and will continue to do so in the future. These operators are joined by Russia and China, and soon by the Republic of Korea, to populate the Equator with sensors that can observe all of the tropical and temperate latitudes at any time according to the demand.

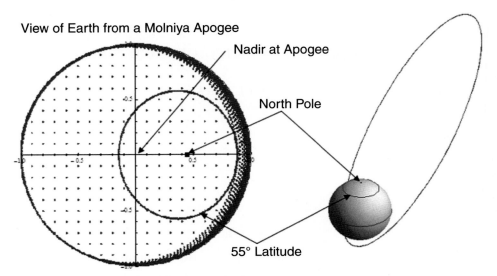

Figure 15.1. At apogee, the Molniya orbit provides complete coverage from the Pole to 55° latitude.

Meteorological phenomena evolve over timescales much shorter than those of interest in geology or agriculture; hence, the use of GEO satellites is of great interest for meteorological remote sensing. A sensor in GEO orbits the Earth at the same rate that the Earth rotates. It is, therefore, capable of sensing the same geographic site with a rapid revisit rate. In this regard, GEO meteorological remote sensing is more akin to movie making than still photography. The great distance from the Earth (>35 780 km) makes GEO remote sensing especially challenging. The solid angle subtended by even kilometer-scale pixels is quite small from GEO. For this reason, applications such as microwave sounding are much more practical from LEO. Like most other remote sensing systems, LEO meteorological systems are deployed in near-polar Sun-synchronous orbits, permitting similar illumination conditions for all overpasses.

GEO and LEO sensors are complementary parts of an overall meteorological remote sensing system of systems. These sensors may be joined in the future by sensors deployed in GPS-like Medium Earth Orbit (MEO) constellations (Gerber *et al.*, 2004). The highly elliptical Molniya orbit is also of interest for coverage of the polar regions (Riishojgaard, 2005). While polar orbiting spacecraft fly over each Pole every orbit, their overpasses are brief, limiting their ability to observe the evolution of weather on short timescales. A classic Molniya orbit has a 12-hour period, an eccentricity $e = 0.74$, and an inclination $i = 63.4°$. When at apogee, the Molniya orbit provides complete coverage from the Pole to 55° latitude, as shown in Fig. 15.1. Because of its high eccentricity, a Molniya satellite would appear to hover nearly above a fixed point for about 8 hours each orbit, followed by a rapid descent to perigee and an equally rapid ascent back to apogee.

15.2 Geolocating meteorological data

Existing LEO and GEO meteorological sensors operate at kilometric scale spatial resolution rather than at the meter-scale resolutions common to many land remote sensing systems (e.g., Landsat and SPOT). Their coverage is also typically limb-to-limb rather than in a narrow ribbon under the ground track. Such differences are important to the respective georegistration problems. In particular, terrain effects are less important for meteorological applications (although, cloud height introduces parallax that can be the basis for stereo imaging).

Geolocating severe weather is obviously of operational interest. There are two geolocation approaches in general use. The first utilizes embedded metadata that describes the latitudes and longitudes for a sparse grid of pixels. Interpolation between grid points enables the user to precisely locate any arbitrary pixel and refining the grid can, in principle, control the interpolation error. The High Resolution Picture Transmission (HRPT) format for the POES Advanced Very High Resolution Radiometer (AVHRR) is an example of this geolocation approach. The second approach is to distribute the data rectified to a fixed grid. Here, there is an invariant relationship between row and column image coordinates and latitude and longitude geographic coordinates defined by a standard. NOAA utilizes this method for GOES Imager data, which is distributed as a GVAR (GOES VARiable) product. Users can easily calculate the geolocations of pixels using simple algorithms defined in the Earth Location Users Guide (ELUG) (NOAA, 2005). The ELUG describes the image geometry for the GOES Imager in its "ideal" configuration, that is, in a perfectly circular geostationary equatorial orbit and oriented with respect to the orbital plane with zero roll, pitch, and yaw angles. An Image Motion Compensation (IMC) system (described below) enables images to be acquired in this geometry, even though these ideal conditions never exist in reality.

The GOES Imager would nominally scan the Earth disk using a bidirectional raster scan. The IMC is calculated in real time in the onboard processor and fed into the scan system to deflect the line-of-sight so as to follow the same trajectory across the face of the Earth that would be followed in the ideal geometry. The IMC can be thought of as a transformation function that depends on spacecraft orbit, attitude, and internal instrument alignments. Figure 15.2 plots the IMC in the North–South axis as a function of East–West scan angle to correct for orbital motion at several positions within a quarter of an orbit. At the node crossing, the correction is purely orbit yaw, while at other points in the orbit, the correction curves to follow the surface of the Earth.

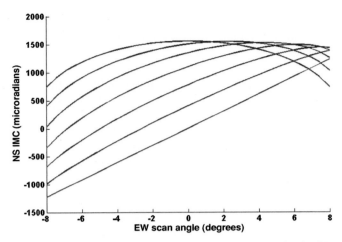

Figure 15.2. Plot of the Image Compensation Motion (IMC) in the North–South axis as a function of East–West scan angle to correct for orbital motion at several positions within a quarter of an orbit.

The discipline of geolocation (image navigation), controlling image navigation errors, and stabilizing sequences of images so that they coregister is referred to as Image Navigation and Registration (INR). INR holds a prominent place in meteorological satellite engineering (Carr, 1992, 2008). The GOES I-M series is the first successful example of INR applied to a three-axis stabilized platform. Systems prior to GOES I-M were spin stabilized. Spinning spacecraft are intrinsically stable platforms for remote sensing due to their gyroscopic stiffness; however, such systems spend barely 5% of their spin cycles viewing the Earth. This is less than ideal, especially for IR sounding, which needs to achieve high signal-to-noise ratios in narrow spectral bands. The INR problem for a three-axis stabilized spacecraft is exceptionally challenging, especially with the IMC concept adopted for GOES I-M (Kamel, 1996). INR is implemented for the GOES Imager and Sounder using an IMC coefficient set that is determined on the ground and uploaded to the spacecraft. The IMC coefficient set is intended to predict the orbit and attitude motion for a 24-hour period into the future. The GOES I-M orbit and attitude are determined from star senses, control points (landmarks), and station-to-satellite ranges in a large multidimensional estimation process. The orbit and attitude solutions are, in turn, propagated ahead and converted into the IMC coefficient set representation. Star sensing operations that support attitude determination are scheduled approximately every 30 minutes between frames, during which the instruments point off Earth near the locations of cataloged stars. Locations of control points are measured in the meteorological images in both visible and IR channels without any special scheduling, enabling orbit determination by means of the INR system.

While extremely clever, the GOES I-M attitude control system is bedeviled by unpredictable low-frequency errors from the Earth sensor used to control roll and pitch attitude (Carr, 1996; Harris *et al.*, 2001). To maintain good image navigation, the residual navigation error is constantly monitored and adjusted as needed (typically several times per day). These adjustments introduce step changes in the IMC attitude model, degrading frame-to-frame registration. The new GOES N-P series improves considerably on GOES I-M by introducing advanced stellar-inertial attitude control. GOES N-P spacecraft use star trackers and a three-axis gyroscope to orient themselves in space rather than an Earth sensor, offering much more accurate and stable spacecraft pointing that improves INR performance (Carson *et al.*, 2008; Carr *et al.*, 2008).

METEOSAT data are distributed in a mode similar to that of GOES according to the High Rate Information Transmission (HRIT) standard. Unlike GOES, the METEOSAT spacecraft are spin stabilized at 100 rpm, providing a stable platform for high-quality INR. METEOSAT Second Generation (MSG) is in many respects the most advanced weather satellite in the world as of this writing. It images the Earth with more spectral bands (12), finer IR resolution (3 km), and with a faster full-disk coverage rate (once per 15 minutes) in comparison to the GOES Imager. The problem of low imaging duty cycle has been solved in large part by the light collection capacity provided by its 50-cm aperture. Engineering studies for METEOSAT Third Generation (MTG) are underway and it seems likely that MTG satellites will be three-axis stabilized. In spite of its success, MSG may be the last of the spin-stabilized weather satellites.

While the MSG platform is intrinsically stable, its INR problem is by no means trivial (Blancke *et al.*, 1997, 1998). Accurate INR is achieved by high-fidelity modeling of internal optical alignments, the spacecraft orbital motion, its spin angular momentum vector, changes in spacecraft inertia tensor – including wobble amplification due to fluids (Blancke and Carr, 1996) – while scanning South-to-North. Model parameters are determined as part of an estimation process, as with GOES, but using control points, IR horizon crossing, and star senses taken during the back-scan. An IMC scheme similar to GOES is infeasible for METEOSAT, so images are rectified to a fixed grid by resampling. The Japanese Multi-functional Transport Satellites, MTSAT-1R and MTSAT-2, systems also utilize resampling in spite of being three-axis stabilized. GOES images are also resampled when the orbital inclination grows large enough to saturate the IMC dynamic range. Resampling on GOES extends the lifetimes of satellites because they may continue to operate with onboard IMC disabled while in high-inclination orbits when there is not enough fuel left to control inclination growth. NOAA currently operates GOES-10 in this mode at 60°W longitude, where it is dedicated to serving South America (Tehranian *et al.*, 2008).

15.3 Control points

Control points are a common feature of INR for both GEO and LEO systems. Control points can either measure the relationship between image features and map features or image features and other image features. The former measures image navigation error ("absolute error") and the latter measures image registration error ("relative error"). While star and tracking data can be used to assess orbit and attitude determination error, they do not directly measure INR performance. Control points directly measure INR performance and, therefore, represent the error signal that an INR system would seek to control. It is natural that they would be used both as part of the INR processing and as validation for such processing; however, the process of gathering control points in a fully automatic mode without human supervision is tricky and the measurement errors are nuanced.

GOES control points are operationally measured using the Replacement Product Monitor (RPM), which is part of the GOES ground system (Madani *et al.*, 2004). The RPM evolved from a prototype system called the GOES Landmark Analysis System (GLAS) using algorithms originally designed for METEOSAT (Carr *et al.*, 1997). The RPM continuously ingests images in real time as they are broadcast to users in GVAR format; therefore, it monitors exactly the product that is being disseminated to the users. INR monitoring by the RPM involves absolute and relative error measurements made automatically without any human supervision. The absolute error measurements are fed back into the operational navigation processing to help create the IMC coefficients for the next day. A similar product, called POLARIS, has been developed for making human supervised control point measurements in POES images. The POLARIS capabilities have been extended several times to now include processing of AVHRR, Advanced Microwave Sounding Unit (AMSU), High-resolution Infrared Sounder (HIRS), and Microwave Humidity Sounder (MHS) images. POLARIS has been routinely used to calibrate the boresight alignments of POES instruments during the Orbital Validation (OV) phase of each new satellite.

Control points for meteorological applications tend to be natural features because image resolutions, typically 1 km to 10 km, are too coarse to resolve man-made features. Land–water boundaries are the most useful natural features for such applications because they are globally mapped, with adequate accuracy in most places, and resolvable in both visible and IR bands (window channels only). The RPM (as well as other systems developed for the METEOSAT and MTSAT programs) utilizes the NOAA Global Self-consistent Hierarchical High-resolution Shoreline (GSHHS) database (Wessel and Smith, 1996). The GSHHS organizes all land–water boundaries into a hierarchical structure, with continents represented by polygons at the highest level. Interior bodies of water are represented as embedded

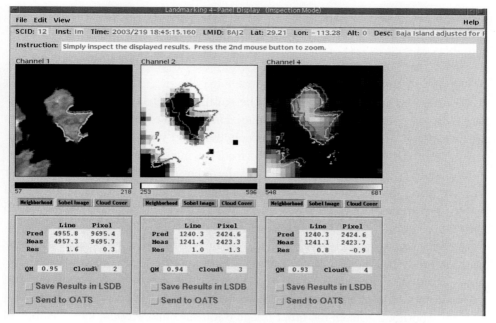

Figure 15.3. Screen capture from a Replacement Product Monitor (RPM) browse tool with one visible (channel 1) and two IR neighborhoods. The chips are shown in two locations, before and after matching with the land–water boundary edges. See color plates section.

polygons; islands within such bodies of water are further represented as embedded polygons at the next hierarchical level, and so forth. The RPM matches edges found in an image "neighborhood" centered about a cataloged land–water feature with the mapped land–water boundary contour transformed into the GOES image geometry ("chip"). Because measurements are made locally within individual neighborhoods and on a small scale in comparison to the overall image dimensions, it may be assumed that the image navigation error is represented by a locally constant translational shift between image and chip. Figure 15.3 shows a screen capture from an RPM browse tool with one visible (channel 1) and two IR neighborhoods. The chips are shown in two locations – first before matching and second after matching with the land–water boundary edges. The shift that is required to best match the chip with the edges is the measured image navigation error.

Subpixel precision for the measured absolute error is achieved by iteratively refining the resolution of the search for a maximum of a similarity metric defined as

$$\rho_{abs}\,(\Delta x,\, \Delta y) = \sum_n E\,(x_n - \Delta x,\, y_n - \Delta y), \qquad (15.1)$$

where $E(x, y)$ is the neighborhood edge image and $\{(x_n, y_n)\}$ are the image coordinates for the chip created from the map. The chip coordinates are floating-point numbers as well as $(\Delta x, \Delta y)$, but $E(x, y)$ is an edge image indexed by integers; therefore, an interpolation scheme (e.g., bilinear) is implicit in Eq. (15.1). The RPM uses the Sobel operator to create the edge image and then masks pixels that have been classified as cloud edges. The similarity metric is essentially a correlation between neighborhood edge image and map shorelines. Other algorithms can also work well, for example, the neighborhood may be super-sampled and cross-correlated with a template created from the map database.

Relative error between images may be inferred by differencing the absolute error measurements made at the same site in each image. It may also be directly measured by cross-correlation of one image with the other. The former approach has proved to be very accurate for measuring the coregistration between different spectral bands, while the latter approach has proved to be better for measuring frame-to-frame motion between images taken at different times in the same spectral band. The cross-correlation algorithm extracts a subset C from one image neighborhood and forms a correlation surface by two-dimensional normalized cross-correlation with the neighborhood N from the other image (Gonzalez and Woods, 1992). Subpixel precision is achieved by interpolation to the maximum of the normalized cross-correlation surface $\rho_{rel}(\Delta x, \Delta y)$; specifically, one assumes that within a pixel of the maximum $(\Delta x_0, \Delta y_0)$, the correlation surface may be represented by a low-order function of the form

$$\rho_{rel}(\Delta x, \Delta y) \cong \rho_0 - (\Delta x - \Delta x_0, \Delta y - \Delta y_0) M (\Delta x - \Delta x_0, \Delta y - \Delta y_0)^T,$$
(15.2)

where M is a positive-definite symmetric matrix. Fitting in the vicinity of the correlation image maximum determines the correlation coefficient ρ_0, the shift $(\Delta x_0, \Delta y_0)$, and the sharpness of the peak (described by M). Similar algorithms are also in operational use for tracking cloud tracers or water vapor features for the purpose of deriving wind field science data products.

Distinguishing between valid and invalid measurements is the greatest challenge to implementing a fully automatic control point measurement system. Invalid measurements can occur because scenes are fully or partially covered by clouds, visible (VIS) illumination is poor, IR temperature contrast is poor, or sharp correlation peaks are not formed because features either lack sufficient topological structure or they possess too much topological structure at scales finer than the image resolution. IR absolute error measurements are particularly treacherous. Typically, land masses are warmer than water bodies during the day and colder at night, with most of the diurnal temperature variation being over land. Thus, high-positive-contrast scenes are plentiful during the day, but one must make do with low-negative-contrast scenes at night. Moreover, the retreat of the thermal boundary between land and sea is affected by littoral meteorology. One often observes small islands to shrink as beaches will cool more rapidly in the evening than the interior land. Overall, reliable IR absolute error measurements are more difficult to achieve during the nighttime and impossible when the scene contrast is changing signs. Figure 15.4 shows VIS and IR (channel 2) landmark production rates for a year of RPM operational experience spanning 2004 and 2005. The diurnal patterns of solar illumination and heating are apparent in the number of valid landmarks produced each hour.

15.4 Cloud masking and quality metrics

Cloud masking is essential for reliable absolute and relative error measurements. The RPM and other systems include algorithms to detect clouds based on multispectral, local spatial coherence, and multitemporal tests (Madani *et al.*, 2004). Such algorithms are similar to those used for scientific assessments of cloud cover, but biased towards false-positive cloud identification error. Cloud fraction, peak similarity metric or peak cross-correlation, (visible) illumination, and IR contrast all individually indicate the quality of an absolute or relative error measurement. In the RPM, these indicators are combined together using a fuzzy-logic algorithm to calculate an overall Quality Metric (QM) used to automatically discriminate between valid and invalid measurements. Absolute error measurements with QM > 0.9 for the visible channel and QM > 0.87 for the IR channels are generally considered valid. The IR standard may be relaxed to permit greater production of nighttime measurements. Relative error measurements with peak correlation >0.96 and masked cloud fraction <0.05 are similarly considered valid. Measurements can be further screened using dynamic models to test for physical reasonableness and consistency checks between absolute and relative measurements. The RPM includes a Kalman filter with a simple dynamic model that tracks navigation error and channel-to-channel alignment states. This model assumes that neither navigation error nor channel-to-channel alignment can change too quickly, providing a basis for identifying and vetoing invalid data as being statistically implausible. The fitting process for orbit and attitude estimation can also identify implausible data according to the statistics of its fit residuals. Consistency checks similarly test the reasonableness of two absolute measurements linked by a relative one. Suppose that A_n and A_{n-1} are two absolute error vectors giving the residual image navigation error in rows and columns for a common landmark site measured in successive images. Further, suppose that $R_{n,n-1}$ is the relative error vector for the same site that has been independently measured between a pair of images. One might suppose

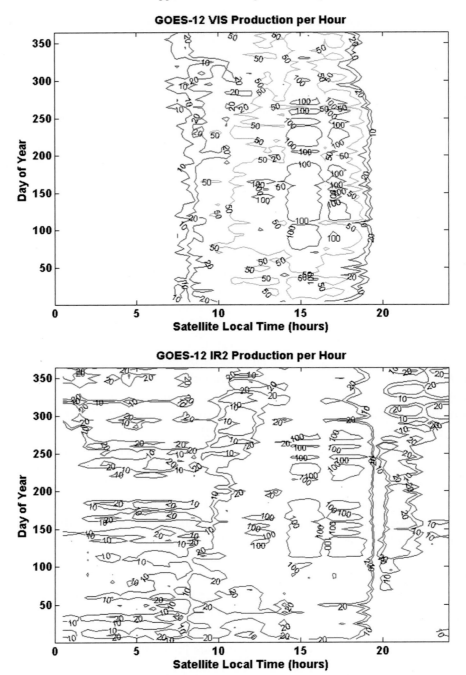

Figure 15.4. VIS and IR (channel 2) landmark production rates for a year of Replacement Product Monitor (RPM) operational experience spanning 2004 and 2005. See color plates section.

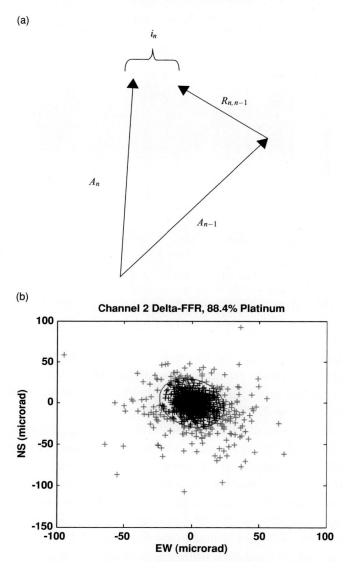

Figure 15.5. (a) The two absolute error vectors A_n and A_{n-1} giving the residual image navigation error in rows and columns for a common landmark site measured in successive images, and (b) plot of the IR channel 2 inconsistency data.

that the vector diagram in Fig. 15.5(a) would close; however, measurement errors will introduce a small inconsistency,

$$i_n = A_n - A_{n-1} - R_{n,n-1}. \tag{15.3}$$

Such inconsistency data can be used to identify the most trustworthy subset of data, those with the least inconsistency, which are called "platinum" landmarks

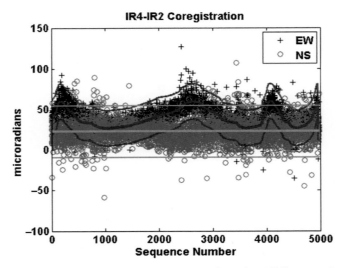

Figure 15.6. Measured channel-to-channel coregistration (differences in absolute errors at the same time and the same site in different bands), covering about three days of GOES-12 in-orbit experience. See color plates section.

(Carr and Madani, 2007). For the IR channel 2 inconsistency data that are plotted in Fig. 15.5(b), a 2σ ellipse identifies the platinum subset. Indeed, such screening is required to reliably assess specification compliance at 3σ levels, defined for the GOES programs as the 99.7th error percentile.

It is difficult to assess the accuracy of the absolute and relative error measurements from the RPM. One is tempted to analyze the statistics of the inconsistency data; however, inconsistency data is immune to errors that are correlated between frames, so it only shows evidence of uncorrelated random errors. Clearly, from Fig. 15.5, the random error component is much less than one pixel because an IR pixel spans 112 microrad. One may also get the sense that the accuracy of the absolute error measurements is much less than a pixel by examining measured channel-to-channel coregistration (differences in absolute errors at the same time and the same site in different bands) as shown in Fig. 15.6. The plot covers about three days of GOES-12 in-orbit experience. IR4-IR2 coregistration data have been plotted versus chronological ordinal number rather than time to avoid bunching up during daylight hours of peak landmark production. A model fits E–W and N–S coregistration components as harmonic functions of time. The fits and their 3σ-residual error bounds have been overplotted. The 3σ-error bounds are clearly less than an IR pixel, showing that this method is very effective for determining channel-to-channel coregistration corrections. However, it is similarly blind to correlated errors in the absolute error measurements. Overall, this sort of analysis

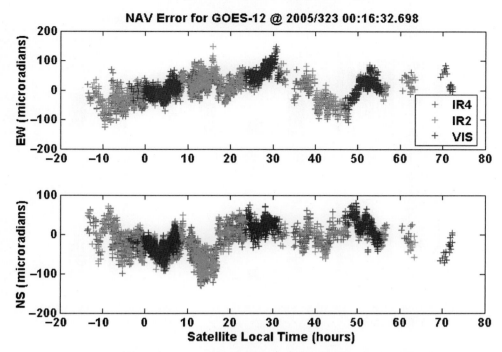

Figure 15.7. Absolute navigation error observed for GOES-12 during operations using platinum quality data from the RPM. See color plates section.

is consistent with random errors that are on the order of 0.15 pixels for relative error measurements and 0.25–0.30 pixels for absolute error measurements, but nothing is learned about systematic errors in the absolute error measurements. In fact, there have been RPM catalogs with evidence of mapping errors at the kilometer scale, so future systems should consider alternatives to the GSHHS digital map.

Figure 15.7 plots the absolute navigation error observed for GOES-12 during operations using platinum quality data from the RPM. Overall statistical metrics of INR performance have been synthesized and plotted in Fig. 15.8 using these measurements together with the relative error measurements. Navigation (NAV) error describes the error in knowledge of the geographic location of pixels, Within Frame Registration (WIFR) error describes nontranslational image distortion, and Frame-to-Frame Registration (FFR) error describes image stability. FFR is computed over 15-minute and 90-minute intervals and is perhaps the most important INR quality metric. It describes the quality of the movies made of the Earth and how well the imaging system has conformed to the paradigm that the clouds can move but the Earth must stay still.

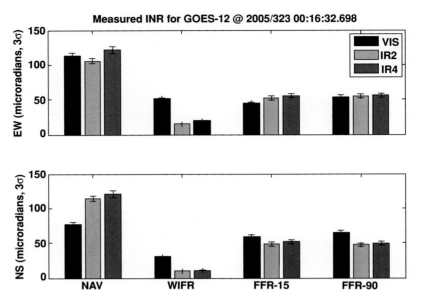

Figure 15.8. Overall statistical metrics of the Image Navigation and Registration (INR) performance.

15.5 Conclusion

Because weather is large scale and highly dynamic, meteorological satellites are deployed into GEO as well as LEO orbits. The high altitude of GEO orbit makes controlling georegistration errors especially challenging. Image Navigation and Registration (INR) systems permit accurate geolocation (navigation), with control and stabilization of the image geometry (registration). Images are generally corrected either with onboard Image Motion Compensation (IMC) or by on-ground resampling. In either case, sequences of frames are produced that form movie loops showing the dynamic evolution of the weather. Absolute geolocation error is important to identify the geographic locations of severe weather. Relative registration between frames is also very important because tracking tracers in movie loops provides quantitative measurements of wind fields. Measurements of stars and control points (landmarks) made through the aperture of the instruments are used in the INR processing. Landmarks provide data on INR performance, and the technology for automatically measuring landmarks with subpixel accuracy is already quite mature.

References

Blancke, B. and Carr, J. L. (1996). Liquid migration and the static stability of spinning spacecraft. *The Journal of the Astronautical Sciences*, **44**(3), pp. 277–292.

Blancke, B., Carr, J. L., Mangolini, M., and Pourcelot, B. (1998). Processing and quality measurement of METEOSAT Second Generation images: The IQGSE software. In *Proceedings of the 49th IAF International Astronautical Congress*, Melbourne, Australia, IAF Paper 96-B601.

Blancke, B., Carr, J., Pomport, F., Rombaut, D., Pourcelot, B., and Mangolini, M. (1997). The MSG image quality ground support equipment. In *Proceedings of the EUMETSAT Meteorological Satellite Data Users' Conference*, Brussels, Belgium. Unpaginated CD-ROM.

Carr, J. L. (1992). Image navigation for geostationary weather satellites. In *Proceedings of the 4th International Space Conference of Pacific-Basin Societies*, Kyoto, Japan, November 1991; *Advances in the Astronautical Sciences*, P. M. Bainum, G. L. May, M. Nagatomo, Y. Ohkami, and Y. Jiachi, eds., **77**, pp. 513–531.

Carr, J. L. (1996). Long-term stability of GOES-8 and -9 attitude control. In *Proceedings of the SPIE Conference on GOES-8 and Beyond*, Denver, CO, Vol. 2812, pp. 709–720.

Carr, J. L. (2008). Twenty-five years of INR. In *Proceedings of the American Astronautical Society F. Landis Markley Astronautics Symposium*, Cambridge, MD, pp. 867–876.

Carr, J. L. and Madani, H. (2007). Measuring image navigation and registration performance at the 3-σ level using platinum quality landmarks. In *Proceedings of the 20th International Symposium on Space Flight Dynamics*, Annapolis, MD, NASA/CP-2007–214158. Unpaginated CD-ROM.

Carr, J., Gibbs, B., Madani, H., and Allen, N. (2008). INR performance of the GOES-13 spacecraft measured in orbit. In *Proceedings of the 31st Annual AAS Guidance and Control Conference*, Breckenridge, CO, Vol. 131, pp. 663–680.

Carr, J. L., Mangolini, M., Pourcelot, B., and Baucom, A. W. (1997). Automated and robust image geometry measurement techniques with applications to meteorological satellite imaging. In *Proceedings of the CESDIS Image Registration Workshop*, NASA Goddard Space Flight Center, Greenbelt, MD, pp. 89–99.

Carson, C., Carr, J. L., and Sayal, C. (2008). GOES-13 end-to-end INR performance verification and post launch testing. In *Proceedings of the American Meteorological Society 5th GOES Users' Conference*, New Orleans, LA, Poster No. Pl. 26.

Gerber, A. J., Baron, R. L., Tralli, D. M., Crison, M., Bajpai, S., and Dittberner, G. (2004). Medium Earth orbit architecture for an integrated environmental satellite system. *Earth Observation Magazine*, **13**(6), pp. 10–15.

Gonzalez, R. C. and Woods, R. E. (1992). *Digital Image Processing*, 3rd edn., Reading, MA: Addison-Wesley.

Harris, J., Carr, J., and Chu, D. (2001). A long-term characterization of GOES I-M attitude errors. In *Proceedings of the NASA Flight Mechanics Symposium*, NASA Goddard Space Flight Center, Greenbelt, MD, pp. 217–228.

Kamel, A. (1996). GOES image navigation and registration system. In *Proceedings of the SPIE Conference on GOES-8 and Beyond*, Denver, CO, Vol. 2812, pp. 766–776.

Madani, H., Carr, J. L., and Schoeser, C. (2004). Image registration using AutoLandmark. In *Proceedings of the IEEE International Geoscience and Remote Sensing Symposium*, Anchorage, Alaska, Vol. VI, pp. 3778–3781.

NOAA (2005). *Earth Location User's Guide (ELUG)*. DRL 504–11, Revision 2 NOAA-GOES/OSD3-1998-015R2UD0, DCN 2.

Riishojgaard, L. P. (2005). High-latitude winds from Molniya orbit: A mission concept for NASA's ESSP program. In *Proceedings of the American Meteorological Society*

Ninth Symposium on Integrated Observing and Assimilation Systems for the Atmosphere, Oceans, and Land Surface, San Diego, CA, Session 6, Paper 604.

Tehranian S., Carr, J. L., Yang, S., Madani, M., Vasanth, S., DiRosario, R., McKenzie, K., Schmit, T., and Swaroop, A. (2008). Remapping GOES Imager instrument data for South American operations: Implementing the XGOHI system. In *Proceedings of the American Meteorological Society 5th GOES Users' Conference*, New Orleans, LA. Unpaginated CD-ROM.

Wessel, P. and Smith, W. H. F. (1996). A global self-consistent, high resolution shoreline database. *Journal of Geophysical Research*, **101**(B4), 8741–8743.

16

Challenges, solutions, and applications of accurate multiangle image registration: Lessons learned from MISR

VELJKO M. JOVANOVIC, DAVID J. DINER, AND ROGER DAVIES

Abstract

A novel approach implemented to meet coregistration/georectification require-ments and continuous data-intensive processing demands of the Multi-angle Imag-ing SpectroRadiometer (MISR) science data system has been in operation since the beginning of on-orbit data acquisition in February 2000. Remote sensing image data are typically only radiometrically and spectrally corrected as a part of standard processing, prior to being distributed to investigators. In the case of MISR, with its unique configuration of nine fixed pushbroom cameras, continuous and autonomous coregistration and geolocation of the data are essential prior to application of any subsequent scientific retrieval algorithm. A fully automated system for continuous orthorectification, including removal of errors related to camera internal geometry, spacecraft attitude data, and surface topography, has been implemented.

The challenges involved in employing such a system range from purely algo-rithmic issues to those related to limitations on computational resources and data volumes. Processing algorithms had to be designed so that $\sim$35 GB of image data per day are orthorectified without interruption and with high fidelity, as verified by an automated quality assessment process. We adopted a processing strategy that distributes the effort between the MISR Science Computing Facility at the Jet Propulsion Laboratory in Pasadena, CA and the Distributed Active Archive Center (DAAC) at the NASA Langley Research Center, Hampton, VA.

Accurate geolocation and coregistration of multiangle, multispectral MISR data is critical for the higher-level science retrieval algorithms. Applications include retrieval of (1) global geometric cloud heights with sufficient accuracy and preci-sion to infer climate trends, (2) height-resolved tropospheric winds with pole-to-pole coverage as a proof-of-concept for design of a future operational system, and (3) stereoscopically derived aerosol plume heights as inputs to chemical transport models. Cloud geometric retrievals, where cloud-top heights and cloud motion vectors are retrieved simultaneously using camera triplets, are especially demanding on image coregistration accuracies. In addition, nonrigid cloud deformations and complex behavior of multilayered cloud surface are currently difficult to handle in an automated image-processing environment. Directions of future research, including 3D reconstructions from optical multiangle imagery, are outlined.

16.1 Introduction

The Multi-angle Imaging SpectroRadiometer (MISR) is a part of the payload for NASA's Terra spacecraft, which was launched in December 1999 (Diner *et al.*, 1998). MISR continuously acquires systematic, global, multiangle, and multispectral imagery in reflected sunlight, with 275-m spatial resolution over a wide range of viewing angles at the surface (nadir to 70° off-nadir). Remote sensing image data are typically only radiometrically and spectrally corrected as a part of standard processing, prior to being distributed to investigators. In the case of MISR, with its unique configuration of nine fixed pushbroom cameras, continuous and autonomous coregistration and geolocation of the data are essential prior to applying any subsequent scientific retrieval algorithms.

In order to meet pixel-level (or better) image registration requirements, errors related to camera internal geometry, spacecraft attitude data, and surface topography have to be taken into account while removing geometric distortions. Traditionally, reduction of the errors associated with the sensor/surface viewing geometry is done in a semi-interactive mode for a targeting sensor where images of relatively small spatial and temporal extent can be controlled by few ground control points and an experienced operator. MISR global processing necessitates a fully automated system for continuous orthorectification and error reduction. The challenges involved in implementing such a system range from purely algorithmic issues to those related to limitations on computational resources and data volumes. Imagery acquired over the wide range of view angles has to be processed simultaneously, and the oblique angles have high sensitivity to pointing errors and geometric distortions. Processing algorithms had to be designed so that ~35 GB of image data per day are orthorectified without interruption and with high fidelity, as verified by an automated quality assessment process (Bothwell *et al.*, 2002). Other challenges are related to the stability of camera internal pointing geometries, varying accuracy of spacecraft attitude data, overall availability of imagery suitable for image matching, and availability of accurate digital elevation models.

To address these challenges, the data production algorithm is based on the integration of digital photogrammetry methods, including the following: (1) Area-based/feature-based image matching, (2) image point intersection, (3) space resection, (4) simultaneous bundle adjustment, and (5) image-to-image registration

(Jovanovic *et al.*, 1998). We adopted a processing strategy that distributes the effort between the MISR Science Computing Facility at the Jet Propulsion Laboratory (JPL) in Pasadena, CA, and the Distributed Active Archive Center (DAAC) at the NASA Langley Research Center, Hampton, VA. This minimizes the amount of off-line processing required at the DAAC, where routine product generation occurs. In-flight geometric calibration activities at the Science Computing Facility generate specialized datasets, which are then used as inputs to standard production at the DAAC.

Accurate geolocation and coregistration of multiangle, multispectral MISR data is critical for the higher-level science retrieval algorithms. Experiment objectives include studies of the abundance and types of tropospheric aerosols and measurements of surface structure (i.e., vegetation canopy architecture and ice sheet roughness). Applications placing the greatest demand on camera coregistration accuracy include retrievals of (1) global geometric cloud heights with sufficient accuracy and precision to infer climate trends, (2) height-resolved tropospheric winds with pole-to-pole coverage as a proof-of-concept for design of a future operational system, and (3) stereoscopically derived aerosol plume heights as inputs to chemical transport models. Cloud geometric retrievals, where cloud-top heights and cloud motion vectors are retrieved simultaneously using camera triplets, are especially demanding on image coregistration accuracies. In addition, nonrigid cloud deformations and complex behavior of multilayered cloud surface are currently difficult to handle in an automated image-processing environment. Directions of future research, including 3D reconstructions from optical multiangle imagery, are outlined.

16.2 MISR mission overview: Imaging event

The MISR instrument consists of nine pushbroom cameras. The cameras are arranged with one camera pointing toward the nadir (designated An), one bank of four cameras pointing in the forward direction (designated Af, Bf, Cf, and Df in order of increasing off-nadir angle), and one bank of four cameras pointing in the aftward direction (using the same convention but designated Aa, Ba, Ca, and Da). Images are acquired with nominal view angles, relative to the surface reference ellipsoid, of $0.0°$, $\pm 26.1°$, $\pm 45.6°$, $\pm 60.0°$, and $\pm 70.5°$ for An, Af/Aa, Bf/Ba, Cf/Ca, and Df/Da, respectively. The instantaneous displacement in the along-track direction between the Df and Da views is about 2800 km (see Fig. 16.1), and it takes about 7 minutes for a ground target to be observed by all nine cameras.

Each camera uses four charge-coupled device line arrays parallel in a single focal plane. The line array contains 1504 photoactive pixels, each 21 μm × 18 μm. Each line array is filtered to provide one of four MISR spectral bands. The spectral

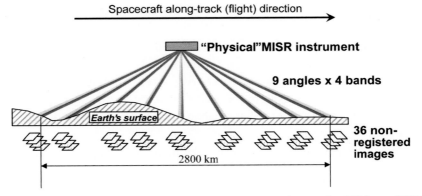

Figure 16.1. MISR imaging event. (Source: Jovanovic *et al.*, 1998, © IEEE, reprinted with permission.)

band shapes are approximately Gaussian, and are centered at 446, 558, 672, and 866 nm in the reflected portion of the solar spectrum.

The letter designation for the cameras (A, B, C, D) indicate different effective focal lengths for the lenses, with A being the shortest (59 mm) and D being the longest (124 mm). Focal length increases with increasing off-nadir viewing angle in order to preserve cross-track sample spacing. To simplify manufacturing, the same optical design was used for the An, Af, and Aa cameras, resulting in slightly different cross-track instantaneous fields of view. The cross-track instantaneous field of view and sample spacing of each pixel is 275 m for all of the off-nadir cameras, and 250 m for the nadir camera. The along-track instantaneous fields of view depend on view angle, ranging from 250 m in the nadir to 707 m at the most oblique angle. Sample spacing in the along-track direction is 275 m in all cameras owing to the 40.8-msec line repeat time and the 6.74-km/sec ground speed of the spacecraft.

16.3 Required georectified and coregistered data products

MISR multiangle multispectral imagery is continuously acquired from pole-to-pole on the day side of each orbit. Higher-level science retrieval algorithms require coregistered and geolocated data to be provided in a form suitable for an automated processing on an orbit-by-orbit basis. We have selected Space Oblique Mercator (Snyder, 1987) as the reference map projection grid, because it is designed for continuous mapping of satellite imagery with minimal distortions. The ground resolution of the map grid is 275 m. We define this segment of ground processing as "georectification," and the derived product as the georectified radiance product (GRP). There are two basic parameters in the GRP depending on the definition of the reflecting surface: (1) Ellipsoid-projected radiance, and (2) terrain-projected

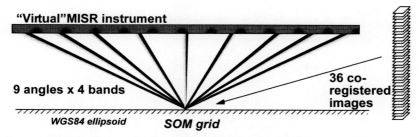

Figure 16.2. Ellipsoid-projected radiance product as it would be acquired from "virtual" MISR. (Source: Jovanovic *et al.*, 1998, © IEEE, reprinted with permission.)

radiance. The ellipsoid-projected radiance is referenced to the surface of the WGS84 ellipsoid (no terrain elevation included) and the terrain-projected radiance is referenced to a Digital Elevation Model (DEM) over land and inland water. An ideal instrument would collect each angular view for the terrain-projected and ellipsoid-projected radiance parameters for a ground point at the same instant, giving the radiance for each band and angle for that ground point (the so-called "virtual" MISR instrument, see Fig. 16.2). The terrain-projected radiance product is in fact an orthorectified map of multiple angle data used for surface retrievals. The ellipsoid-projected radiance product can be thought of as approximations of multiangle data reprojected to epipolar geometry suitable for subsequent cloud height/wind retrievals.

The spatial horizontal accuracy requirements are driven by the needs of the geophysical parameters retrieval algorithms, especially those designed to simultaneously derive cloud-top heights and flow field (wind) data (Moroney *et al.*, 2002a; 2002b). The initial goal is to achieve an uncertainty better than $\pm140\,\text{m}\,(1\sigma)$ regarding both the absolute geolocation of nadir camera and coregistration between all nine cameras. The 140 m corresponds to about half the size of (1) the instrument ground sampling distance, (2) the nadir camera instantaneous field of view, and (3) the GRP ground resolution. It should be pointed out that this requirement assumes no bias in coregistration between nine cameras. It has been estimated (Davies *et al.*, 2007) and confirmed with the real mission data that even a small coregistration bias of one tenth of the pixel (i.e., 27.5 m) will increase the *root mean square* (RMS) error in retrieved wind velocity vectors up to several meters per second.

16.4 Production algorithm

In response to the specific spatial accuracy requirements, together with the need for autonomous and continuous production capabilities, we implemented a processing

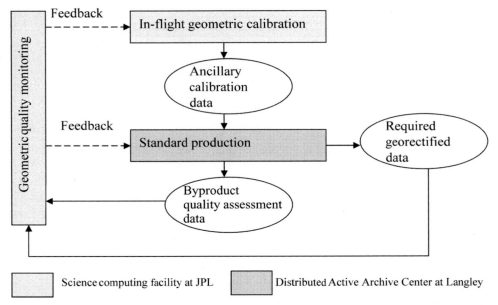

Figure 16.3. Partitioning georectification and coregistration operations. (Source: Jovanovic *et al.*, 2007, © Elsevier science, reprinted with permission.)

strategy which partitions the effort between the MISR Science Computing Facility and the EOS DAAC in a way that minimizes the amount of processing required at the latter location (see Fig. 16.3).

Activities at the Science Computing Facility lower the computational need at the DAAC by precalculating certain datasets early in the mission and staging them for ongoing use, in a manner that avoids much calculation during routine ground processing. These datasets include the camera geometric model, reference orbit imagery, and projection parameters. Their preparation, completed in stages at the beginning of the mission, was highly computationally intensive, involving techniques such as ray casting, simultaneous bundle adjustment, and the matching of imagery from different camera angles.

Consequently, routine processing of MISR data at the DAAC, the characteristics of which are dominated by the very high data volume, is optimized to require only less computationally intensive work, such as matching of imagery from the same camera angle (not different camera angles) with no need for ray casting nor a high-resolution digital elevation model.

In addition, a geometric quality monitoring system was implemented to complement the production system, in order to verify geometric performance on a global basis over specific time periods. This system was particularly effective in identifying a problem with the pointing stability of one of the most oblique cameras (Da) as well as verifying the performance of the other cameras relative to the requirements of the MISR wind and height retrievals.

In-flight geometric calibration In-flight geometric calibration activities have been designed to produce specialized datasets, which are then used as inputs to standard production. These datasets not only reduce the overall processing load but also assure the required georectification accuracy, taking into account various error sources. The errors can be categorized into three groups: (1) Static pointing errors, (2) dynamic pointing errors, and (3) errors associated with the topography of the projection surface. The camera geometric model (CGM) has been designed and completed with the aim of removing static pointing errors. In addition, projection parameters (PP) and reference orbit imagery (ROI) are created to provide for removals of: (1) Parallax errors due to surface topography, and (2) any remaining errors in the camera pointing geometry, including errors in the spacecraft navigation and attitude.

16.4.1 Camera Geometric Model (CGM)

The CGM dataset is designed to deal with static pointing errors. It consists of a set of parameters used in a mathematical expression that gives the pointing direction of an arbitrary pixel to the spacecraft attitude frame of reference. These parameters represent the geometry of the camera system and account for distortions from an ideal optical system (Korechoff *et al.*, 1996). There are nine sets of parameters corresponding to nine MISR cameras. The mathematical expression relating line and sample (l, s) coordinates of an image band in one of MISR cameras to the vector $\hat{r}_{scs}$ in the spacecraft coordinate system can be written as

$$\hat{r}_{scs} = T_{si} \cdot T_{ic} \cdot T_{cd} \cdot \begin{bmatrix} -(k + (l - \lfloor l + 0.5 \rfloor)d_x) \\ f \sum_{i=0}^{5} \alpha_i (s - c_y)^i \\ f \end{bmatrix}, \qquad (16.1)$$

where

T_{si} is a rotation matrix, function of the angles between spacecraft and instrument coordinate system;

T_{ic} is a rotation matrix, function of the angles between instrument and camera coordinate system;

T_{cd} is a rotation matrix, function of the angles between camera and detector coordinate system;

k is the offset of a particular band from the intersection between z-axis of the detector coordinate system (the normal to the focal plane) and a reference position in the focal plane;

c_y is the pixel number (i.e., boresight pixel) corresponding to the x-axis of the detector coordinate system;

f is the effective focal length;

$\alpha_i, i = 0, 1, 2, 3, 4, 5$ are coefficients of a fifth-order polynomial to account for nonlinear distortions of the field angle in the cross-track direction.

The in-flight calibration of the CGM is done on a camera-by-camera basis utilizing a set of globally distributed ground control points (GCPs). The MISR GCP database contains 120 individual points distributed across all latitudes, the majority of which are in the USA. About 50 points are equally distributed across Russia, Africa, and South America. The remaining points are located in Australia and New Zealand. A single MISR ground control point is a collection of nine geolocated image patches of a well-defined and easily identifiable ground feature. The database construction involved two stages. Firstly, the acquisition and production of terrain-corrected Landsat TM scenes over the desired ground locations. The second is the extraction of image chips from the TM imagery and update of the GCPO database. The selection of candidate ground location and TM scenes was carried out in collaboration with MODIS and EROS Data Center (Bailey *et al.*, 1997). Seasonally invariant features (e.g., man-made objects, coastlines) were the first choice for GCPs. Once the Landsat TM scenes were selected, EROS Data Center was responsible for precise geometric processing and terrain correction of these scenes prior to its distribution to MODIS and MISR teams. This terrain-corrected imagery is then used as the input to ray-casting simulation software. The software replicates MISR viewing geometry producing nine images (corresponding to nine MISR cameras), which are then used for the extraction of smaller image chips. This warping of TM imagery is necessary in order to obtain image chips with the best possible chances to be identified in corresponding MISR imagery. The corresponding geodetic coordinates define the location of a particular ground feature. If $(X_{ctr}, Y_{ctr}, Z_{ctr})$ are the geodetic coordinates of a particular ground feature, then Eq. (16.1) can be expanded as follows to describe the ray between a ground point and the image of that point, as seen by a MISR camera:

$$T_{gc} \begin{bmatrix} X_{ctr} \\ Y_{ctr} \\ Z_{ctr} \end{bmatrix} = \widehat{P}_{gci} + \lambda T_{gs} \widehat{r}_{scs}. \tag{16.2}$$

The variables $\widehat{P}_{gci}$ and T_{gs} are functions of spacecraft ephemeris and attitude data representing the position and transformation between spacecraft and geocentric inertial coordinates system at the time of imaging. T_{gc} is the transformation between conventional terrestrial and geocentric inertial coordinate systems, and λ is a scale factor. Equation (16.2), known as the collinearity condition, is used for least-squares estimation (Mikhail, 1976) of certain camera model parameters, that is, space resection (Paderes *et al.*, 1989). A large number of GCP observations over a period of time were made so that static errors of the individual cameras were isolated from slowly varying errors in the navigation data or any other source. The process of accessing, identifying, and measuring GCP is fully automatic. It is built on a combination of area-based cross-correlation and least-squares image matching, providing an accuracy of 1/8 pixel (1σ) (Förstner, 1982; Ackermann, 1984).

16.4.2 Projection parameters (PP) and reference orbit imagery (ROI)

In order to routinely deal with slowly varying pointing errors, and facilitate rapid and automated geometric processing and quality assessments, a specialized set of files was produced at the beginning of the mission and delivered as input to the standard production system. A total of 233 by 9 pairs of PP and ROI files have been created, one pair for each of the 233 unique Terra orbit paths multiplied by 9 MISR cameras. A PP file provides geolocation information for a corresponding ROI file on a pixel-by-pixel basis for an entire orbit path. This geolocation information is referenced to the selected Space Oblique Mercator map projection grid. A corresponding ROI file consists of selected cloud-free MISR imagery, extracted and mosaicked from four to five orbit passes over the same orbital path. The ROI is produced using rigorous photogrammetric methods to accurately fit geolocation data contained in the corresponding PP file pair. The process of creating PP and ROI pairs can be thought as similar to regular orthorectification of time-dependent imagery. A major difference is that the acquired imagery (ROI) is geolocated through PP but not resampled to the map projection grid. It is preserved in its as-acquired viewing geometry (i.e., image coordinate space).

The coupled PP and ROI files provide two major benefits to standard georectification processing. First, expensive computation required to account for topographic displacement is performed only once, off-line. The resulting information is saved in a file and utilized during on-line processing throughout the mission. This is possible because of the small orbit-to-orbit variations at the same location within an orbit path, adding relatively small changes to the topographic displacements that can be accounted for in a separate process during georectification. Second, nonresampled but geolocated MISR imagery is used as ground control information. MISR imagery with close to the same viewing geometry provides for a high success rate of the least-square area-based image matching performed in standard processing during image-to-image registration.

The creation of PP file involves a series of steps dealing with sensor-to-ground and ground-to-sensor projection algorithms based on nominal spacecraft position and orientation, and calibrated CGM parameters. A global DEM of nominally 100-meter resolution is used to define the terrain-projection surface (Logan, 1999). A sensor-to-ground ray casting algorithm is applied to find locations in the map projection grid which cannot be seen by a particular MISR view angle because it is topographically obscured by the surrounding terrain. As a second step, the location in the MISR imagery where the center of each map grid pixel is seen is determined for each camera angle. This is done through ground-to-image projection where an iterative root-finding method is used to solve for the image coordinates in Eq. (16.2).

With the goal of assuring an accurate fit between geolocation data in the PP and imagery in the ROI, a process of accurate determination of viewing geometry to account for all error sources was carried out. This involves a method called simultaneous bundle adjustment which takes advantage of the following MISR characteristics: (1) At a single instant of time MISR "sees" nine different, widely separated, targets on the ground, and (2) a single location on the ground is seen at nine different instants of time. Slowly varying position errors are modeled as time-dependent linear variations for the entire orbit while attitude errors are modeled as time-dependent spline curves. These error models are included in Eq. (16.2), so that the terms $\hat{P}_{gci}$ and T_{gs} are modified. As such, Eq. (16.2) leads to an overdetermined system of equations given that a sufficient number of tie-points has been extracted from multiple MISR images.

A tie-point refers to the conjugate image feature locations of the same ground point across multiple images viewed from various angles. Based on initial conjugate image locations determined using known MISR navigation data, interest point features (Förstner and Gülch, 1987) are detected independently on all nine local conjugate image patches extracted from MISR imagery. A feature-based matching scheme, namely consistent labeling with forward check (Castleman, 1979), is used to match conjugate interest points as improved tie-points. In particular, a relational matching algorithm was implemented so that the problem of mapping of one set of features with another is treated as a consistent labeling problem. The algorithm for solving the consistent labeling problem is based on the forward checking tree search (Haralick and Shapiro, 1993, Vol. II, p. 379), modified to expand along a potential branch according to its total benefit (Zong, 2002). The accuracy of relational-based feature matching is about two pixels due to both image distortions and the relaxed geometric constraints. Area-based matchers then refine the accuracy to subpixels.

In addition to the tie-points and GCPs we also included the global DEM as an additional constraint to the simultaneous bundle adjustment. The system of equations underlying simultaneous bundle adjustment cannot be solved exactly, so the equality should be taken in a least-square sense, weighted by the appropriate covariance matrix. In particular, we let the ground location of the points as well as the certain ephemeris and attitude parameters vary, scaled by their respective covariance matrixes. The solution is then obtained using a standard Levenberg-Marquardt method (Dennis and Schnabel, 1983).

Once the accurate viewing geometry is known for the data to be used in creation of the ROI, images of interest are resampled to look as acquired under conditions of nominal navigation, thereby fully complementing the geolocation data in the PP file. As the final step in the creation of ROI we used the radiometric camera-by-camera cloud mask (RCCM) derived through a series of thresholding tests (Diner *et al.*, 1999a; Yang *et al.*, 2007) to assemble a mosaicked image from multiple passes in order to maximize cloud-free regions.

16.4.3 Standard production

With the geometric calibration datasets as input, the georectification and coregistration approaches during standard processing are significantly simplified. The major information implicitly contained in these datasets is error-free navigation and attitude data, georeferenced, with surface topography relative to the various geometries of the nine MISR cameras. This information is routinely exploited through a hybrid image registration algorithm. In particular, a central part of the autonomous and continuous georectification is a recursive image registration between ROI and acquired MISR imagery which consists of the following steps: (1) Image point intersection, i.e., a backward projection function used to provide an initial location of the conjugate points (Paderes *et al.*, 1989), (2) image matching for the precise identification of the tie-points (Ackermann, 1984), and (3) generation of the transformation (mapping) function between two images (see Fig. 16.4).

The registration method is adaptive with regard to the character and size of image displacement, in order to minimize the processing load. The adaptive nature of the algorithm is attained by recursively dividing images into subregions until the required registration accuracy is achieved. An assessment of the registration accuracy which is tested against a user-defined threshold is obtained as a part of the least-square estimation of the mapping function. The mapping function associated with a subregion is a simple modification of the affine transform appropriate to model distortions between two images due to Earth curvature, coupled with small differences in the navigation data:

$$
\begin{aligned}
L_{new} &= L_{ref} + K_1(S_{ref} - S_{center}) + K_2(S_{ref} - S_{center})^2 + K_3, \\
S_{new} &= K_4(S_{ref} - S_{center}) + K_5(S_{ref} - S_{center}) + K_6,
\end{aligned}
\tag{16.3}
$$

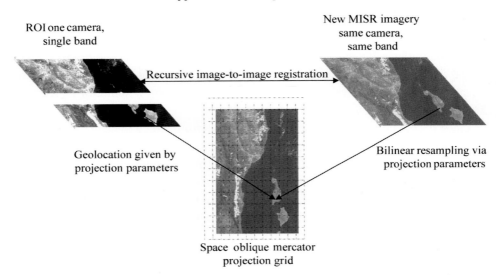

ROI one camera,
single band

New MISR imagery
same camera,
same band

Recursive image-to-image registration

Geolocation given by
projection parameters

Bilinear resampling via
projection parameters

Space oblique mercator
projection grid

Figure 16.4. Georectification algorithm as implemented for standard production.

where (L_{new}, S_{new}) are image coordinates in a newly acquired image of the same points located in the ROI at the location (L_{ref}, S_{ref}). Once a mapping between the two images is established, the last processing step is the assignment of the appropriate radiance value to the grid points of the Space Oblique Mercator map. This is done using bilinear interpolation. It should be noted that data obtained from image-to-image registration for the terrain-projected product are also used for the ellipsoid-projected product via slightly modified transformation functions.

As implemented, this simplified georectification and coregistration algorithm performed very well for eight out of the nine MISR cameras. The number and distribution of tie-points identified across an orbit segment during image-to-image registration are the important factors influencing the quality of the map-projected data. In order to improve tie-point distribution, a cloud-mask is included and the search area includes polar water regions where large ice features can be found as good tie-point candidates. In cases where a large region of data lacks conjugate points, use of information obtained through the registration of the closest subregion is applied. The idea is to correct for slowly varying parameters through the use of a Kalman filter built while processing previous subregions. However, we discovered that the pointing stability of one of the most oblique cameras (Da) was significantly worse than the stability of the other eight. The nominal algorithm as designed with only forward processing through the data was not adequate for the unexpected stability of the Da camera. Consequently, an update to the nominal algorithm has been made. This update is based on a second pass through the data so that tie-points collected over the entire orbit can be jointly used to best define the ROI-to-new image mapping prior to resampling. The georectification and coregistration performance measures prior to and after this update are included in the next section.

16.5 Operational results

This section summarizes operational results, including those related to calibration and those from the automated quality monitoring system. We focus on camera pointing stability and final georectification and coregistration accuracies.

16.5.1 Camera geometric model, PP, and ROI

The first operational results that addressed georectification and coregistration requirements were related to the performance of the camera geometric model (CGM) with its prelaunch calibration (Jovanovic *et al.*, 2002). Geolocation errors measured over a globally distributed set of GCPs (Bailey *et al.*, 1997) were used as input to the calibration algorithm. After several iterations, a final estimate of the CGM parameters was generated and included into operations in April, 2002.

Table 16.1 summarizes the calibration and quality assessment results. Results of the calibration adjustments for roll, pitch, and yaw angle (which define the orientation between the camera and spacecraft frames of reference) obtained using observations over a one-month time period are given. Validity of these newly calibrated parameters, as evaluated using observations of GCPs during a longer, six-month time period, is demonstrated with geolocation mean and RMS errors shown in the two last columns of Table 16.1. As expected, RMS errors are increasingly larger for more oblique cameras due to the fact that potential random pointing errors have a larger effect on those more oblique view angles. However, the mean error for Da is significantly larger than for any other camera, indicating a potential issue with either its calibration data, the mechanical stability of that camera, or some timing idiosyncrasy in the flight software. Even after careful analysis and recalibration a significant gelocation bias remained with the data from this camera. Further analyses and the results of the second-pass update are addressed in subsequent subsections.

Variability of the pointing errors for the other eight cameras over a longer time period performed as expected, and is mostly attributed to the small pointing errors contained in the spacecraft attitude data. As explained earlier, one of the reasons for the utilization of (PP, ROI) file pairs is to deal with this type of error. Simultaneous bundle adjustment, used as a part of PP and ROI creation, models remaining pointing errors as a cubic spline function in order to ensure accurate coregistration for all the images used in producing PP and ROI. Figure 16.5 illustrates the typical magnitude of pitch angle corrections as obtained for one day side of an orbit.

Table 16.1 *Results of the in-flight camera geometric model calibration and quality assessment*

Camera (as defined in Subsection 16.2.1)	In-flight geometric calibration results					Camera geometric model quality assessment		
	Number of observations	Corrections to pointing (arcsec)				Number of observations	Errors (units: 1 pixel = 275 m)	
		Roll	Pitch	Yaw			Mean	RMS
Df	43	327	5	−326		279	−0.16	1.00
Cf	63	323	8	86		428	0.09	0.81
Bf	61	520	3	343		506	−0.05	0.65
Af	61	526	−13	661		514	0.01	0.54
An	56	274	−2	774		473	0.06	0.48
Aa	73	−67	−9	767		627	0.07	0.52
Ba	65	−214	−13	557		620	0.03	0.57
Ca	57	−920	−95	972		596	0.02	0.66
Da	54	−1030	−106	757		507	0.89	0.99

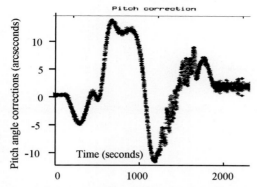

Figure 16.5. Pitch angle corrections modeled as a cubic spline function within simultaneous bundle adjustment for one orbit pass.

In addition to facilitating standard production, the (PP, ROI) pairs also provide for the qualitative assessment measurements used to evaluate (1) the pointing stability of all nine cameras, and (2) geolocation and coregistration errors in the final product. Pointing stability and errors in the final product on a global basis are presented in the next section.

16.5.2 Final validation results

The reference orbit imagery (ROI) provides a stable "ground truth" and is used as an integral part of standard production. The nine MISR camera images are compared to the ROIs in order to estimate coefficients of a two-dimensional transformation describing the fit between images for a predefined length of a labeled image block. These coefficients, called image coordinate corrections (ICC), were subsequently used in the processing software to optimally "warp" new images to match the reference. The ICC were also collected and summarized in order to give insight into pointing stability over desired time periods during the mission.

Figure 16.6 summarizes the quality assessment data that illustrate the pointing stability and coregistration performance for the period of eight years. It shows the plot of mean offsets in fractional pixels (ICC) taken every 500 orbits. As illustrated, pointing stability is fairly good for eight out of nine cameras. Some variations of approximately a one-year time period are considered acceptable and controllable by nominal standard processing. They appear to correspond to spacecraft inclination maneuver events, which were more prevalent at the beginning of the mission prior to the spacecraft reaching a stable orbit. This is not the case for the Da camera, which exhibits independent and very irregular pointing changes.

For the purpose of validation, the final geolocated products were scanned with a version of the image-to-image matching program that evaluated and reported

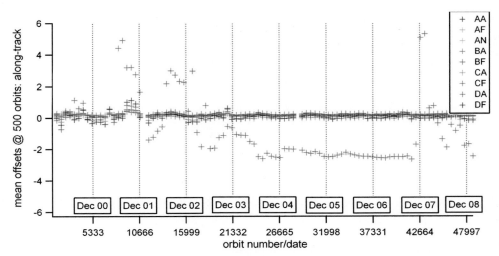

Figure 16.6. Global geometric performance for the nine MISR cameras; stability of pointing geometry shown with the mean values at 500 orbits plotted against orbit number. See color plates section.

coregistration errors for each camera relative to the nadir (An) using 275-meter resolution, red band images. The algorithm performed numerous image-matching operations and computed the mean, fractional pixel offset errors per fixed image segment size of 512 lines (called a block) for both along-track and across-track directions. In the following figures, we report only errors in the along-track direction. As can be seen in Fig. 16.7(a), in which errors are displayed for a four-year time period within the time range in Fig. 16.6, most of the static pointing biases are removed and the RMS errors are within the limits of the automatic image-matching algorithm. However, it is clear that nominal implementation of the standard production algorithm was inadequate to fully take into account the pointing instability of the Da camera.

The update to the standard processing, based on the second pass through the data, has been operational since May 2005. It is applied not only to the Da but also to Df, Ba, and Bf cameras in order to enhance accuracy of the simultaneous retrieval of cloud height and motion vectors, an application described in the following section.

As in Fig. 16.7(a), Fig. 16.7(b) shows the global quality of the georectified product by summarizing coregistration errors (in fractional pixels). However, this figure uses data obtained from the final version of the product. The reprocessed data available from years 2000–2008, at the time this chapter was written, show obvious improvement when compared with Fig. 16.6. Da camera performance is no longer significantly different from other cameras, and virtually all of the mean errors overlap within a ±0.1 pixel range.

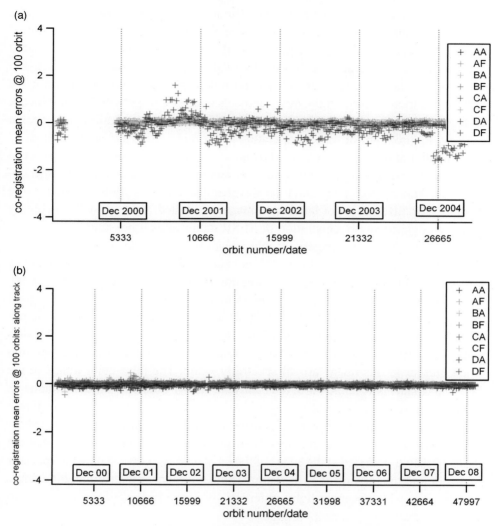

Figure 16.7. Global geometric performance for the nine MISR cameras; product accuracy given as mean coregistration error at 100 orbits plotted against orbit number: (a) Prior to update of standard production algorithm, and (b) final version product. See color plates section. (Source: Jovanovic *et al.*, 2007, © Elsevier science, reprinted with permission.)

16.6 Applications

16.6.1 Surface properties retrievals

Mapping of variations in bidirectional reflectance from pixel to pixel would not be possible without the great degree of care put into multiangle image registration, as described in the previous sections. Among the applications of such imagery is the retrieval of aerosol properties over land. The MISR aerosol algorithm models the shape of the surface bidirectional reflectance as a linear sum of angular empirical orthogonal functions derived directly from the image data, making use of spatial contrast and angular variation in the observed signal to separate the surface and atmospheric signals. This approach has been demonstrated to be highly effective even in situations where bright, dusty aerosols overlay a bright desert (Martonchik *et al.*, 2004; Diner *et al.*, 2005; Kahn *et al.*, 2005), which has traditionally been a challenging problem for sensors operating in the mid-visible and near-infrared.

Color digital imagery is generated by allocating the data from different spectral bands to the red, green, and blue planes of a color display, and when the spectral bands used are at red, green, and blue wavelengths, the imagery is known as "true color"; for a different set of bands the imagery is called "false color." The highly accurate angle-to-angle registration of MISR data makes it possible to generate a new type of false-color imagery, in which data from a single spectral band (e.g., the red band, collected globally at 275-m resolution) but at three different angles are used. When this is done, variations in color across the image serve as a proxy for changes in the angular shape of the bidirectional reflectance pattern. A good example was obtained on February 24, 2000, as MISR flew over Hudson and James Bays, Canada. The left-hand panel of Fig. 16.8 is a true-color image from MISR's nadir (An) camera. The bays are frozen at that time of year, and the ice, snow, and cloud present in this image are all spectrally white. The false-color image in the right-hand panel is a composite of red band data taken by the Ba, An, and Bf cameras displayed in red, green, and blue colors, respectively. In this rendering, different portions of the scene are now readily distinguishable. Low stratiform clouds show up in purple due to similar amounts of forward and backscatter at the B camera angles. Fast (smooth) ice at the edges of the bays scatters with a somewhat specular character into the forward-viewing camera, and thus appear with a light blue tinge. Rougher ice, which scatters light preferentially in the backward direction due to facets and shadowing, appears orange.

Based on the Hudson/James Bay example, Nolin *et al.* (2002) developed a proxy for surface roughness, consisting of a Normalized Difference Angular Index (NDAI) that uses a combination of forward and backward scattered radiation. NDAI is defined as *(back − fore)/(back + fore)*, where *back* and *fore* correspond to bidirectional reflectance factors at the 60° viewing angles of MISR that happen to be viewing predominantly backward and forward scattering, respectively. NDAI imagery for the area around the Jakobshavn Glacier in Western Greenland shows an increase in roughness associated with the formation of melt ponds between spring and fall. NDAI-derived surface roughness correlates well with Airborne Topographic Mapper (ATM) laser altimetry data (Nolin *et al.*, 2002; Diner *et al.*, 2005). Classification of sea ice types using surface roughness derived from MISR

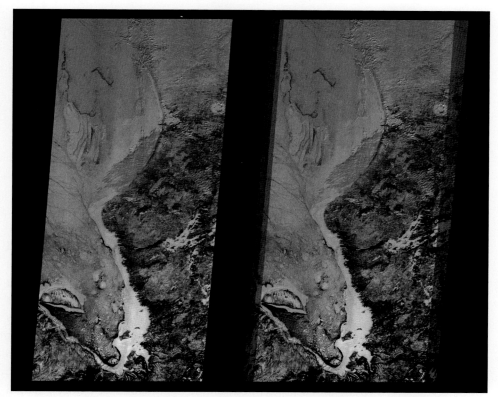

Figure 16.8. True color (left) and multiangle false color (right) images of the areas around Hudson and James Bays, Canada, generated from MISR data collected on February 24, 2000 during Terra orbit 995. These data were acquired shortly after the opening of the instrument's cover. See color plates section.

shows great potential for clear discrimination of new and older ice types. In a preliminary study by Nolin *et al.* (2002) over the Beaufort Sea, ice types within areas classified by the National Ice Center (NIC) as first-year ice were distinguishable by MISR but not discernible by synthetic aperture radar. In an Antarctic study, Nolin *et al.* (2002) used MISR data to characterize "blue ice" surfaces on the basis of their surface roughness. These regions of net mass loss are thought to be indicators of changes in ice sheet ablation and accumulation patterns (Bintanja and Reijmer, 2001). Multispectral methods are not able to distinguish them from the spectrally similar crevassed blue ice exposed on active glaciers. However, multiangle data can readily accomplish this distinction based on their textural signatures.

Other surface applications making use of coregistered MISR data include retrieval of vegetation structure. A variety of approaches has been employed for exploring the relationship between structure and MISR bidirectional reflectances, including parametric surface models (Widlowski *et al.*, 2004), 3D radiative transfer (Hu *et al.*, 2003), geometric optics (Chopping *et al.*, 2006), and neural networks (Braswell *et al.*, 2003). MISR data show that only the presence of vertically elongated foliage clumps (tree crowns) of medium-to-high densities generates "bell-shaped" angular anisotropy patterns (higher nadir reflectance relative to off-nadir), whereas sparse tree coverage and closed vegetation canopies have "bowl-shaped" anisotropies (lower nadir reflectance relative to off-nadir) (Pinty *et al.*, 2002). Information on additional surface applications is contained in Diner *et al.* (2005).

16.6.2 Cloud applications

16.6.2.1 Introduction

The coregistration of multiangle views is particular challenging when it is applied to the remote sensing of cloud properties. There is no DEM relevant to clouds, which by their nature have heights that range freely and unpredictably throughout the troposphere, and may occasionally be present in the stratosphere or even in the mesosphere. Many cloud tops have heights about 1 km above the surface, whereas cloud tops above 18 km are extremely rare. An altitude error of only 1 km, however, results in a horizontal disparity between the Df and Da views of about 5.7 km, so that knowledge of the actual cloud-top height is essential if all viewing angles are to be coregistered to the cloud top without undue degradation of MISR's spatial resolution.

Clouds may at times be multilayered or semi-transparent, so that the choice of reference level can be problematic. In an operational context, MISR coregisters its multiangle measurements to a dynamic reference level defined on a 2.2-km grid. This is termed the Reflecting Layer Reference Altitude (RLRA), and is defined to be the level corresponding to the maximum observed contrast in reflected solar radiation, assuming this can be found. Thus, thin cirrus above thick low cloud typically results in an RLRA being placed at the lower cloud deck. This is consistent with the need to coregister multiangle radiances at higher spatial resolution when their horizontal variability is greater.

Another complication is cloud motion. During the time that elapses between the Df and Da views (about 7 minutes), for example, a cloud in the lower troposphere will typically move about 4 km, and one in the upper troposphere typically twice as far. The actual speed and direction is of course highly variable, but tends to be well correlated over mesoscale distances (taken to be about 70 km in the MISR operational processing).

As may be gathered from these comments, the goal of operational coregistration of cloud reflectivity is to provide an efficient automated technique that works adequately for all cloud types irrespective of global location, and this involves compromise. Specific individual cloud scenes may have characteristics that permit a more rigorous coregistration, but these inevitably require manual inspection and the luxury of an iterative approach. Nonetheless, the automated approach produces cloud properties that have excellent statistical robustness, given the vast amount of global data that is made available by adopting such an approach.

16.6.2.2 *Operational coregistration of multiangle measurements of clouds*

The nadir view is registered to the reference ellipsoid, irrespective of cloud height or motion. The remaining eight viewing directions are initially coregistered to the reference surface ellipsoid using cloud-free control points as described earlier. This is achieved at subpixel accuracy and is assumed to be relatively static for a given orbit, with the coregistration of each orbit being dynamically reassessed independently of other orbits. For cloudy scenes, the effects of cloud altitude and cloud motion then appear as apparent disparities from one viewing angle to the next.

A variety of empirical pattern-matching algorithms have been developed to assess the apparent disparity between any pair of MISR viewing directions. These form the working core of the operational approach, and are described more thoroughly in Diner *et al.* (1999b), Moroney *et al.* (2002a; 2002b), and Muller *et al.* (2002). Briefly, the main techniques are termed *NestedMax*, *M2*, and *M3*.

NestedMax is designed to retrieve, very efficiently, sparse matches of the brightest cloud reflectivities at the feature level. It compares sequences of points in two image pairs by noting their relative brightness, finding the local maxima using inequality logic, and operating iteratively in nested sense. Ambiguous matches are rejected, leaving a sparse set of quickly found feature matches that can be used directly or to optimize area search patterns with other techniques.

M2 (a multipoint matcher using means) and *M3* (a multipoint matcher using medians) are closely related area-matchers that provide high coverage. They typically operate on a patch size of 10×6 high-resolution (i.e., 275-m) MISR pixels, with reference and comparison patches being shifted relative to each other to find the pixel disparities across-track and along-track that provide the optimum match.

There are two main phases to the operational use of these algorithms. *NestedMax*, *M2*, and *M3* are applied to four pairs of views, as described below, over mesoscale (70-km) regions to determine the consensus height-resolved wind field for each region, and this is used to correct the height classifications of the second phase, as well as being an output product in its own right. The second phase is the application of *M2* and *M3* to pairs of near-nadir views (cameras An, Af, and Aa) to provide disparities on a 1.1-km grid, which are then used to create the RLRAs that are used

Table 16.2 *Retrieval errors due to systematic along-track errors in coregistration to the surface ellipsoid*

Component	Sensitivity to a systematic 1-pixel along-track error in the B camera coregistration	Sensitivity to a systematic 1-pixel along-track error in the D camera coregistration
Along-track wind error	15.5 m/s/pixel	5.6 m/s/pixel
Across-track wind error	3.4 m/s/pixel	1.4 m/s/pixel
Height assignment error	1159 m/pixel	519 m/pixel

for coregistration purposes, as well as a stereoscopic-derived cloud mask (SDCM) at the 1.1-km resolution.

16.6.2.3 Height-resolved winds

The main problem with interpreting the disparity between a single image pair of clouds from a polar orbiter is that the effect of an along-track wind component appears almost identical to that of cloud height. For an Af/An image pair from MISR, for example, the apparent disparity due to a 1-m/s difference in the along-track wind component is equivalent to that due to a difference in cloud-top height of 94 m. By taking a triplet of images using appropriate cameras, it is, however, possible to separate the two effects, as examined in detail by Horváth and Davies (2001) and described by Zong (2002), yielding both the wind component and its height. Optimal camera combinations for this turn out to be the Df-Bf-An and Da-Ba-An triplets.

Since the wind retrieval is quite sensitive to errors in the image matching process, many samples are taken from each mesoscale region in extent of 70.4-km × 70.4-km squares and the results analyzed for the modal disparities, which can be retrieved with subpixel accuracy, as discussed by Davies *et al.* (2007). The random component of the error in image matching over the mesoscale region is thus largely eliminated, with the main assumption being that the wind field is essentially constant over the region at the cloud-top altitude. While generally true, this assumption is clearly violated in cases of strong horizontal wind shear, as at frontal boundaries, or when cloud heights are highly variable within the region. Such cases give a poor estimate of a unique modal value, and are usually identifiable from the statistical distribution. Systematic errors, however, cannot be detected as easily, and this is where the need for accurate orbital coregistration of the relevant cameras to the surface ellipsoid is most critical. Using the above triplets, good coregistration of the Bf, Ba, Df, and Da cameras to the An camera is required. As shown in Davies *et al.* (2007), it is the along-track coregistration of the B cameras that is the most important, noting that the Bf and Ba images are effectively used twice each. Table 16.2

summarizes the errors in the along-track and across-track wind components, and in their height assignment, due to systematic 1-pixel (i.e., 275-m) along-track errors in the coregistration of either a D or a B camera.

Fortunately, the B camera coregistration error is typically far less than 1 pixel. However, an independent check still needs to be applied to ensure quality control of the resulting winds. This is accomplished by using both forward and aft looking triplets to provide two relatively independent estimates of the height-resolved mesoscale wind vector. If the two estimates differ by a predetermined amount then the regional retrieval is rejected. Statistics for the entire orbit are also accumulated separately for the forward and aft retrievals, providing additional identification of systematic coregistration errors. The statistical distribution of differences between the forward and aft retrievals that pass the quality control, taken over many orbits, then provides a robust internal measure of the wind quality retrieval. Davies *et al.* (2007) show that this results in a typical operational coverage of 67%, with a root mean square (RMS) speed difference of 2.4 m/s, an RMS directional difference of 17°, and an RMS difference in height assignment of 290 m. Smaller RMS differences are also possible by using even tighter quality control, but at the cost of reduced coverage.

16.6.2.4 Cloud heights and Reflecting Layer Reference Altitude (RLRA) assignment

Irrespective of the wind retrieval, the near-nadir views provide apparent disparities that are interpreted as cloud-top heights at high spatial resolution. These are obtained initially on a 1.1-km grid (i.e., 4×4 pixels) and form the basis of the SDCM, for which a cloud-top altitude sufficiently above the terrain height indicates the presence of cloud. Without the benefit of statistical sampling, the application of the stereo matchers to produce these cloud heights remains quantized at the pixel level. For a typical near-nadir image pair, the retrieved cloud height is thus similarly quantized in steps of about 560 m. This is usually the main source of error for the stereo-derived cloud heights at 1.1-km resolution, as has been investigated using a comparison with higher-resolution ASTER data by Seiz *et al.* (2006). A version of the cloud heights corrected for the effects of the mesoscale wind is also produced, and these heights are more accurate on average, but their RMS error remains dominated by the quantization limit.

Four adjacent 1.1-km heights are then combined to produce the RLRA on a 2.2-km grid, for the purposes of reprojecting all nine viewing angles to a common reference level. Here, the maximum of the four heights is assigned to the RLRA, since this is most useful for estimating the local albedos (discussed below). For a small percentage (<10%) of cloudy scenes, an operationally determined RLRA cannot be found, due either to lack of usable contrast, or to too complex a variation in cloud heights within the initial 1.1-km regions.

16.6.2.5 Cloud albedos

For a given target, the nine radiances measured by MISR span much of the total range of possible viewing zenith angles in a single azimuthal plane (to a good approximation). These radiances can be integrated directly over solid angle to estimate the total irradiance reflected into the upward hemisphere above a flat surface, which here is taken to be the RLRA. More sophisticated techniques of integration can also be used, adopting statistical or deterministic angular models of reflectivity that may be scene-dependent, the main purpose of which is to account for unmeasured variations in reflectivity as a function of azimuthal angle.

The difficulty with this approach for clouds is that the reflectivity typically changes rapidly with the horizontal distance, as does the morphology of the cloud top. If the top is flat over large areas, having a constant well-defined RLRA, the effect of local coregistration errors would tend to cancel out and the concept of a cloud-top albedo would be relatively straightforward. As the cloud becomes structured on smaller scales, so that the RLRA changes from one 2.2-km region to the next, the obliquely directed radiances quickly become obscured going from low to high regions, and the conventional definition of albedo breaks down. MISR defines albedos on two scales, one at the local 2.2-km level corresponding to an individual RLRA, and the other at a coarser 35.2-km level, being the combination of 16×16 local values. The local albedo exploits the coregistration of the multiangle measurements to a single RLRA in order to best apply the angular model used for integration. This includes measures of the degree of obscuration due to neighboring RLRAs that are higher, so that on later analysis this effect can be accounted for, depending on the application. The coarser-scale albedos are directly useful for radiation budget studies, and account for the angular corrections inherent in the local albedos (by addition), as well as accounting for the solid angle contributions from the remaining radiances that emerge from the sides of RLRA columns of different heights. The main consequence of registration accuracy for the albedo product is thus at the local albedo level, as it affects the choice of angular model.

16.7 Conclusion

A novel approach implemented to meet coregistration/georectification require-ments and continuous data-intensive processing demands of the MISR science data system has been in operation since the beginning of on-orbit data acquisition in February of 2000. As a complement to the production cycle, a global monitor-ing system continuously acquires and summarizes performance assessment data.

Higher-level scientific applications, including retrieval of abundance and types of tropospheric aerosols (especially over land), various measurements of surface structure, and remote sensing of cloud properties are highly dependable on accurate registration of MISR multiangle data with a wide range of viewing angles. As it turned out, the initially implemented algorithm was producing data of sufficient coregistration accuracy to meet the requirements of all scientific applications other than sensitive simultaneous retrievals of cloud-top height and cloud-motion vectors. Additionally, it has been determined that pointing stability of one of the nine MISR cameras is significantly reduced when compared to the other eight cameras. Consequently an update to the algorithm was implemented and made operational in May of 2005 so that by October of 2008 all of the data from the beginning of the mission, including newly acquired, are at a consistent quality level appropriate for all subsequent application processing. In terms of cloud property retrieval requirements we are confident that coregistration RMS errors are around desirable values, ranging from 0.2 to 0.5 of a pixel for the B and D cameras, respectively. Most importantly, there is no noticeable coregistration bias (i.e., within ± 0.05 of a pixel) for all cameras and data since the beginning of the mission.

Validation of retrieved cloud and other atmospheric properties with the tuning and optimizing of the underlying algorithm is still in progress. For example, we would like to take advantage of increased computational capacity at the Langley DAAC and improve resolution and coverage of the global cloud product. At the same time, a specialized version of the algorithm focusing on hurricane properties in support of regional studies is now feasible. However, there are no anticipated future updates to the current operational coregistration/georectification and geometric quality monitoring system. The operational software, ancillary data, and georectifed products are maintained in support of the ongoing mission, as well as in preparation for any future mission with similar multiangle imaging configuration and mapping requirements.

References

Ackermann, F. (1984). Digital image correlation: Performance and potential application in photogrammetry. *The Photogrammetry Record*, **11**(64), pp. 429–439.

Bailey, G. B., Carneggie, D., Kieffer, H., Storey, J. C., Jovanovic, V. M., and Wolfe, R. E. (1997). *Ground Control Points for Calibration and Correction of EOS ASTER, MODIS, MISR and Landsat 7 ETM+ Data*. SWAMP GCP Working Group Final Report, USGS, EROS Data Center, Sioux Falls, SD.

Bintanja, R. and Reijmer, C. H. (2001). Meteorological conditions over Antarctic blue-ice areas and their influence on the local surface mass balance. *Journal of Glaciology*, **47**(156), 37–50.

Bothwell, G., Hansen, E. G., Vargo, R. E., and Miller, K. C. (2002). The MISR science data system: Its products, tools, and performance. *IEEE Transactions on Geoscience and Remote Sensing*, **40**(7), 1467–1476.

Braswell, B. H., Hagen, S. C., Salas, W. A., and Frolking, S. E. (2003). A multivariable approach for mapping sub-pixel land cover distributions using MISR and MODIS: Application in the Brazilian Amazon. *Remote Sensing of Environment*, **87**, 243–256.

Castleman, K. R. (1979). *Digital Image Processing*. Englewood Cliffs, NJ: Prentice-Hall.

Chopping, M. J., Su, L. H., Laliberte, A., Rango, A., Peters, D. P. C., and Martonchik, J. V. (2006). Mapping woody plant cover in desert grasslands using canopy reflectance modeling and MISR data. *Geophysical Research Letters*, **33**, L17402, doi: 10.129/2006 GL027148.

Davies, R., Horváth, A., Moroney, C., Zhang, B., and Zhu, Y. (2007). Cloud motion vectors from MISR using sub-pixel enhancements. *Remote Sensing of Environment*, MISR Special Issue, **107**, 194–199.

Dennis, J. E. and Schnabel, R. B. (1983). *Numerical Methods for Unconstrained Optimization and Nonlinear Equations*. Englewood Cliffs, NJ: Prentice-Hall.

Diner, D. J., Beckert, J. C., Reilly, T. H., Bruegge, C. J., Conel, J. E., Kahn, R. A., Martonchick, J. V., Ackerman, T. P., Davies, R., Gerstl, S. A. W., Gordon, H. R., Muller, J. P., Myneni, R., Sellers, P. J., Pinty, B., and Verstraete, M. M. (1998). Multi-angle Imaging SpectroRadiometer (MISR) instrument description and experiment overview. *IEEE Transactions on Geoscience and Remote Sensing*, **36**(4), 1072–1087.

Diner, D. J., Braswell, B. H., Davies, R., Gobron, N., Hu, J., Jin, Y., Kahn, R. A., Knyazikhin, Y., Loeb, N., Muller, J.-P., Nolin, A. W., Pinty, B., Schaaf, C. B., Seiz, G., and Stroeve, J. (2005). The value of multiangle measurements for retrieving structurally and radiatively consistent properties of clouds, aerosols, and surfaces. *Remote Sensing of Environment*, **97**(4), 495–518.

Diner, D. J., Clothiaux, E., and Di Girolamo, L. (1999a). *Level 1 Cloud Detection Algorithm Theoretical Basis Document*. JPL Tech. Doc. D-11397, Rev. B. Jet Propulsion Laboratory, California Institute of Technology, Pasadena, CA.

Diner, D. J., Davies, R., DiGirolamo, L., Moroney, C., Muller, J.-P., Paradise, S., Wenkert, S., and Zong, J. (1999b). *Level 2 Cloud Detection and Classification Algorithm Theoretical Basis Document*. JPL Tech. Doc. D-11399, Rev. D. Jet Propulsion Laboratory, California Institute of Technology, Pasadena, CA.

Förstner, W. (1982). On the geometric precision of digital correlation. In *Proceedings of the Commission III International Society for Photogrammetry and Remote Sensing*, Helsinki, Finland, Vol. XXIV, pp. 176–189.

Förstner, W. and Gülch, E. (1987). A fast operator for detection and precise location of distinct points, corners and centers of circular features. In *Proceedings of the International Society for Photogrammetry and Remote Sensing Intercommission Workshop on Fast Processing of Photogrammetric Data*, Interlaken, Switzerland, pp. 281–305.

Haralick, R. M. and Shapiro, L. G. (1993). *Computer and Robot Vision*. Reading, MA: Addison-Welsey.

Horváth, Á. and Davies, R. (2001). Feasibility and error analysis of cloud motion wind extraction from near-simultaneous multiangle MISR measurements. *Journal of Atmospheric Oceanic Technology*, **18**, 591–608.

Hu, J., Tan, B., Shabanov, N., Crean, K. A., Martonchik, J. V., Diner, D. J., Knyazikhin, Y., and Myneni, R. B. (2003). Performance of the MISR LAI and FPAR algorithm: A case study in Africa. *Remote Sensing of Environment*, **88**, 324–340.

Jovanovic, V. M., Bull, M., Smyth, M. M., and Zong J. (2002). MISR in-flight camera geometric model calibration and achieved georectification performances. *IEEE Transactions on Geoscience and Remote Sensing*, **40**(7), 1512–1519.

Jovanovic, V. M., Moroney, C., and Nelson, D. (2007). Multi-angle geometric processing for globally geo-located and co-registered MISR image data. *Remote Sensing of Environment*, **107**(1–2), 22–32.

Jovanovic, V. M., Smyth, M. M., Zong, J., Ando, R., and Bothwell, G. W. (1998). MISR photogrammetric data reduction for geophysical retrievals. *IEEE Transactions on Geoscience and Remote Sensing*, **36**(4), 1290–1301.

Kahn, R. A., Gaitley, B. J., Crean, K. A., Diner, D. J., Martonchik, J. V., and Holben, B. N. (2005). MISR global aerosol optical depth validation based on two years of coincident AERONET observations. *Journal of Geophysical Research*, **110**, D10S04, doi: 10.129/2004JD004706.

Korechoff, R. P., Jovanovic, V. M., Hochberg, E. B., Kirby, D. M., and Sepulveda, C. A. (1996). Distortion calibration of the MISR linear detector arrays. In *Proceedings of the SPIE Conference on Earth Observing Systems*, Denver, CO, Vol. 2820, pp. 174–183.

Logan, T. L. (1999). *EOS/AM-1 Digital Elevation Model (DEM) Data Sets: DEM and DEM Auxiliary Datasets in Support of the EOS/Terra Platform*. JPL Tech. Doc. D-013508. Jet Propulsion Laboratory, California Institute of Technology, Pasadena, CA.

Martonchik, J. V., Diner, D. J., Kahn, R., Gaitley, B., and Holben, B. N. (2004). Comparison of MISR and AERONET aerosol optical depths over desert sites. *Geophysical Research Letters*, **31**, L16102, doi:10.1029/2004GL019807.

Mikhail, E. M. (1976). *Observations and Least Squares*. New York: Harper & Row.

Moroney, C., Davies, R., and Muller, J. P. (2002a). Operational retrieval of cloud-top heights using MISR. *IEEE Transactions on Geoscience and Remote Sensing*, **40**(7), 1532–1540.

Moroney, C., Horváth, A., and Davies, R. (2002b). Use of stereo-matching to coregister multiangle data from MISR. *IEEE Transactions on Geoscience and Remote Sensing*, **40**(7), 1541–1546.

Muller, J. P., Mandanayake, A., Moroney, C., Davies, R., Diner, D. J., and Paradise, S. (2002). MISR stereoscopic image matchers: Techniques and results. *IEEE Transactions on Geoscience and Remote Sensing*, **40**(7), 1547–1559.

Nolin, A. W., Fetterer, F. M., and Scambos, T. A. (2002). Surface roughness characterizations of sea ice and ice sheets: Case studies with MISR data. *IEEE Transactions on Geoscience and Remote Sensing*, **40**(7), 1605–1615.

Paderes, F. C., Mikhail, E. M., and Fagerman, J. A. (1989). Batch and on-line evaluation of stereo SPOT imagery. In *Proceedings of the Annual American Society of Photogrammetry and Remote Sensing and the American Congress on Surveying and Mapping Convention*, Baltimore, MD, Vol. 3, pp. 31–40.

Pinty, B., Widlowski, J.-L., Gobron, N., Verstraete, M. M., and Diner, D. J. (2002). Uniqueness of multiangular measurements–part I: An indicator of subpixel surface heterogeneity from MISR. *IEEE Transactions on Geoscience and Remote Sensing*, **40**(7), 1560–1573.

Seiz, G., Davies, R., and Grün, A. (2006). Stereo cloud-top height retrieval with ASTER and MISR. *International Journal of Remote Sensing*, **27**(9), 1839–1853.

Snyder, J. P. (1987). *Map Projections – A Working Manual*. United States Geological Survey Professional Paper 1395, U.S. Government Printing Office, Washington, DC.

Widlowski, J.-L., Pinty, B., Gobron, N., Verstraete, M. M., Diner, D. J., and Davis, A. B. (2004). Canopy structure parameters derived from multi-angular remote sensing data for terrestrial carbon studies. *Climatic Change*, **67**(2–3), 403–415.

Yang, Y., Di Girolamo, L., and Mazzoni, D. (2007). Selection of the automated thresholding algorithm for the Multi-angle Imaging SpectroRadiometer camera-by-camera cloud mask over land. *Remote Sensing of Environment*, MISR Special Issue, **107**(1–2), 159–171.

Zong, J. (2002). Multi-image tie-point detection applied to multi-angle imagery from MISR. In *Proceedings of the Commission III International Society of Photogrammetry and Remote Sensing Symposium on Photogrammetric Computer Vision*, Graz, Austria, p. A-424.

Zong, J., Davies, R., Muller, J. P., and Diner, D. J. (2002). Photogrammetric retrieval of cloud advection and top height from the Multi-angle Imaging Spectro-Radiometer (MISR). *Photogrammetric Engineering & Remote Sensing*, **68**(8), 821–829.

17

Automated AVHRR image navigation

WILLIAM J. EMERY, R. IAN CROCKER, AND DANIEL G. BALDWIN

Abstract

To enable automated (without human intervention) AVHRR (Advanced Very High Resolution Radiometer) image navigation, a base image is defined and the *maximum cross-correlation* (MCC) method is used to automatically compute the satellite attitude parameters required to geometrically correct images to this base image. The auto attitude corrections are shown to be more accurate than the traditional linear translation methods and provide a significant improvement in geolocation accuracy over two other AVHRR image navigation methods. Geolocation accuracies are given for near-real-time use of this method for operational applications using daily imagery off the U.S. East and West Coasts. A further application of the attitude corrections is demonstrated whereby attitude corrections computed over land can be carried forward in the satellite's orbit to accurately navigate imagery over the open ocean where map reference points are not available.

17.1 Introduction

The accurate georegistration of satellite imagery typically requires the application of an orbital model to predict the location of the spacecraft, as well as an instrument pointing model to determine the geolocation of the sensor field of view (FOV) (Rosborough *et al.*, 1994). The implementation of these two models is straightforward and easily automated. However, the obtained registration accuracy is dependent on the accuracies of the timing of the data and the spacecraft attitude (roll, pitch, and yaw) (Rosborough *et al.*, 1994; Baldwin and Emery, 1994). Although the sources of the timing and orientation errors are separate, it is possible to compensate for the timing error by including it in the pitch correction. Determination of the spacecraft attitude corrections has traditionally required human intervention to identify registration offsets using known ground control points (GCPs). This process can be time consuming and tedious, especially when processing large volumes of satellite data. Frequently, time and effort constraints preclude making these corrections, and the associated registration errors are ignored. This chapter discusses an automated approach for computing and applying the timing/attitude corrections for Advanced Very High Resolution Radiometer (AVHRR) image navigation. This approach provides accurate georegistrations without the time-consuming efforts inherent in manual corrections. Section 17.2 describes the technique and presents examples of the results. Then, in Section 17.3, the obtained georegistration accuracies are compared against two other methods of AVHRR georegistration. Finally, an example is discussed which demonstrates an extended application of the procedure for computing attitude corrections over the open ocean where GCPs are not available.

17.2 Image navigation methodology

The automatic image navigation technique (hereafter referred to as AUTONAV) consists of three basic elements: An accurately registered base image, a roughly registered target image, and a pattern tracking algorithm called the *maximum cross-correlation* (MCC) method (Emery *et al.*, 1986). The base image, which only needs to be constructed once, is accurately registered using a standard manual correction technique. The required characteristics of the base image and the manual correction technique are discussed in detail in Subsection 17.2.2. The target image is first roughly registered using only the orbital and instrument models with no attitude corrections. The MCC method is then used to detect displacements between the base image and the target image, and these displacements are subsequently used to compute the roll, pitch, and yaw attitude corrections necessary to accurately register the target image.

17.2.1 The maximum cross-correlation (MCC) method

The MCC method is an automated technique that calculates the displacements of data features between sequential images (Ninnis *et al.*, 1986; Emery *et al.*, 1992; Kelly and Strub, 1992; Bowen *et al.*, 2002; Wilkin *et al.*, 2002). The procedure, illustrated in Fig. 17.1, cross-correlates a template subwindow of data from the first image with all possible similarly sized subwindows of data that fall within a search window (dashed window in Fig. 17.1) of the second image. The location of the second-image subwindow that produces the highest cross-correlation with the template subwindow indicates the most likely displacement of the data feature. The origin of the displacement vector is centered in the template subwindow and the terminus is centered in the second-image subwindow. This procedure is repeated over a specified grid spanning the image dimensions to generate a displacement

First Image   Second Image

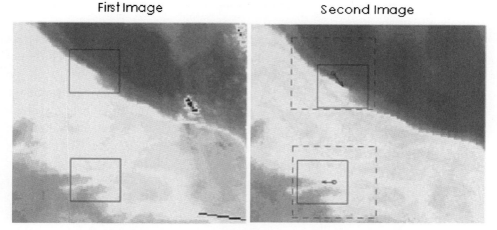

Figure 17.1. The MCC method illustrated for two grid locations. The solid boxes
in the first image are the template subwindows; this is the data feature to search
for within the dashed search windows in the second image. The second-image
subwindows that provide the highest cross-correlations suggest the most likely
displacements of the data features (solid boxes). See color plates section.

vector field for the image pair. When used for image georegistration, the MCC
method is applied over land to define the displacements needed to correct the target
image to fit the base image. The nonlinear nature of attitude errors frequently results
in displacements which are nonuniform over the image. An example illustrating
this effect is presented in Section 17.3.

17.2.2 Creating the base image and cloud filtering

Constructing an accurately registered base image is a crucial step towards the
successful use of the AUTONAV algorithm. The base image must be manually
registered to the exact same grid as the target images and must be as cloud free as
possible. The typical steps to creating an accurately georegistered base image from
AVHRR data are as follows: (1) The orbital and instrument models are applied to
the data and a roughly registered image is created, (2) the image is then compared
to an overlay of known geographic map features (coastlines, rivers, etc.) and geographic offsets between the map features and the corresponding image features are
visually determined, (3) the offsets are used to calculate a roll, pitch, and yaw, and
(4) Step 1 is repeated using the attitude parameters, resulting in an attitude corrected
image. As was mentioned in the Introduction, any timing errors are accounted for
in the pitch correction. Since the geographic offsets are subjectively determined by
visual estimation, the attitude parameters may require some additional tweaking

to remove any residual errors and ensure that the image features exactly coincide with their map counterparts.

A second requirement is that the radiance distributions of the base and target images be similar. The widely varying illumination conditions that exist for different orbits prevent the use of the reflected channels for the AUTONAV algorithm. While the thermal channels are better suited, care must be taken in selecting the base image since differential heating and cooling over land can result in thermal patterns which could wrongly be interpreted as registration displacements. Experience has shown that thermal differences between the base and target image can be minimized by having daytime and nighttime base images for each season. Processing a year's worth of data would thus require a minimum of eight base images for a particular location.

The base image must be as cloud free as possible. Cloudy pixels can result in erroneous displacements and must be identified and excluded from the MCC computation. The AUTONAV algorithm has three means for eliminating the effects of clouds. The first is a simple threshold method where pixels having brightness temperatures less than a user specified value are ignored in the cross-correlation computation. The threshold value should be larger than the brightness temperatures of the clouds present in the images. While this technique will identify most of the cloudy pixels, it is likely that a few will pass through and be used to compute cross-correlations. Since the MCC method computes displacements by correlating data patterns between two images, each displacement vector has a correlation coefficient which quantifies how well a pattern was matched. Displacements contaminated by clouds usually have low cross-correlations, and eliminating all vectors with cross-correlations lower than the 95% confidence level ($\sim$0.7) is the second means of cloud filtering. The third step, known as coherency filtering, makes use of the fact that displacements due to misregistration are fairly coherent for local regions, while those due to clouds are not. The coherency filter removes vectors that have different lengths and directions than their immediate neighbors. The three cloud filtering processes result in a set of vectors that represent the registration offsets of the target image without the influence of clouds.

17.2.3 Calculating the attitude corrections

As previously mentioned, AUTONAV is an algorithm which, given a roughly navigated target image and a precisely navigated base image, will return a precisely navigated target image. The AUTONAV algorithm consists of a series of steps that can be summarized as follows: (1) Cloud filtering of the target image, (2) the generation of MCC displacement vectors, (3) quality control filtering of the vectors (correlation threshold and coherency filters), (4) creation of optimally sorted pairs

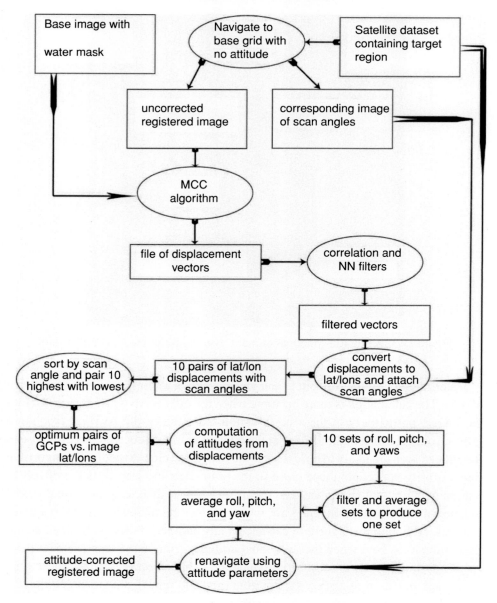

Figure 17.2. The processing steps of the AUTONAV algorithm.

of displacement vectors, (5) conversion of the image displacements to latitude and longitude offsets, (6) computation of attitude parameters from the displacement pairs, (7) calculation of a mean roll, pitch, and yaw, and (8) navigation of the target image using the average attitude parameters. The AUTONAV algorithm is summarized in Fig. 17.2, which shows the various stages of processing.

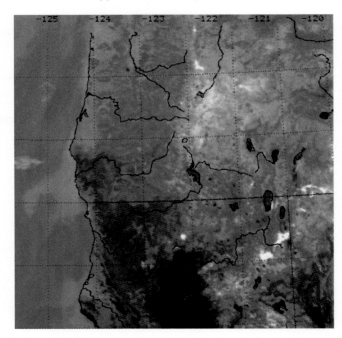

Figure 17.3. A nighttime thermal infrared base image for the U.S. West Coast.

Determination of the roll, pitch, and yaw requires a minimum of two geographic offsets. The calculation is more accurate when the offsets are widely spaced across the image, particularly in the scan direction. Also, since the AVHRR resolution grows from 1.1 km at nadir to 7 km at the edge of scan, offsets which occur close to the edge of scan are likely to be less accurate than those close to nadir. The algorithm filters offsets to exclude those close to the scan edge and also sorts the offsets to maintain an optimal width between the offset pairs. It then uses the orbital model to determine the roll, pitch, and yaw, which correct the offsets using least-squares analysis. Although the algorithm can accept any number of offsets, we have found the results to be best when the individual attitudes from many displacement pairs (width-optimized) are averaged to produce the final correction values. These final values are then used by the sensor pointing model to renavigate the target image.

17.2.4 Applying the attitude corrections

This section demonstrates the typical georegistration performance of the AUTONAV algorithm. The base image used for this example, shown in Fig. 17.3, is a nighttime infrared image of the U.S. West Coast. This image was selected due to its lack of cloud contamination; a necessary requirement to provide the maximum

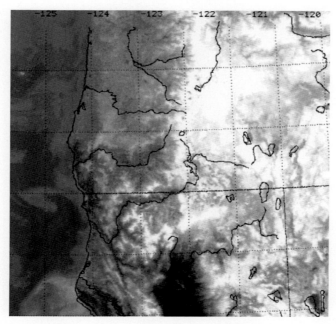

Figure 17.4. An AVHRR thermal infrared target image to be used for AUTONAV georegistration. Note the misalignment of lakes, rivers, and coastline.

number of image reference points. As described in Subsection 17.2.2, this image was roughly navigated using an orbital and instrument pointing model, and then accurately navigated by hand to ensure the most precise image registration possible. Note the excellent agreement between the image and map coastlines, rivers, and lakes across the entire image.

A thermal infrared target image that was roughly georegistered without attitude corrections is shown in Fig. 17.4. Note the misalignment of the map and image coastline, rivers, lakes, and other prominent landmarks. Figure 17.5 shows this same image with the MCC displacement vectors after the correlation threshold and coherency filters have been applied. Notice that most of the vectors are oriented in a southwest direction, but they are not all uniformly directed and there are subtle differences in each group of vectors. These vectors are used by AUTONAV to compute the attitude corrections. The final corrected image is shown in Fig. 17.6 where it is evident that the image features align with the map reference base.

One alternative to attitude correction has been to linearly translate the image to align image features with GCPs. This method is commonly called "nudging" and can be automated by using the average MCC displacements as the nudge factors. While nudging is computationally faster than attitude correction, it is not capable of accurately correcting for either roll or yaw. A different image which has been

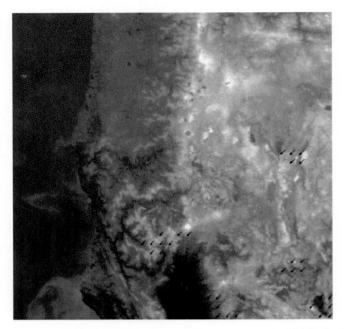

Figure 17.5. The AVHRR target image from Fig. 17.4 with the MCC displacement vectors in cloud-free regions.

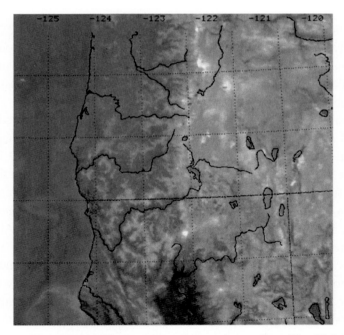

Figure 17.6. The AUTONAV georegistered target image from Fig. 17.4. Note the alignment of map and image features.

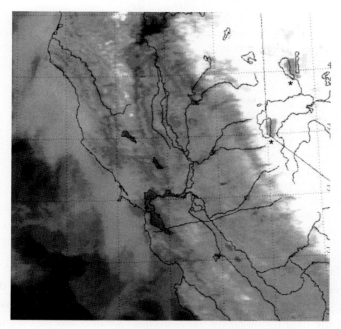

Figure 17.7. An example of a different target image after applying only nudging corrections; examples of remaining misregistration errors are indicated by a * below them.

corrected only by nudging is displayed in Fig. 17.7. Although the coastline and coastal lakes are well registered, the inland lakes towards the eastern boundary of the image are poorly registered. This is a typical result when only nudging is used. Although the example of Fig. 17.7 is from a different image, a comparison between Fig. 17.7 and the attitude corrected image of Fig. 17.6 clearly shows that the attitude corrected image has accurate georegistrations throughout the image.

17.2.5 AUTONAV georegistration accuracy

The AUTONAV algorithm is currently being used by researchers at the Colorado Center for Astrodynamics Research (CCAR) to compute near-real-time ocean currents for both the East and West Coasts of the U.S. The currents are calculated using the MCC technique to track thermal ocean patterns between sequential AVHRR images (Bowen *et al.*, 2002). The coastal imagery used contains enough land to enable accurate automatic attitude corrections using the AUTONAV algorithm. To establish the quantitative georegistration accuracy when the algorithm is used for automated processing of large quantities of data, we have randomly selected ten images from the East Coast CCAR dataset and manually inspected them to determine the georegistration errors. A geographic map was overlaid on the images and

Table 17.1 *AUTONAV registration errors*

Satellite pass identification	X-offset (km)	Y-offset (km)
3 Sept 2007, 6:34 GMT, NOAA-18	−1	0
3 Sept 2007, 10:24 GMT, NOAA-15	0	0
4 Sept 2007, 21:20 GMT, NOAA-15	−1	−1
5 Sept 2007, 15:34 GMT, NOAA-17	1	−1
6 Sept 2007, 15:11 GMT, NOAA-17	0	−1
7 Sept 2007, 14:48 GMT, NOAA-17	1	−1
7 Sept 2007, 17:20 GMT, NOAA-18	2	−3
8 Sept 2007, 2:08 GMT, NOAA-17	0	0
8 Sept 2007, 10:07 GMT, NOAA-15	0	0
9 Sept 2007, 7:13 GMT, NOAA-18	0	0

then magnified so that the registration errors could be recorded. Table 17.1 lists the offsets in both the East–West (X-offset) and North–South (Y-offset) directions for each image. The satellite pass identification listed in Table 17.1 contains the date and time of the data followed by the National Oceanic and Atmospheric Administration (NOAA) satellite number. The average absolute X-offset and Y-offset for these 10 images is 0.6 km and 0.7 km, respectively, and is caused mainly by one poorly navigated image. Most images are no more than 1 km off, and four of the ten images are perfectly registered. These numbers suggest that the AUTONAV technique can be implemented to yield well-navigated AVHRR images for various applications in both operational and scientific research.

17.3 Comparison with other georegistration techniques

In this section we compare the georegistration results of the AUTONAV system with those from three other sources. The first comparison is done using archived AVHRR data from a NOAA-11 orbit in 1991. The first set of georegistered images was generated using the latitude and longitude values which are computed by the NOAA National Environmental Satellite, Data, and Information Service (NOAA/NESDIS) and embedded in the NOAA Level 1b data. Although these geolocation values are only given for every eighth satellite FOV on the scan line, they can be interpolated to every FOV and then resampled to create a registered image.

The effect of attitude on registration error can be seen in Fig. 17.8(a), which depicts an NOAA-11 image that has been georegistered using the NOAA/NESDIS Level 1b latitudes and longitudes. The 5-km offset between geographic features in the map overlay and the image are mainly the result of a 2-milliradian roll. The

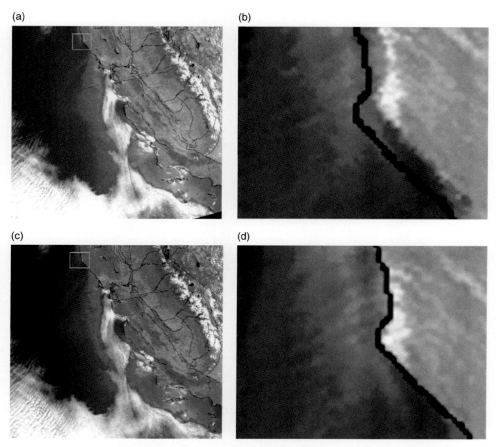

Figure 17.8. (a) NOAA-11 channel 4 AVHRR image georegistered using NOAA/NESDIS Level 1b data; the 5-km offset from the map is due mostly to a 2-milliradian roll error, (b) white rectangle portion of (a), (c) same image in (a) georegistered using AUTONAV, and (d) white rectangle in (c).

registration error can be seen more clearly in Fig. 17.8(b), which is a close-up of the delineated region in Fig. 17.8(a). Note the offset between the map and image coastline.

Figures 17.8(c) and 17.8(d) show the same data after it has been registered using the AUTONAV system, which automatically computed and corrected for the attitude, resulting in pixel-level agreement with the map overlay. Notice the excellent agreement between the map and image coastline, especially when compared to Fig. 17.8(b). It should be noted that the images shown in Figs. 17.8(a)–(d) are close to the center of the swath, which is the nadir position of the satellite's orbit. As mentioned earlier, the resolution of the AVHRR instrument is 1.1 km at nadir and grows to approximately 7 km towards the edge of scan. This can result in

(a)   (b)

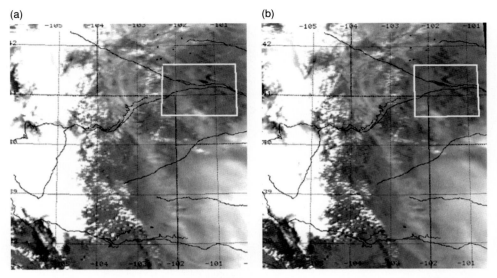

Figure 17.9. (a) An edge-of-scan AVHRR image georegistered using NOAA/
NESDIS Level 1b data; the image lake is offset about 30 km to the right of
the map lake shown in the white rectangle, and (b) image in (a) georegistered
using AUTONAV.

attitude-induced registration errors which are greater at the edge of scan than at
nadir. This effect is evident in the images depicted in Figs. 17.9(a) and 17.9(b),
which show the same dataset as Fig. 17.8, except the images were made close to
the edge of scan. Figure 17.9(a) was registered using the NOAA/NESDIS Level 1b
geolocations, which result in a 30-km offset between the image and map overlay.
This offset is clearly seen in a comparison between the map and image locations of
the river and lake in the white rectangular region. The dark shape of the somewhat
linear lake is offset to the right of the lake on the map.

The georegistration of the image shown in Fig. 17.9(b) was done with the
AUTONAV algorithm using the same attitude corrections computed for the image
in Fig. 17.8(c). The geographic features within the white rectangle are now very
consistent between the map overlay and the satellite image. The correspondence
between the map overlay and the geographic image features in Fig. 17.9(b) demonstrates the accuracy of the AUTONAV geolocation toward the edge of scan.

The previous examples used Level 1b data from a 1991 NOAA-11 orbit. In
more recent years, NOAA has improved the accuracy of the latitude and longitude
values for the Level 1b data. The timing of the data has improved and some
attitude information has been implemented in the geolocation calculations. The
attitude values are, however, kept constant for an extended period of time, and
are not computed for each orbit. The figures that follow show images that have

been registered using recent NOAA-18 Level 1b geolocations from two different sources. As before, the region is close to nadir. The geolocation accuracies of these images are then compared with images generated from the AUTONAV system. The two sources of Level 1b geolocations are the standard values produced by NOAA using their attitude parameters, and latitudes and longitudes obtained from the European Organization for the Exploitation of Meteorological Satellites Advanced ATOVS Processing Package (EUMETSAT AAPP) software, which uses attitude parameters obtained from Moreno and Meliá (1993).

It should be noted that although AAPP can compute latitudes and longitudes from Level 1b AVHRR data, the software is not capable of producing a georegistered AVHRR image. Figures 17.10(a) and 17.10(b) respectively show georegistered images generated using the NOAA and AAPP latitudes and longitudes. The improvement in NOAA's registration accuracy resulting from their inclusion of attitude corrections is evident when the image and map offsets in Fig. 17.10(a) are compared with those of the older NOAA-11 images shown in Figs. 17.8(a) and 17.8(b).

The delineated areas in Figs. 17.10(a) and 17.10(b) are enlarged and presented as Figs. 17.10(c) and 17.10(d), respectively, to more clearly depict the georegistration errors. The NOAA geolocations result in a 4-km offset, while those from the AAPP software produce a 3-km offset. The corresponding images obtained from the AUTONAV software are depicted in Figs. 17.10(e) (the full image) and 17.10(f) (the close-up of the delineated region). As in the earlier NOAA-11 results, the registration obtained from the AUTONAV system is accurate to within 1 km.

Since all of these images (NOAA, AAPP, and AUTONAV) were attitude corrected, the registration errors do not amplify towards the edge of scan, as was the case for the nonattitude corrected images from NOAA-11 shown in Figs. 17.9(a) and 17.9(b).

17.4 AUTONAV georegistration over the open ocean

An added benefit of the AUTONAV navigation is the potential to apply attitude corrections computed from one part of the satellite's orbit to another part of the orbit which does not contain sufficient land surface features to perform the attitude corrections. Emery *et al.* (2003) presented an example where attitude parameters computed from images of Puerto Rico were then used to navigate an image of Cape Cod at a later time in the same orbit. This navigation was found to be accurate to within 1 pixel.

As an added demonstration in the present paper, we use attitude parameters computed over the U.S. West Coast to navigate images of Hawaii two orbits later. The AUTONAV georegistered image of Hawaii with no attitude corrections is

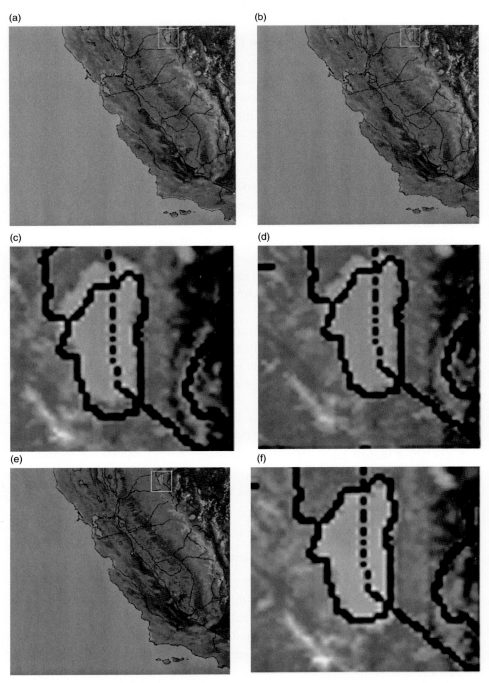

Figure 17.10. (a) NOAA-18 channel 4 AVHRR image georegistered using NOAA/NESDIS attitude-corrected Level 1b data, (b) image in (a) georegistered using the AAPP latitudes and longitudes, (c) white rectangle in (a) georegistered with constant-attitude-corrected NOAA/NESDIS Level 1b data, (d) white rectangle in (b) georegistered with the AAPP software, (e) image of (a) georegistered with AUTONAV, and (f) white rectangle in (e) georegistered with AUTONAV.

(a)   (b)

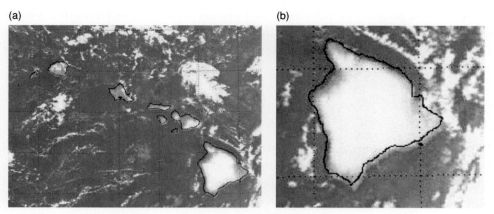

Figure 17.11. Scenes navigated using AUTONAV without attitude correction: (a) Hawaiian Islands, and (b) Big Island of Hawaii.

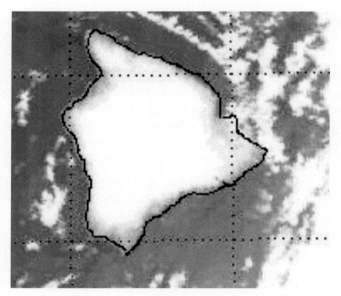

Figure 17.12. Big Island of Hawaii navigated using AUTONAV with attitude correction.

presented here in Fig. 17.11(a), and a close-up image of the Big Island of Hawaii is shown in Fig. 17.11(b). It can be seen that the geolocation is off by more than 4–5 pixels in this 1-km resolution image. This is particularly evident on the southeast shore of the Big Island, which shows the highly reflective land offset from the map coastline. By applying the attitude parameters computed almost 3 hours earlier (2 orbits prior) over the U.S. West Coast we achieve the geolocation shown in Fig. 17.12, which shows only the Big Island of Hawaii. The improvement is clear, and now the image agrees with the map overlay. Small geolocation errors can be seen at the island's northernmost point and along the western coast; however, these errors are much smaller than those seen in Fig. 17.11(b) where attitude corrections were not applied.

17.5 Conclusion

By constructing a precise base image for the appropriate season and time of day, it is possible to automatically and accurately navigate satellite imagery without the need for human intervention, significantly increasing the speed and efficiency of image geolocation. This is primarily of importance when a large number of satellite images require navigation. In this case, an automated approach makes it possible to realistically consider processing hundreds or even thousands of images for any one application, a near impossible task if human intervention is required to calculate attitude parameter corrections.

The AUTONAV image navigation approach uses the MCC method to compute displacements between a target image and an accurately navigated base image. These displacements are used to compute the satellite attitude corrections, which are then applied to accurately renavigate the image. Manual inspection of ten randomly selected images indicates that the AUTONAV technique is capable of FOV-level georegistration accuracy when used in an automated fashion to process large volumes of satellite imagery. Comparisons between two other sources of AVHRR geolocations, those of NOAA Level 1b data and of AAPP, demonstrate that the AUTONAV method yields more accurate georegistrations than either of these other methods. An added benefit of AUTONAV is that attitude corrections computed for one part of an orbit that contains land can be applied to later orbits for different regions that may not contain any land features. This approach was tested by using the attitude corrections computed from an image of the U.S. West Coast to accurately georegister an image of the Hawaiian Islands two orbits later.

Acknowledgements

The authors would like to acknowledge the financial support of the NASA Physical Ocean Program and its support of the TOPEX/Poseidon and Jason altimeter programs through the Jet Propulsion Laboratory. The automated image navigation procedure was developed to meet the demand for the routine and automated processing of a large number of AVHRR images for our Jason Science Team project. The authors would like to thank Drs. Lee Fu and Eric Lindstrom for their continued support of these research efforts.

References

Baldwin, D. and Emery, W. J. (1994). Spacecraft attitude variations of NOAA-11 inferred from one-year of AVHRR imagery. *International Journal of Remote Sensing*, **16**, 531–548.

Bowen, M. M., Emery, W. J., Wilkin, J. L., Tildesley, P. C., Barton, I. J., and Knewtson, R. (2002). Extracting multiyear surface currents form sequential thermal imagery using the maximum cross correlation technique. *Journal of Atmospheric and Oceanic Technology*, **19**(10), 1665–1676.

Emery, W. J., Baldwin, D., and Matthews, D. (2003). Maximum cross correlation automatic satellite image navigation and attitude corrections for open-ocean image navigation. *IEEE Transactions on Geoscience and Remote Sensing* **41**(1), 33–42.

Emery, W. J., Fowler, C. W., and Clayson, C. A. (1992). Satellite-image-derived Gulf Stream currents compared with numerical model results. *Journal of Atmospheric and Oceanic Technology*, **9**(3), 286–304.

Emery, W. J., Thomas, A. C., Collins, M. J., Crawford, W. R., and Mackas, D. L. (1986). An objective procedure to compute advection from sequential infrared satellite images. *Journal of Geophysical Research*, **91**(12), 12 865–12 879.

Kelly, K. A. and Strub, P. T. (1992). Comparison of velocity estimates from AVHRR in the coastal transition zone. *Journal of Geophysical Research – Oceans*, **97**(C6), 9653–9668.

Moreno, J. F. and Meliá, J. (1993). A method for accurate geometric correction of NOAA AVHRR HRPT data. *IEEE Transactions on Geoscience and Remote Sensing*, **31**, 204–226.

Ninnis, R. M., Emery, W. J., and Collins, M. J. (1986). Automated extraction of sea ice motion from AVHRR imagery. *Journal of Geophysical Research*, **91**(10), 10 725–10 734.

Rosborough, G. R., Baldwin, D., and Emery, W. J. (1994). Precise AVHRR image navigation. *IEEE Transactions on Geoscience and Remote Sensing*, **32**, 644–657.

Wilkin, J. L., Bowen, M. M., and Emery, W. J. (2002). Mapping mesoscale currents by optimal interpolation of satellite radiometer and altimeter data. *Ocean Dynamics*, **52**(3), 95–103.

18

Landsat image geocorrection and registration

JAMES C. STOREY

Abstract

As the primary archive for the data acquired by the Landsat series of spacecraft, the U.S. Geological Survey Center for Earth Resources Observation and Science (EROS) is responsible for the processing systems that capture, correct, and distribute Landsat image data products. The Landsat ground system includes product generation components that apply the radiometric and geometric processing necessary to convert the digital detector samples acquired by the sensor to top-of-atmosphere radiance measurements referenced to an Earth-fixed coordinate system. These product generation systems implement geometric correction algorithms that use the instrument and spacecraft support data, provided in the Landsat data stream, to construct a model of the geometric relationship between the acquired image data samples and an Earth-fixed ground reference system. These support data include onboard measurements of instrument timing, spacecraft attitude (orientation) and jitter, and estimates of the spacecraft position and velocity derived from ground tracking data. The accuracy of the basic systematic geometric registration model is limited by uncertainties in these supporting data. In particular, the spacecraft position and attitude data typically contain residual biases on the order of tens to hundreds of meters. Fortunately, these errors are usually slowly varying, allowing for the registration of multiple Landsat images using simple, low-order techniques. When higher-accuracy products are required, additional capabilities of the Landsat ground system are employed to perform more precise geometric correction. This higher-level processing requires ancillary data sources, such as ground control points, to remove the residual pointing and position biases, and digital elevation data, to correct for the effects of terrain height on the Landsat viewing geometry. These ancillary data are used to derive corrections to the systematic geometric model that yield image products registered to a ground reference system to subpixel accuracy.

18.1 Introduction

The Landsat program has launched and operated medium-resolution multispectral imaging satellite systems since 1972. As of this writing (2009), the Landsat-5 and Landsat-7 missions continue to collect operational remote sensing data. As the primary archive for the data acquired by the Landsat series of spacecraft, the U.S. Geological Survey (USGS) Center for Earth Resources Observation and Science (EROS) is responsible for the processing systems that capture, correct, and distribute Landsat image data products. The Landsat ground system includes product generation components that apply the radiometric and geometric processing necessary to convert the digital detector samples acquired by the sensor to top-of-atmosphere radiance measurements referenced to an Earth-fixed coordinate system. These product generation systems provide products with varying levels of registration accuracy, depending upon the application requirements.

18.2 Landsat overview

Landsat-7 and its Enhanced Thematic Mapper Plus (ETM+) sensor, launched in 1999, is the most recent in the Landsat series of multispectral spaceborne Earth remote sensing satellites that began with Landsat-1 in 1972. The Landsat-7 ETM+ provides medium-resolution (15-m panchromatic, 30-m multispectral, 60-m thermal) imagery of Earth's land areas from a near-polar, Sun-synchronous orbit (NASA, 1998). Landsat-7 provides seasonal global coverage of the Earth's land areas (up to ±82° latitude) using an onboard solid-state recorder to capture data for later playback over a Landsat ground station. The Landsat-5 Thematic Mapper (TM) has been operating since 1984 and continues to provide valuable data through direct downlink to an international network of ground stations. Landsat-7 and Landsat-5 use the same 16-day repeat cycle orbit, with an 8-day offset between the spacecraft. Both missions acquire nadir-viewing imagery on the Worldwide Reference System (WRS-2) so that the data are spatially coincident to facilitate multitemporal image registration and analysis. The relationship between consecutive Landsat orbits and the 185-km Landsat swath is shown in Fig. 18.1.

The cross-track whiskbroom sensor technology used in the ETM+ is similar to the TM instruments flown on both Landsat-4 and -5. The TM/ETM+ instrument detectors are aligned in parallel rows on two separate focal planes: the primary focal plane, containing the visible and near-infrared (VNIR) bands 1–4 and (in the ETM+) the panchromatic band 8, and the cold focal plane containing the short-wave infrared (SWIR) and thermal bands 5, 6, and 7. The primary focal plane is illuminated by the scanning mirror, primary mirror, secondary mirror, and scan line corrector mirror. In addition to these optical elements, the cold focal plane optical

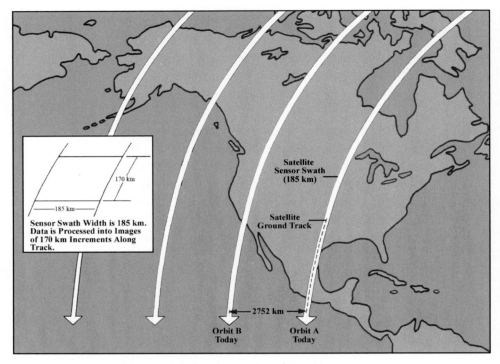

Figure 18.1. Landsat orbit and swathing pattern.

train includes the relay folding mirror and the spherical relay mirror. The scan mirror provides a nearly linear cross-track scan motion that covers a 185-km-wide swath on the ground. The scan line corrector (SLC) compensates for the forward motion of the spacecraft and allows the scan mirror to produce usable data in both scan directions. In May 2003, the ETM+ suffered a failure in the scan line corrector mechanism, leading to data with a disrupted scanning pattern since that time. The resulting scan-to-scan gaps have accentuated the need for multitemporal image registration so that multiple "SLC-off" images can be used to provide complete coverage (Storey *et al.*, 2005).

An image dataset is built as a time sequence of cross-track scans, acquired as the spacecraft flies over the target area. Each scan covers approximately 480 m in the along-track direction. A full scene is defined to contain 375 scans, which makes the ground coverage approximately 185 km by 180 km. Variations in scan timing lead to imperfect alignment between adjacent scans without geometric correction. This can be seen as discontinuities in the image features of Fig. 18.2, which shows a portion of an uncorrected ETM+ Level 0 image. Telemetry describing the spacecraft position, velocity, and attitude, as well as measurements of the instrument's high-frequency jitter and scan mirror motion is packaged with the

Figure 18.2. Uncorrected Level 0 Landsat-7 ETM+ image.

image data in the downlink, for use in subsequent correction processing. Basic Level 1 systematic geocorrection processing is required to correct for these instrument and spacecraft effects in order to provide a spatially consistent image suitable for subsequent analysis or geometric refinement.

The basic level of image registration provided by the Landsat ground system is thus a systematic registration to World Geodetic System of 1984 (WGS84) Earth-fixed coordinates, derived using instrument and spacecraft support data transmitted from the spacecraft or provided by the Landsat flight operations team. The resulting products are subject to the accuracy limitations associated with these support data and any terrain-induced image displacements. Higher levels of absolute registration accuracy are achieved by including information from ancillary ground control and digital elevation data. For applications in which relative image registration accuracy is of paramount importance, reference imagery is used as control to refine the relative registration of multiple scenes. In this case, the registered set of images retains the absolute ground registration accuracy of the reference image.

18.3 Landsat image systematic geocorrection

The Landsat product generation systems implement geometric correction algorithms that use the instrument and spacecraft support data, provided in the Landsat data stream, to construct a model of the geometric relationship between the acquired image data samples and an Earth-fixed ground reference system. These support data include onboard measurements of instrument timing, spacecraft attitude (orientation) and jitter, and estimates of the spacecraft position and velocity derived from ground tracking data. A rigorous model of the sensor imaging geometry is constructed using these instrument and spacecraft data in conjunction with a set of geometric calibration parameters derived and maintained by the Image Assessment System (IAS), a component of the Landsat ground system dedicated to calibration and data validation (Storey *et al.*, 1999).

The accuracy of the basic systematic geometric registration model is limited by uncertainties in the supporting data. In particular, the spacecraft position and attitude data typically contain residual biases on the order of tens to hundreds of meters (Lee *et al.*, 2004). Fortunately, these errors are usually slowly varying, allowing for the registration of multiple Landsat images using simple, low-order techniques. Ephemeris errors can also be reduced by using definitive ephemeris data, derived by the Landsat flight operations team subsequent to image acquisition, in place of the predicted ephemeris broadcast with the instrument data stream.

Neither the Landsat-5 nor Landsat-7 spacecraft carry an onboard navigation system (e.g., the Global Positioning System (GPS)) to provide real-time position and velocity information. Instead, predicted ephemeris data are computed and uploaded daily by the Landsat flight operations team to support spacecraft operations. These predicted data are included in the payload correction data (PCD) stream transmitted to the ground with the Landsat image data. Each day, the Landsat flight operations team performs an orbit determination process that integrates the tracking data from the Landsat ground station and, when available, from the Tracking and Data Relay Satellite System (TDRSS) satellites to obtain the "definitive" best estimates of the spacecraft position and velocity for a 61-hour period spanning the previous two days and the first 13 hours of the current day. The orbit determination process is also used to generate the orbit predictions that are uploaded to the spacecraft. In addition to being, on average, more accurate than the PCD ephemeris, the definitive ephemeris data are also more consistent. The routine use of definitive ephemeris greatly reduces the frequency of, though it does not eliminate, the occurrence of unusually large geopositional errors ($>250\,$m), and makes pointing knowledge, rather than spacecraft position, the dominant source of error in systematically geocorrected Landsat data products (Dykstra and Storey, 2004).

The Landsat-7 system was designed to produce geometrically corrected image products accurate to $250\,$m (1σ) or better without the use of ground control (NASA, 1998). Prelaunch measurements of the ETM+ to Landsat-7 spacecraft alignment were believed to be accurate to approximately 4 arc-minutes, whereas alignment knowledge accurate to 24 arc-seconds or better was required to meet this

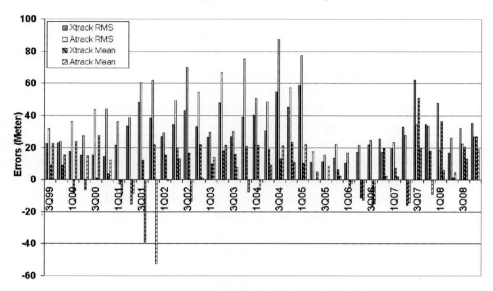

Figure 18.3. Landsat-7 absolute registration accuracy statistics by calendar quarter.

specification, according to the geodetic error budget (USGS, 2006). The IAS on-orbit sensor alignment capability was developed to refine the prelaunch knowledge of the relationship between the ETM+ and Landsat-7 attitude control system sufficiently to achieve product geodetic accuracy better than 250 m. Fortunately, the Landsat-7 attitude determination system has exceeded its accuracy requirements. Coupled with the use of definitive ephemeris, this has enabled the Landsat-7 mission to achieve absolute ground registration (excluding the effects of terrain) accuracy on the order of 55 m (1σ) (Lee *et al.*, 2004). Lifetime *root mean square* (RMS) and mean absolute ground registration error statistics, compiled by calendar quarter, are shown in Fig. 18.3. The improvement in accuracy that began in early 2005 has been attributed to turning off a problematic gyroscope in the spacecraft's inertial measurement unit. Accuracy was degraded somewhat when the ETM+ was switched into bumper mode in April 2007. This backup operating mode, used by the Landsat-5 TM since 2002, compensates for scan mirror bumper wear and extends the useful mission life, but provides poorer knowledge of the sensor absolute line-of-sight. Lacking a dedicated image assessment capability similar to the Landsat-7 IAS, and – until recently – definitive ephemeris processing, Landsat-5 products have been an order of magnitude less accurate.

The sensor alignment calibration procedure examines the apparent pointing errors, derived using ground control points (GCPs), in sequences of scenes acquired at different times, to look for pointing biases that could be attributed to systematic alignment errors. It is important to use scenes acquired at different times in this process so that any spacecraft attitude determination errors, which are typically highly correlated over time periods of seconds to minutes, will be random and averaged out. If systematic biases are detected, they are used to compute corrections to the ETM+ to Landsat-7 sensor alignment matrix, and an update to the Landsat-7 calibration parameter file is issued for use by the Landsat product generation systems. This sensor alignment calibration procedure was carried out as part of the initial on-orbit verification for Landsat-7 (Storey and Choate, 2000) and has been repeated routinely since then. Only one on-orbit alignment calibration operation has been performed for Landsat-5 – in late 2004, coincident with the introduction of the definitive ephemeris processing capability for that mission. More robust software tools were subsequently designed to support routine alignment calibrations, and facilitate a retrospective calibration campaign for the full mission history.

Data users often place greater importance on the relative registration between multitemporal images of the same area than on the images' absolute registration to a ground coordinate system. Given the consistent acquisition geometry employed by the Landsat sensors, geometric effects due to target area topography are largely common from acquisition to acquisition. Thus, explicit terrain correction is not always required to provide image-to-image registration accuracy sufficient for many applications. Higher-order corrections are generally required if registration to data from other sensors or to external reference data is necessary.

18.4 Landsat image precise geocorrection

When higher-accuracy products are required, additional capabilities of the Landsat ground system are employed to perform more precise geometric correction. This higher-level processing requires ancillary data sources, including ground control points, to remove the residual pointing and position biases, and digital elevation data, to correct for the effects of terrain height on the Landsat viewing geometry. These ancillary data are used to derive corrections to the systematic geometric model that yield image products registered to a ground reference system to subpixel accuracy.

The Landsat-7 precision correction process begins with the generation of a systematically corrected image as described above. This image is input to an automated correlation procedure that matches it to control-point image chips, extracted from a library, to measure the locations of known ground control points in the systematic image. The ground control points are derived from a variety of sources but two control point sources of particular note are the USGS digital orthophoto quadrangle (DOQ) data and the Global Land Survey (GLS) 2000 dataset.

The DOQ data are high-resolution (1-m) aerial images that have been digitized and orthorectified to an accuracy of 33.3 feet (CE90) (USGS, 1996). This is equivalent to a one-sigma accuracy of approximately 4.8 m. Each image covers a ground area corresponding to one fourth of a USGS 7.5-min topographic quadrangle map. These so-called quarter quad images are mosaicked to cover a full Landsat scene, and reduced to 15-m resolution using a Gaussian pyramid algorithm (USGS, 2006). Image windows extracted from these DOQ-derived images are paired with latitude/longitude coordinates computed from the DOQ header information and height coordinates derived from the associated USGS digital elevation data. This process creates ground control points that consist of an image chip and the associated geodetic position information.

Since DOQ data are only available within the United States, a similar procedure can be applied to the global GLS 2000 dataset when control is required elsewhere. The GLS 2000 data are a collection of orthorectified Landsat-7 scenes covering the Earth's land areas visible from the Landsat orbit. These scenes were originally created to be accurate to 50-m RMS (Tucker *et al.*, 2004) as part of the GeoCover project. The GeoCover data were subsequently readjusted and reprocessed using more accurate elevation data from the Shuttle Radar Topography Mission (SRTM) to improve the accuracy of the resulting GLS data (Gutman *et al.*, 2008). The GLS data are available from the USGS Landsat website.

When building control from the GLS data, control point image windows are extracted directly from the GLS images. This control point selection and extraction process is directed by an automated process that uses a multiscale variant of the Moravec interest operator (Moravec, 1979) to identify image points that should provide strong correlation targets. Although the absolute geolocation accuracy of the GLS data is not substantially better than that of systematically corrected Landsat-7 products, the use of GLS-derived control has the advantage of providing a consistent geometric reference for multitemporal image registration. This is the primary reason that the GLS control is used for standard Landsat product generation.

Whatever the source of the ground control information, the assembly of the ground control point library is an offline process that is performed one time for each Landsat WRS-2 scene area. When a newly acquired Landsat scene is being corrected, the control point image windows falling within the scene area are extracted from the library and correlated with the systematically corrected image. Several different control point matching methods are available for use, for example, methods based on the *normalized gray-scale correlation* (see Chapters 4 and 5) and *mutual information* (MI) (see Chapter 6), which are most common, as well as methods based on *robust feature matching* (see Chapter 8). The normalized gray-scale correlation method is the default selection due to its speed and simplicity. The mutual information method is typically used when the systematic image has missing data (image gaps). The failure of the ETM+ scan line corrector mechanism has made this a characteristic of images acquired since May 31, 2003. This will be discussed in more detail later.

The normalized gray-scale correlation algorithm is implemented in the frequency domain for processing efficiency by correlating the control point chips of 64×64 pixel with search areas of size 128×128 pixel (or larger). The IAS implementation uses a peak finding algorithm to achieve subpixel precision in locating the match point (USGS, 2006). The peak finding algorithm fits a two-dimensional second-order polynomial to the 3×3 set of correlation coefficients surrounding the peak of the discrete correlation surface. The best-fit coefficients are then used to analytically compute the location of the polynomial maximum, providing a subpixel estimate of the correlation peak. This peak-fitting approach reduces the ± 0.5-pixel quantization errors inherent in simply selecting the line/sample offset associated with the maximum coefficient computed by the correlation algorithm, but can introduce offset-dependent errors as large as 0.05 pixels (Dvornychenko, 1983).

Each measured ground control point systematic image location is mapped backwards through the geometric model to determine the corresponding input Level 0 image location, from which the time of observation of the control point is computed. The observation time and the geometric model are used to determine the spacecraft position associated with the measured control point. This information and the differences between the systematic image locations and the known ground locations of the control points are input to an iterative least-squares procedure that computes corrections to the spacecraft ephemeris and attitude to best fit the ground control. Correction terms are computed for the spacecraft position, velocity, attitude, and attitude rate. To account for errors in the control point correlation process, the solution procedure includes automated outlier detection and removal. This procedure uses a t-distribution test to isolate points that are inconsistent with the control point set.

The spacecraft position, velocity, attitude, and attitude rate corrections computed by the precision correction process are stored in an updated copy of the geometric model. This precision model is then used to repeat the image geocorrection procedures to create a precision-corrected output image. It is important to note that the original systematic image used for control point mensuration is discarded. A new "first generation" precision image is created directly from the unresampled input data using the updated precision model.

If an elevation model of the area covered by a Landsat-7 scene is available, it can be used to correct for the geometric effects of terrain displacement. This correction is computed and applied during image resampling. Since the ETM+ instrument is nadir pointing with a very narrow along-track (cross-scan) field of view, the terrain offset primarily affects the cross-track (along-scan) direction. The terrain correction logic in the image resampling process uses the input terrain model and the spacecraft geometric model to compute a terrain correction table that provides the along-scan terrain offset correction as a function of target elevation and distance from nadir. Each input image scan is then examined to determine the along-scan sample location corresponding to nadir. The terrain offset table and the list of scan nadir sample locations are precomputed and used as lookup tables during image resampling to determine the along-scan offset to apply to each pixel based on its elevation and distance from nadir.

The IAS measures the internal accuracy of Landsat-7 precision and terrain-corrected images as it monitors the calibration of the ETM+ scanning system. Along- and across-scan deviations are measured by comparing each ETM+ scan with the DOQ-derived reference imagery over selected calibration test sites. Although the main purpose of this operation is to accurately characterize scan mirror behavior, it also provides a measure of image internal accuracy. The overall measured accuracy of 6.9-m along-track (90% linear error) and 6.5-m across-track (90% linear error) compares favorably with the 12-m (90% linear error) Landsat image registration requirement.

18.5 Landsat-7 ETM+ scan line corrector failure

On May 31, 2003 the Landsat-7 ETM+ experienced a failure in its scan line corrector (SLC) mechanism. When operating nominally, the SLC deflects the ETM+ line-of-sight in the along-track direction to ensure that the sensor's bidirectional cross-track scanning pattern provides continuous coverage of the full Landsat swath (NASA, 1998). Without this compensation, the resulting zigzag scanning pattern exhibits wedge-shaped scan-to-scan gaps, alternating with scan-to-scan overlap, of increasing magnitude away from nadir. Figure 18.4 shows a diagram depicting the change in the ETM+ scanning pattern with and without the scan line corrector. The ETM+ has continued to acquire data with the SLC powered off, leading to images that are missing approximately 22% of the normal scene area, but that are otherwise of the same radiometric and geometric quality as images collected prior to the failure. Figure 18.5 shows an example of the impact of the SLC failure on ETM+ imagery near the edge of a scene, where the scan gaps are the largest.

The USGS EROS Center developed the infrastructure to implement a production capability for multiscene (same path/row) gap-filled products, in an effort to improve the usability of ETM+ data acquired after the SLC failure (Storey *et al.*, 2005). The random placement of the scan gaps from one acquisition to the next leads to variations in the amount of gap-fill that is achieved when combining the data from a pair of SLC-off scenes. Thus, the gap-fill production process allows for

Normal Scanning

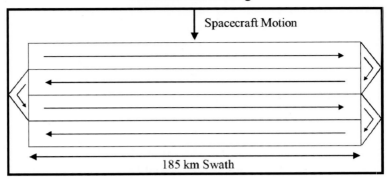

SLC-Off Scanning

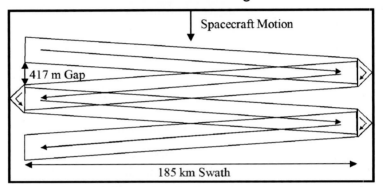

Figure 18.4. ETM+ normal versus SLC-off scanning pattern.

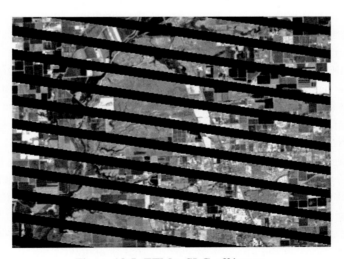

Figure 18.5. ETM+ SLC-off image.

a minimum of two and a maximum of five scenes to be combined to provide complete coverage of the scene area. One of these scenes may be a prefailure SLC-on scene, though the longer the time lag between "primary" and "fill" scenes, the more likely it is that temporal changes will lead to scan gaps filled with inappropriate data. Implicit in this process is the need to geometrically register all of the input scenes so that they can be appropriately merged to fill the scan gaps.

18.6 SLC-off image-to-image registration

One of the input scenes is selected as a reference for the registration procedure. The registration procedure was found to work best if an SLC-on scene was used as the reference image. Using a complete (gap-less) image as the reference maximizes the number of common pixels available for image matching, especially near the edges of the scene. If the user selects an SLC-on scene for inclusion in the gap-fill process, it is used as the geometric reference. Otherwise, a cloud-free SLC-on scene is automatically selected and used for the image registration processing. All input scenes are first systematically corrected. Tie-points are automatically extracted from the reference scene, and stored in a temporary control point library. This temporary library is then used to precision-correct each of the other scenes, using the same methods described above, with one important difference. Instead of using the normalized gray-scale correlation procedure typically chosen for precision correction processing, the MI-based image matching method (see Chapter 6) was found to provide better performance, though at a considerable computational cost, when working with images having missing data. This methodology was more easily adapted to ignore the missing scan gap pixels, as it was straightforward to implement the algorithm so as to exclude fill pixels from the histogram analysis. This gap-insensitive MI algorithm is the preferred image matching method when working with SLC-off data.

Initially, *partial volume interpolation* (PVI) (Maes *et al.*, 1997) was used to achieve subpixel precision in the mutual information image matching procedure. This was abandoned in favor of the second-order polynomial peak-fitting method described above, in order to avoid the difficulties that the data gaps introduced into the PVI algorithm. The maximum peak-fitting errors were found (through experimentation) to be somewhat larger (up to 0.08 pixels) when used with the MI algorithm than was the case with normalized gray-scale correlation.

18.7 Landsat band-to-band registration

The design of the Thematic Mapper class of sensors is well suited to achieving highly accurate registration between the Landsat spectral bands. The whiskbroom design features a small focal plane with a modest number of detectors (100 in the TM, 136 in the ETM+) that are swept across the target area by the scan mirror. In this architecture, all of the spectral bands sample the same ground area within a period of 2.1 msec. This means that all bands observe the Earth from essentially the same location, making the band registration insensitive to terrain parallax effects. This near-simultaneity in band sampling also makes the TM/ETM+ band registration relatively insensitive to variations in platform pointing, since the platform is generally stable over these short time intervals. Landsat band registration accuracy is limited mainly by scan mirror stability, primary to cold focal plane alignment, and instrument jitter.

The TM/ETM+ spectral bands are treated individually in the Landsat geometric model. Differences in focal plane location and sample timing are explicitly modeled in the geometric correction process. No image matching techniques are used during geocorrection processing, but band-to-band image correlation is used to measure band registration performance and estimate refinements to the geometric calibration parameters that define the placement of the spectral bands on the sensor focal planes.

The band-to-band registration accuracy characterization procedure correlates an array of tie-points spanning the scene, across all possible unique band-pair combinations. In this process, the 15-m resolution ETM+ panchromatic band is reduced to 30-m resolution, using a Gaussian pyramid procedure, for comparison to the multispectral bands. Similarly, the 30-m bands (including the reduced panchromatic band) are reduced to 60-m resolution for comparison to the thermal band. Normalized gray-scale correlation is used to perform the image matching in SLC-on scenes, whereas the MI method is used in SLC-off data. Summary statistics are computed for each band by combining the valid correlation results for all band combinations that include the subject band. Band-average registration accuracy results for Landsat-7 are plotted in Fig. 18.6, with SLC-on and SLC-off results shown separately. The apparently poorer results for band 6 (thermal) are attributed more to measurement difficulties than actual misregistration. Note that band registration performance improved slightly when the SLC was turned off. Apparently, halting the SLC eliminated one source of instability in the scanning process.

The band-pair measurements can also be used to estimate corrections to the focal plane band location parameters. This was necessary shortly after the launch of Landsat-7 when a small systematic offset was observed between the primary focal plane VNIR bands and the cold focal plane SWIR/LWIR bands (Storey and Choate, 2000). This offset was attributed to an increase in the instrument temperature once it reached its operational duty cycle. The focal plane calibration for the cold focal plane bands was updated and the characterization and calibration procedure was iterated until no systematic offset was evident. This iteration overcomes the effects of the small biases introduced by the subpixel correlation refinement technique.

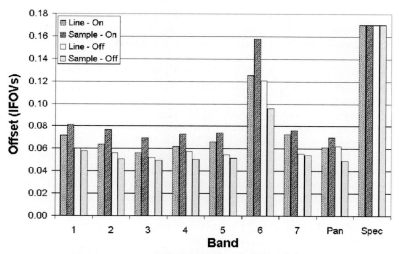

Figure 18.6. Landsat-7 RMS band-to-band registration accuracy.

18.8 Summary

The operational Landsat image registration procedures rely primarily on model-based rather than image-based techniques. Image matching is used in the image correction process only for measuring and removing low-order model biases during precision control point correction and for registering multidate SLC-off images during gap-fill processing. The absolute registration of systematic products and band-to-band registration both rely exclusively on the inherent accuracy of the Landsat geometric model. This approach has the advantage of being insensitive to scene content. Image matching is used extensively in the calibration and characterization procedures to measure system performance and estimate model parameters, but in these cases the target scenes can be carefully chosen to avoid clouds or other problem image features. The level of accuracy achieved in Landsat-7 product generation using this model-based registration approach requires highly accurate supporting data, a stable sensor, and robust models.

References

Dvornychenko, V. N. (1983). Bounds on (deterministic) correlation functions with applications to registration. *IEEE Transactions on Pattern Analysis and Machine Intelligence*, 5(2), 206–213.

Dykstra, J. D. and Storey, J. C. (2004). Landsat-7 definitive ephemeris: An independent verification of geocover-ortho 1990 positional accuracy. In *Proceedings of the American Society for Photogrammetry and Remote Sensing 2004 Annual Conference*, Denver, CO, unpaginated CD-ROM.

Gutman, G., Byrnes, R., Masek, J., Covington, S., Justice, C., Franks, S., and Headley, R. (2008). Towards monitoring land-cover and land-use changes at a global scale: The Global Land Survey 2005. *Photogrammetric Engineering & Remote Sensing*, **74**(1), 6–10.

Lee, D. S., Storey, J., Choate, M. J., and Hayes, R. W. (2004). Four years of Landsat-7 on-orbit geometric calibration and performance. *IEEE Transactions on Geoscience and Remote Sensing*, **42**(12), 2786–2795.

Maes, F., Collignon, A., Vandermeulen, D., Marchal, G., and Suetens, P. (1997). Multimodality image registration by maximization of mutual information. *IEEE Transactions on Medical Imaging*, **16**(2), 187–198.

Moravec, H. P. (1979). Visual mapping by a robot rover. In *Proceedings of the Sixth International Joint conference on Artificial Intelligence*, Tokyo, Japan, pp. 599–601.

NASA, National Aeronautics and Space Administration (1998). *Landsat 7 Science Data Users Handbook*. Landsat Project Science Office, NASA Goddard Space Flight Center, Greenbelt, MD, http://landsathandbook.gsfc.nasa.gov/handbook.html.

Storey, J. C. and Choate, M. J. (2000). Landsat 7 on-orbit geometric calibration and performance. In *Proceedings of the SPIE Conference on Algorithms for Multispectral, Hyperspectral, and Ultraspectral Imagery VI*, Orlando, FL, Vol. 4049, pp. 143–154.

Storey, J. C., Morfitt, R. A., and Thorson, P. R. (1999). Image processing on the Landsat 7 Image Assessment System. In *Proceedings of the 1999 American Society for Photogrammetry and Remote Sensing Annual Conference*, Portland, OR, pp. 743–758.

Storey, J., Scaramuzza, P., Schmidt, G., and Barsi, J. (2005). Landsat 7 scan line corrector-off gap-filled product development. In *Proceedings of the 2005 American Society for Photogrammetry and Remote Sensing Pecora 16 Conference on Global Priorities in Land Remote Sensing*, Sioux Falls, SD, unpaginated CD-ROM.

Tucker, C. J., Grant, D. M., and Dykstra, J. D. (2004). NASA's global orthorectified Landsat data set. *Photogrammetric Engineering & Remote Sensing*, **70**(3), 313–322.

USGS, U.S. Geological Survey (1996). *Standards for Digital Orthophotos, Part 2: Specifications*. USGS National Mapping Program Technical Instructions, http://rockyweb.cr.usgs.gov/nmpstds/acrodocs/doq/2DOQ1296.PDF.

USGS, U.S. Geological Survey (2006). *Landsat 7 (L7) Image Assessment System (IAS) Geometric Algorithm Theoretical Basis Document (ATBD)*, USGS Document LS-IAS-01, Version 1, http://landsat.usgs.gov/documents/LS-IAS-01_Geometric_ATBD.pdf.

19

Automatic and precise orthorectification of SPOT images

SIMON BAILLARIN, AURÉLIE BOUILLON, AND MARC BERNARD

Abstract

Following the SPOT-5 launch, Spot Image and the French National Geographic Institute (IGN) have designed a high-accuracy worldwide database called Reference3D^{TM} using data from the High Resolution Stereoscopic (HRS) SPOT-5 instrument. This database consists of three information layers: A Digital Elevation Model (DEM) at 50-m resolution, ortho-images at 5-m resolution, and quality masks with a circular horizontal accuracy better than 16 m for 90% of the points and an elevation accuracy better than 10 m for 90% of the points. A new system (called ANDORRE) was also developed to archive the Reference3D^{TM} database and to automatically produce orthorectified images using Reference3D^{TM} data. ANDORRE takes advantage of Reference3D^{TM} horizontal and vertical accuracies to automatically register and rectify an image from any SPOT satellite in every area where Reference3D^{TM} data are available. This chapter presents the automatic orthorectification algorithm, developed under the CNES (French Space Agency) leadership, that is composed of three main steps: (1) Generation of a reference image (in focal plane geometry) using Reference3D^{TM} orthoimage and DEM layers, (2) modeling of the misregistration between the reference image and the SPOT image to be processed, and (3) resampling of the image into a cartographic reference frame. It also describes the geometric performance measured on operational cases involving different landscapes, DEM data, and image resolutions. Timing measurements show that the rectification of a $24\,000 \times 24\,000$ image can be performed in less than an hour.

19.1 Introduction

SPOT-5 is the latest satellite of the SPOT (Satellite Pour l'Observation de la Terre) family and has both enhanced resolution and location accuracy, as well as a

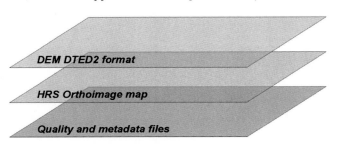

Figure 19.1. Reference3D™ tile.

stereoscopic capacity provided by the HRS (High Resolution Stereoscopic) instrument. It was launched in May 2002 to ensure data continuity with the previous SPOT satellites. This instrument is dedicated to the generation of the worldwide database, Reference3D™ (Bouillon *et al.*, 2006), composed of the three following layers (see Fig. 19.1): (1) A DEM, Level 2 standard (at an approximate 30-m resolution), following the Digital Terrain Elevation Data (DTED), (2) an orthoimage at 5-m resolution, and (3) quality masks. The quality masks consist of 8 layers of information relative to the DEM accuracy (such as water, snow or cloud presence, altitude validity, etc.). Reference3D™ offers high planimetric and altimetric accuracy with a circular horizontal accuracy better than 16 m for 90% of the points and an elevation accuracy better than 10 m for 90% of the points (depending on slopes) (Airault *et al.*, 2003).

This database concept, combined with a correlation method, enables one to automatically register any SPOT image at locations where Reference3™ DEM data are available, thus benefiting from high location accuracy. A preliminary study and prototyping, conducted by the French National Geographic Institute (IGN), demonstrated the feasibility of such an automatic orthorectification process. This process was validated after SPOT-5 launch on real data, and an operational version was developed under the CNES (French Space Agency) leadership and integrated by Spot Image in its ground segment.

The algorithmic core, called TARIFA (French acronym for Automatic Image Rectification and Fusion Processing), is activated by the ANDORRE (French acronym for Atelier Numérique D'ORthoREctification, or digital orthorectification shop) operational system that ensures tasking and production control (Mangolini and Cunin, 2002); see Fig. 19.2. This system has been in operation since March 2005 and is able to process 50 images a day in a fully automatic mode (see ANDORRE reference).

This chapter first presents the algorithm's principle and performance. Then, specific test cases are presented that demonstrate the robustness of the algorithm on different landscapes, DEM data, and image resolutions.

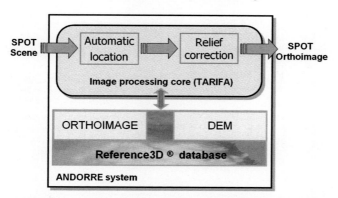

Figure 19.2. Reference3D^{TM} database and ANDORRE. (Source: Baillarin *et al.*, 2005, reprinted with permission from the International Society for Photogramme-try and Remote Sensing (ISPRS).)

19.2 Description of processing method

The automatic rectification process is based on the improvement of the SPOT geometric model using automatic tie-point generation between Reference3D^{TM} data and SPOT images. The algorithm is divided into four main stages:

(1) Generation of a simulated image.
(2) Determination of the correction model by multiresolution image matching.
(3) Processing of the rectification grid.
(4) Resampling of the SPOT scene.

19.2.1 First stage: Generation of a simulated image

This first stage consists in simulating a landscape image as seen by the satellite out of the Reference3D^{TM} orthoimage and DEM layers. This method is described in general in (Datcu and Seidel, 1994). In the case of SPOT, the scene parameters are arrived at utilizing the sensor geometric model to compute a resampling grid from the ground geometry to the raw scene geometry by a ray tracing method based on the DEM variations. The orthoimage layer extracted from Reference3D^{TM} is resampled with this grid to obtain the simulated image (Baillarin *et al.*, 2004) (see Fig. 19.3).

The use of a DEM layer fully compatible with the orthoimage layer ensures that relief effects are well taken into account in the simulated image. Remaining misregistration between the simulated image and the SPOT scene is only due to imprecision of the SPOT location model (Bouillon *et al.*, 2003).

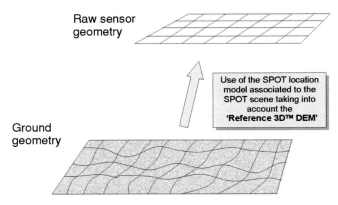

Figure 19.3. Computing simulated resampling grid. (Source: Baillarin *et al.*, 2005, reprinted with permission from the International Society for Photogrammetry and Remote Sensing (ISPRS).)

19.2.2 Second stage: Determination of the correction model by multiresolution image matching

The purpose of the second stage is to obtain a correction model which will allow matching the SPOT scene with the simulated image. Remaining distortions (after taking into account the sensor model and terrain knowledge) can be captured by a low-order polynomial correction model (degree 1 or 2) using a multiresolution image matching strategy from coarse to fine zoom level. The initial zoom level is computed in accordance with the raw location accuracy of the SPOT scene in order to have an initial search window (for matching) of size less than 5 pixels. For example, for a SPOT-4 10-m mode, the initial image un-zoom ratio is $\times$ 1/64. A ratio of two was chosen between successive zoom levels.

At each zoom level, an automatic image matching is performed by maximizing a normalized correlation coefficient as a similarity criterion. The main parameters of the process are the correlation window size and the correlation validity threshold. The resulting matching grid is filtered out in order to select uniformly distributed and most reliable tie-points. The collected tie-points are then used to estimate the correction model of the SPOT scene for the current zoom level. The model estimation process includes a statistical filtering of outlier points with some K standard deviation filter.

The model estimation depends on the zoom level; a bias is estimated only for the first zoom levels whose number is defined by a parameter. Usually a bias is estimated up to 80-m resolution (Phase 1), then the polynomial model is estimated (Phase 2), and at the full resolution, the matching is iterated several times in order to refine the correction model (Phase 3 or refining phase).

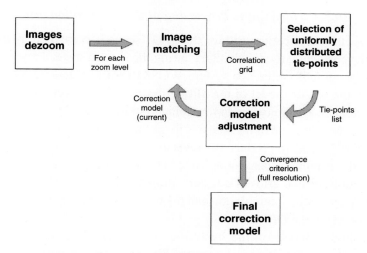

Figure 19.4. Correction model estimation by multiresolution image matching. (Source: Baillarin *et al.*, 2005, reprinted with permission from the International Society for Photogrammetry and Remote Sensing (ISPRS).)

At each zoom level, the model previously estimated is taken into account in the matching process to predict the differences between the two images and keep a 5-pixel search window size. At the full resolution level (refining phase), the convergence criterion is defined as the difference between the current and previous model, and is computed for a grid of control points carefully selected in the image. The model estimation is complete when the difference becomes lower than a convergence threshold (usually <50 cm).

Some specific strategic steps are also used to optimize the processing:

(1) Reference3D™ masks are utilized so that correlation is not computed over water, snow, or cloud areas.
(2) In the case of a multispectral SPOT scene, several bands are combined to maximize the similarity with the simulated image.
(3) The lowest zoom level corresponds to the lowest resolution between Reference3D™ (5 m) and the SPOT scene in order to avoid some useless oversampling.
(4) The matching set of parameters are adapted for each phase of the multiresolution processing.
(5) A densification processing described in the next section is performed in case of an insufficient number of tie-points. The process still fails if the number of tie-points is too small at the finest possible matching step or after the statistical filtering.
(6) The correction model is computed on data strips in order to ensure good registration performance even in the case of a subscene rectification.

Figure 19.4 summarizes the second stage.

19.2.3 Third stage: Processing of the rectification grid

In this stage, the rectification grid for the SPOT scene is computed by the combination of the correction model obtained during the second stage, a reverse location grid in the geographic reference system WGS84 (World Geodetic System 1984) obtained by the sensor model of the SPOT image, and a conversion grid from the WGS84 reference system to the cartographic system required for output.

The main characteristic of this stage is to take advantage of the full resolution of Reference3D^{TM} DEM to produce the reverse location grid with the best accuracy possible and a minimum computing time. The sensor model is used to compute reverse location for two grids of constant altitude H_1 and H_2. Those two grids are then filled up by spatial interpolation and the final dense grid is obtained by altitude interpolation using the DEM.

This method is based on the hypothesis that the reverse location is linearly dependent on both altitude and planimetry. To ensure this linearity, H_1 and H_2 are chosen as the minimum and maximum altitude value on the DEM over the SPOT scene area.

19.2.4 Fourth stage: Resampling of the SPOT scene

The last stage consists of a resampling of the SPOT scene using the rectification grid processed at the previous stage. In addition to the image rectification, all the tie-points finally used to compute the correction model are converted into ground control points using their Reference3D^{TM} location. These points are archived in case of further processing of the same image or the processing of another image from the same data strip.

19.3 Operational validation of the processing

A first validation of the prototype orthorectification system was done step by step in order to tune and optimize the algorithm (Baillarin *et al.*, 2004). Then, an operational validation of the system was achieved on various test cases. In the following paragraph, the robustness of the algorithm is analyzed through a few well-chosen complex cases.

Tuning of the image matching algorithm and tie-point filtering method

Different problems were chosen based on low similarity between images. Figure 19.5 presents an example of seasonal variations over rice field landscapes in China.

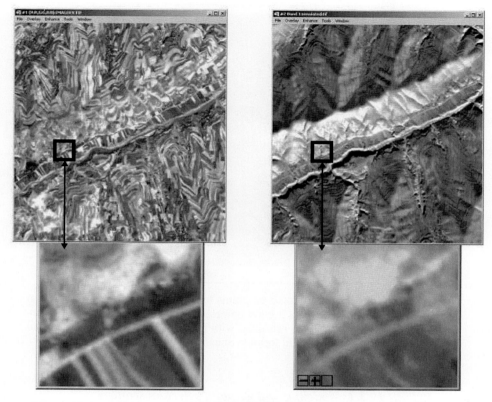

Figure 19.5. Image to process (left) and reference image (right), acquired in Summer and Winter, respectively, and zoomed-in images of corresponding highlighted regions. (Source: Baillarin *et al.*, 2005, reprinted with permission from the International Society for Photogrammetry and Remote Sensing (ISPRS).)

In such cases, the image matching algorithm based on the computation of correlation coefficients cannot select a sufficient number of valid tie-points between the two images. The correlation coefficients are smaller than in other cases and do not reach the validity threshold (80%). Therefore, in order to find enough tie-points in all cases, a lower validity threshold must be selected (e.g., 70% instead of 80%).

However, in some cases, this lower validity threshold may lead to "false correlations" and erroneous tie-points. This problem is mainly observed in images with low radiometric ranges. These invalid tie-points are filtered during the refining phase to avoid the risk that they might decrease the accuracy of the correction model.

Indeed, during the refining phase, the search window size of the image matching algorithm is reduced (from 5 to 3 pixels) taking into account the prediction model obtained in the preceding phase. This eliminates aberrant points.

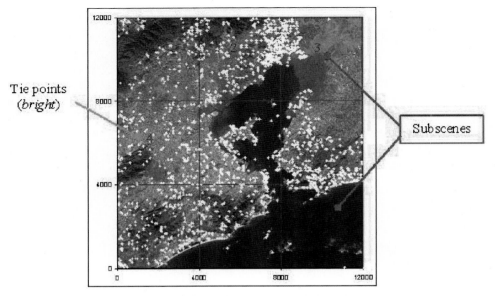

Figure 19.6. Uniformly distributed tie-points by subscenes in a 2.5-m resolution SPOT-5 image (displayed with pseudo natural colors). See color plates section. (Source: Baillarin *et al.*, 2005, reprinted with permission from the International Society for Photogrammetry and Remote Sensing (ISPRS).)

19.3.1 Modeling and tie-point selection strategy

In order to ensure a good quality model over the entire image, a specific tie-point selection strategy has been developed. The idea is to obtain a uniform distribution of tie-points in the entire image. The image is divided into several areas (usually 9) called "subscenes." A given number of points is required for each subscene. If, after image matching and filtering, the resulting number of tie-points is not sufficient, the image matching is performed on a higher density of points. This process is called *densification*. If more valid tie-points than required are found, a random selection is performed. The densification is an iterative operation that can be carried out several times on several subscenes.

Figure 19.6 represents the 9 subscenes on a SPOT-5 image and the resulting tie-point distribution. The subscenes with low texture have been densified (especially subscenes 3, 8, and 9). The densification process found between 150 and 250 tie-points by subscenes.

However, the densification of a subscene can fail if the number of tie-points remains insufficient after several iterations (usually 10). The subscene is then eliminated from the tie-point selection algorithm. If too many subscenes are eliminated, the overall process fails.

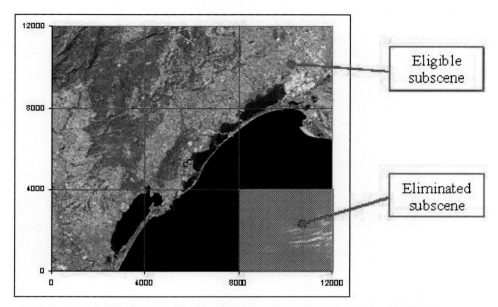

Figure 19.7. Eligible and eliminated subscenes in a 10-m resolution, multispectral SPOT-5 image of a coast landscape. See color plates section. (Source: Baillarin *et al.*, 2005, reprinted with permission from the International Society for Photogrammetry and Remote Sensing (ISPRS).)

Moreover, the Reference3D^{TM} database contains water and coastline mask information that can be used to guide the densification process. Thus, this strategy coupled with Reference3D^{TM} masks allows the processing of images with island and sea landscapes. The subscenes totally located in water areas are directly eliminated by the first tie-point selection step. (For example, subscene 9 is eliminated in Fig. 19.7).

19.3.2 Measuring rectification accuracy

A final estimation of the orthorectified product accuracy is done by measuring the misregistration between the SPOT rectified image and Reference3D^{TM} orthoimage on a random set of 1000 control points (points that have not been used in the computation of the model). The process has been tested over various landscape types (desert, forest, sea coasts, urban areas, etc.), various DEM types, and various SPOT images (from 20-m resolution SPOT-2 to 2.5-m resolution SPOT-5 images). For each processed image (about 100 different images), the residual circular error measured in meters for 80% of the control points was lower than 2 m. Results are shown in Fig. 19.8.

Table 19.1 *Computing time benchmarked on a 1.3-GHz processor for a 24 000 × 24 000 image*

Stage	Time (min)
Simulated image	12
Model estimation	8
Rectification grid	3
Final resampling	25
TOTAL	48

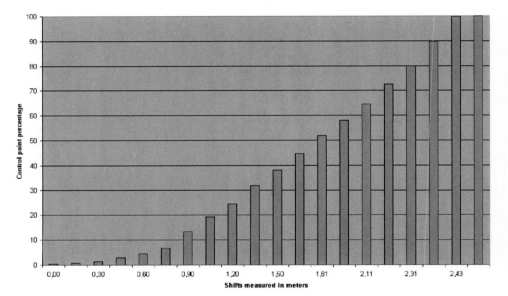

Figure 19.8. Rectification accuracy. (Source: Baillarin *et al.*, 2005, reprinted with permission from the International Society for Photogrammetry and Remote Sensing (ISPRS).)

Moreover, the computing time performances meet the requirements to ensure an intensive operational use of this system; see Table 19.1.

19.4 Conclusion

The results presented in this chapter demonstrate that an automatic production of orthoimages is made possible by using the high horizontal and altimetric accuracy of the SPOT-5 system. With Reference3D^{TM}, manual ground control points are no longer needed to generate orthoimages.

The ANDORRE software with the TARIFA algorithmic core is operational at the Spot Image main processing center in Toulouse and at the China Remote-Sensing

Satellite Ground Station (RSGS) in Beijing. The software is available and ready to be integrated in any SPOT processing center or SPOT Terminal (ANDORRE system: Mangolini and Cunin, 2002).

Since this system was developed, thousands of images have been successfully processed and larger amounts of images are being processed. Furthermore, for each of the SPOT satellite types, only one set of parameters is sufficient to process every image. Since the beginning of the production, parameter adjustments were needed only in a few cases; in no case has the process failed. Moreover, by using this system, the delivery time of orthoimages has been improved from several days to a few hours.

Although initially limited to SPOT imagery, because of the high success rate of the registration method on these images, some studies are in progress to adapt the system for other future sensors, including PLEIADES High Resolution (HR) satellites and the SPOT family successors.

References

Airault, S., Gigord, P., Breton, E., Bouillon, A., Gachet, R., and Bernard, M. (2003). Reference3D™ location performance review and prospects. In *Proceedings of the International Society for Photogrammetry and Remote Sensing Workshop on High Resolution Mapping from Space*, Hannover, Germany, CD-ROM, 5 pp., pagination unknown.

ANDORRE: *Instant automatic orthorectification*. http://www.spotimage.fr/andorre.html.

Baillarin, S., Bouillon, A., Bernard, M., and Chikhi, M. (2005). Using a three dimensional spatial database to orthorectify automatically remote sensing images. In *Proceedings of the International Society for Photogrammetry and Remote Sensing Workshop on Service and Application of Special Data Infrastructure*, Hangzhou, China, XXXVI (4/W6), pp. 89–94.

Baillarin, S., Gleyzes, J. P., Bouillon, A., Breton, E., Cunin, L., Vesco, C., and Delvit, J. M. (2004). Validation of an automatic image ortho-rectification processing. In *Proceedings of the IEEE International Geoscience and Remote Sensing Symposium*, Anchorage, AK, Vol. 2, pp. 1398–1401.

Bouillon, A., Bernard, M., Gigord, P., Orsoni, A., Rudowski, V., and Baudoin, A. (2006). SPOT 5 HRS geometric performances: Using block adjustement as a key issue to improve quality in DEM generation. *ISPRS Journal of Photogrammetry and Remote Sensing*, **60**(3), 134–146.

Bouillon, A., Breton, E., De Lussy, F., and Gachet, R. (2003). SPOT5 geometric image quality. In *Proceedings of the IEEE International Geoscience and Remote Sensing Symposium*, Toulouse, France, Vol. 1, pp. 303–305.

Datcu, M. and Seidel, K. (1994). Fractal and multiresolution techniques for the understanding of Geo-information. In *Proceedings of the Joint European Union/European Association of Remote Sensing Laboratories Expert Meeting on Fractals in Geoscience and Remote Sensing*, Ispra, Italy, pp. 56–87.

Mangolini, M. and Cunin, L. (2002). ANDORRE: Une innovation majeure pour la production automatisée d'ortho-images, *Géo Evenement 2002*, Paris, France.

20

Geometry of the VEGETATION sensor

SYLVIA SYLVANDER

Abstract

VEGETATION data are used mainly for multiple date applications. This implies high-accuracy geometric requirements, especially for multitemporal registration. To comply with these requirements, VEGETATION1 image location is improved by a systematic use of a database of ground control points (GCPs), based on VEGETATION1 chips extracted worldwide. Due to this systematic processing, VEGETATION1 has shown a very accurate and stable geometric performance, ever since its launch onboard SPOT-4 in March of 1998. VEGETATION2, launched onboard SPOT-5 in May of 2002, now complements VEGETATION1 for operational production. It was, therefore, essential to ensure geometric continuity between both sensors. Onboard SPOT-5, a stellar sensor provides high-accuracy satellite attitude, thus enabling high accuracy in the absolute image location. Consequently, it is not necessary to improve VEGETATION2 image geolocation using GCPs. However, it is necessary to perform a fine geometric calibration of VEGETATION2 cameras. This calibration is performed during the commissioning phase using GCPs from the VEGETATION1 database. The monitoring of VEGETATION2 has shown that due to this calibration its geometric performance became at least as accurate and stable as that of VEGETATION1, and that geometric continuity between both sensors was guaranteed. This chapter describes the VEGETATION1 and VEGETATION2 sensors, the method used to build the VEGETATION database of GCPs, as well as the operational method used to register VEGETATION data.

20.1 Introduction

The SPOT-4 (Satellite Pour l'Observation de la Terre 4) satellite, launched in March 1998, provides data continuity for high-resolution SPOT data users through its High Resolution in the Visible and Infra-Red (HRVIR) instruments. These upgraded instruments offer extended capabilities with regard to SPOT-3 High Resolution Visible (HRV) instruments, namely a short-wave infrared band and an onboard registration of the 10-m and 20-m resolution imaging channels.

The SPOT-5 satellite is the latest satellite of the SPOT family, launched in May 2002 to ensure data continuity with the previous SPOT satellites, with enhanced 2.5-m resolution, enhanced location accuracy (due to a star tracker), and stereoscopic capacity provided by the High Resolution Stereoscopic (HRS) instrument. This instrument is dedicated to the generation of Digital Terrain Models (Gleyzes *et al.*, 2003).

In addition of their basic mission, SPOT-4 and SPOT-5 also include an innovative payload based on a large field-of-view imaging instrument (101°, swath width of about 2250 km) for the monitoring of terrestrial vegetation at global and local scales. This payload, named VEGETATION1 on SPOT-4 and VEGETATION2 on SPOT-5, is made independent of the host satellite by having its own recording, transmission, and management equipment. It is designed to provide images of high radiometric accuracy in four spectral bands; red, near-infrared (NIR) and short-wave infrared (SWIR) bands identical to those of the HRVIR instruments, as well as a blue band dedicated to atmospheric measurements. The spatial resolution is about 1.15 km at nadir in both directions, with minimum variations for off-nadir observation (spatial resolution up to 1.7 km in the across-track direction), due to the combination of a telecentric objective with a CCD (charge-coupled device) linear array (Henry, 1999; Barba *et al.*, 1994). The VEGETATION system also comprises a complete ground segment for payload programming, image telemetry receiving, data processing, and product distribution (Pulitini *et al.*, 1994).

The VEGETATION program is an international project between France, Belgium, and Sweden. The main use of VEGETATION data is the monitoring of vegetation evolution for agriculture (e.g., crop monitoring), forestry, or the environment (e.g., net primary production or human settlement). These applications imply the use of a set of images covering a large time period, e.g., a complete growing season for agriculture, year-to-year comparison for deforestation, several years for global change studies, etc. Moreover, the final product generated by the VEGETATION Ground Processing Center (CTIV) in Belgium is a 1-km resolution synthesis product, obtained by merging all the data acquired during 10-day periods. This stresses the need for a very accurate multitemporal registration that is better than 500 m for 95% of the points, with an objective of 300 m, which is much better than the 1-km resolution of the product (Sylvander *et al.*, 2000).

The object of this chapter is to present the methods used to have VEGETATION1 and VEGETATION2 geometric image quality comply with the specifications, i.e., the generation of a global database of ground control points (GCPs) using a space triangulation method and its application to both systematically improve the location accuracy of VEGETATION1 images and process a fine geometric calibration of VEGETATION2 cameras. The geometric accuracy of both sensors will be described in the following section.

20.2 Generation of global database of GCPs

20.2.1 User requirements

From a system point of view, the requirement from multitemporal registration is the most demanding. Assessments performed before the launch on VEGETATION1 had demonstrated that, as far as geometric accuracy was concerned, all the user requirements could be met except that of multitemporal registration. Without any complementary correction, the multitemporal registration depends directly on the absolute location. Preflight assessment predicted a root mean square (RMS) absolute location accuracy around 1 km (not far from the 800-m accuracy effectively measured in orbit), and so gave no hope of reaching the required multitemporal registration accuracy. This assessment justified the decision of improving the VEGETATION1 image location by a systematic use of ground control points as part of the ground processing performed at CTIV.

The first idea was to utilize a coastline database. However, none of the existing databases were accurate enough to reach the final 500-m requirement (at 2σ). Next, the use of SPOT/HRV chips was studied. However, because of their limited Earth coverage (no cloud-free images or no useable maps are available for some parts of the globe), and because of their cost, it turned out that it would be very difficult to create such a database. As a result, it was decided to build the GCP database from well-located VEGETATION1 chips, the location of these chips being obtained by a space triangulation method.

20.2.2 Space triangulation principle

The global GCP database was generated using a space triangulation method, and dedicated tools were developed for the monitoring of VEGETATION image quality.

The space triangulation method is based on a physical description of the image acquisition process. It optimizes some physical correction parameters of a set of images in such a way that it limits the number of unknowns that need to be determined, and improves the relative location between the images. It requires the following three types of information:

(1) *Physical description of the image acquisition process, that is, a function of three attitude biases*: At the orbit scale, some viewing parameters are unknown with a large uncertainty. These parameters are mainly the absolute attitude data of the satellite. A

time error, an error on the along-track position of the satellite or a pitch angle error have very comparable effects on the image and cannot be separated. Similarly, an error on the across-track position of the satellite and a roll angle error are also correlated. Consequently, we chose three attitude biases to represent the unknowns of the geometric model: A pitch bias, a roll bias, and a yaw bias. These three biases are propagated to the whole orbit due to the relative position of the satellite along its orbit provided by the gyrometers.

(2) *Some initial ground control points* (extracted from SPOT HRV; see next section) that provide absolute locations of the VEGETATION images from which the GCP database is extracted.

(3) *Tie-points*, extracted from overlapping areas in a semiautomatic way to provide relative location between images. (An operator selects a certain detail in the reference image and the homologous point in the second image is estimated by automatic correlation.)

The generation of the GCP database by space triangulation requires a sufficient amount of VEGETATION1 images so that:

(1) It provides a global and continuous coverage of all the continental areas.
(2) Tie-points extracted from the overlapping areas ensure a no-gap coverage of these areas.

For each VEGETATION1 image, a correlation-based matching is performed using potential initial GCPs on the one hand, and tie-points in overlapping areas with neighboring images, on the other hand.

The space triangulation process can then simultaneously estimate:

(1) Real ground location of the tie-points, which can then be used as GCPs for the database to be generated.
(2) Unknowns of the geometric models (i.e., the three attitude biases).

20.2.3 Initial ground control points

About 100 SPOT HRV chips were extracted from SPOT Quick Look images by the Swedish Space Corporation Satellitbild. They were chosen as regularly as possible over the globe (see Fig. 20.1). They were then geolocated with maps and rectified in a cartographic projection at a resolution of 800 m. Finally, each of these chips was validated, estimating their location accuracy at 100 m.

20.2.4 Selection of VEGETATION images

The VEGETATION GCP database was then generated progressively, as shown in Figs. 20.2 and 20.3, from about 200 orbits acquired between April and August 1998. The database was further populated in February 2000, using about 60 new

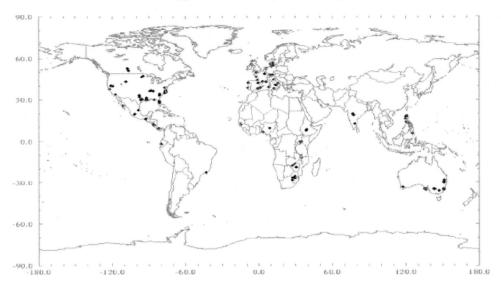

Figure 20.1. Initial ground control points. (Image: Copyright CNES. Reprinted with permission.)

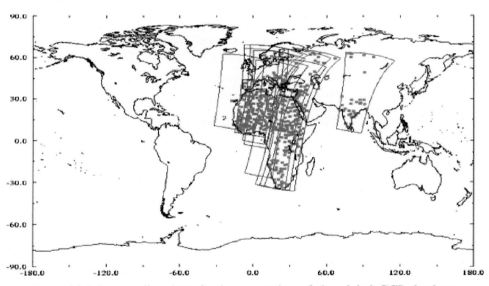

Figure 20.2. Intermediary step in the generation of the global GCP database. (Image: Copyright CNES. Reprinted with permission.)

orbits acquired between October 1998 and February 1999. About 5900 tie-points were utilized altogether. The same point is seen four times on average, so that each tie-point is represented by chips from different seasons and at different angular directions. The location accuracy of these tie-points was estimated at 100-m RMS, through point residuals analysis.

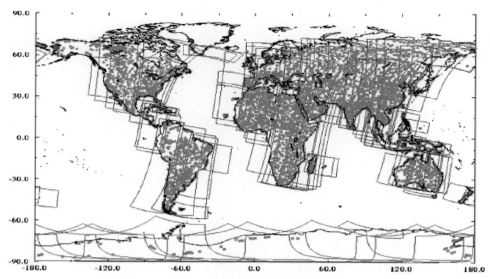

Figure 20.3. Further intermediary step in the generation of the global GCP database. (Image: Copyright CNES. Reprinted with permission.)

20.3 Use of the GCP database for VEGETATION1

20.3.1 VEGETATION1 systematic geometric registration

20.3.1.1 Extraction of a GCP database for the CTIV

A large part of the global GCP database was extracted and supplied to the processing center CTIV in order to allow them to correct any VEGETATION1 image. This CTIV GCP database was supplied to the CTIV before the beginning of the image distribution in March 1999. It was then further populated in February 2000 (see Subsection 20.2.4). CTIV's final database was composed of about 3650 regularly distributed GCPs (see Fig. 20.4), with an average of four VEGETATION1 chips from different seasons and different angular directions per point. The location accuracy of the ground control points was estimated at 100-m RMS.

20.3.1.2 Principle of the geometric registration

VEGETATION1 was used for land monitoring until VEGETATION2 took its place in February 2003. VEGETATION1 was then dedicated to ocean monitoring. VEG-ETATION1 images were systematically registered during the period March 1999 and January 2003. This geometric registration aimed at improving the quality of the viewing parameters (using the GCPs of the CTIV database), and consequently, improving also the absolute location and the multitemporal registration accuracy.

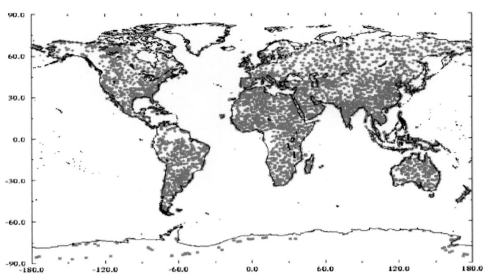

Figure 20.4. The CTIV GCP database. (Image: Copyright CNES. Reprinted with permission.)

As described in Subsection 20.2.2, the GCPs were used to estimate three attitude biases per VEGETATION1 orbit, which modeled the uncertainties of the viewing parameters.

To estimate the three biases of a VEGETATION1 orbit, a relationship was established by correlation between GCPs and their homologous points in the VEGETATION1 images constituting this orbit. The goal of the estimation was to find the values of the attitude biases that best matched the observed differences between the location of the GCPs and the ground location of their homologous points, using the initial viewing parameters.

The CTIV GCP database was used in a semiautomatic way: An operator selected GCPs and the biases were then estimated by an automatic correlation on these points. For each GCP, all the chips (from different seasons and different angular directions) were considered in the correlation process, in order to find the best chip, i.e., the one with the highest correlation coefficient. Few GCPs were theoretically needed to correct a VEGETATION1 image because of the good initial geometric quality. Nevertheless, if possible, a large number of GCPs were selected to be able to monitor them and improve the quality of the GCP database.

To estimate properly the three attitude biases, it was necessary to obtain evenly distributed GCPs over the VEGETATION images; and since the images were often partly covered by clouds, the CTIV needed a very dense GCP database.

20.3.2 VEGETATION1 geometric accuracy

20.3.2.1 Absolute location accuracy

The VEGETATION1 geometric quality was strictly monitored by CNES, the French National Space Agency. In particular, the absolute location accuracy and the multitemporal registration accuracy were regularly measured using checking points, called reference sites, extracted specially from the global VEGETATION GCP database (and not used by the CTIV for image registration). For each VEGETATION1 image, after registration by the CTIV GCP processing, a correlation measure estimated the distance between each reference site and the homologous point in the images overlapping this reference site. This distance represents the absolute location accuracy of the image. An average of 15 images per site was monitored every month using this method.

For VEGETATION1, we used 15 reference sites, regularly distributed over the globe. The absolute location accuracy was estimated from March 2000 to March 2003 at 190-m RMS, i.e., better than 375 m for 95% of the measurements (Sylvander *et al.*, 2003). These measurements showed the high quality of the absolute location accuracy of VEGETATION images that is in full compliance with the initial requirements.

20.3.2.2 Multitemporal registration

From the previous results, we can deduce the multitemporal registration accuracy, which is defined as the distance between the locations of the same point in two different images. We obtain an accuracy of 220-m RMS, i.e., better than 450 m for 95% of the measurements (Sylvander *et al.*, 2003). Consequently, we can also conclude that the multitemporal registration accuracy is very good compared to the pixel resolution, and also fully compliant with the user requirements.

20.4 Use of the GCP database for VEGETATION2

20.4.1 VEGETATION2 geometric calibration

20.4.1.1 Geometric calibration principle

VEGETATION2 was launched onboard SPOT-5 in May 2002. In order to ensure the continuity of VEGETATION1 operational production, it was essential to ensure continuity between the two sensors, i.e., VEGETATION2 had to comply with the same demanding requirements as VEGETATION1. But unlike VEGETATION1, the accuracy of a VEGETATION2 image location does not need to be improved using GCPs, because a stellar sensor onboard SPOT-5 provides a very good estimation of the satellite attitude. This implies that the absolute location of the images is very accurate, provided that a fine geometric calibration of the VEGE-TATION2 cameras is performed.

As a matter of fact, initial location accuracy measurements, performed during the first two months of the in-orbit commissioning phase (i.e., the phase during which operation and system performance are being validated in a real environment), showed a global accuracy of 510-m RMS, distributed as followed:

(1) Across-track; an average of +250 m with a standard deviation of 220 m.
(2) Along-track; an average of −355 m with a standard deviation of 155 m.

Such results meant that a geometric calibration was necessary to improve the accuracy.

On the other hand, correlation measurements between spectral bands showed that they were correctly coregistered to each other. Therefore, the geometric calibration of VEGETATION2 cameras consists essentially in the calibration of the B3 reference camera alignment biases. This process was carried out using GCPs from the VEGETATION1 database. The method consists of calculating, over a long period of time, the satellite pointing average error, using the space triangulation tool: GCPs are identified in each VEGETATION2 image, and then three pointing biases per image are calculated, using a least-squares method. The alignment biases correspond to the average of these biases over all the images.

20.4.1.2 Geometric calibration results

The alignment biases calibration was processed in July 2002, using 85 VEGE-TATION2 images, acquired in June and July 2002. The images selected were distributed over the globe as regularly as possible. The values of the computed biases were:

- 330 microrad for pitch, with a standard deviation of 65 microrad.
- 215 microrad for roll, with a standard deviation of 60 microrad.
- −50 microrad for yaw, with a standard deviation of 95 microrad.

The standard deviation values were very low, which suggested a very stable geometry over the considered time period, and confirmed the pertinence of a geometric calibration. It also showed the high quality of the VEGETATION1 GCP database.

20.4.1.3 Geometric calibration validation

The validation of the alignment biases calibration may be carried out using two methods:

(1) Use of our space triangulation tool, i.e., pointing biases are calculated on new VEGE-TATION2 images; the average of these biases must be close to 0 if the satellite geometry is stable.
(2) Measurement of absolute location accuracy on VEGETATION2 images acquired after calibration results are taken into account; the results of this method are presented in Subsection 20.4.2.

The validation of the geometric calibration by the space triangulation used 40 VEGETATION2 images, acquired in July and August 2002. The calculation of three pointing biases per image showed an average of 20 microradians for pitch, 15 microradians for roll, and 5 microradians for yaw, with a standard deviation of about 45 microradians. These values validate the geometric calibration. They also confirm very good geometric stability.

20.4.2 VEGETATION2 geometric accuracy

20.4.2.1 Absolute location accuracy

After taking into account the geometric calibration results, the VEGETATION2 absolute location accuracy was measured using 36 reference sites well distributed over the globe. The measurement principle is the same as for VEGETATION1, except that the VEGETATION2 images are not registered using GCPs by the CTIV (Sylvander *et al.*, 2003). A statistical calculation performed on 1665 measurements from images acquired between August and December 2002 showed:

- Across-track: an average of $+45$ m with a standard deviation of 120 m.
- Along-track: an average of -55 m with a standard deviation of 100 m.
- A global accuracy of 170-m RMS, better than 345 m for 95% of the measurements.

These results confirmed the validation of the geometric calibration. We can also conclude that the VEGETATION2 absolute location accuracy is very good and fully compliant with the specifications.

20.4.2.2 Multitemporal registration

From the previous absolute location measurements, we can deduce the multitemporal registration accuracy. We obtain an accuracy of 155-m RMS, i.e., better than 320 m for 95% of the measurements (Sylvander *et al.*, 2003). Again, this accuracy is very good, better than VEGETATION1's accuracy, and fully compliant with the specifications.

20.4.2.3 VEGETATION1/VEGETATION2 continuity

VEGETATION1 and VEGETATION2 absolute location measurements were processed in August 2002, using 15 reference sites shared by both sensors. These measurements showed the following VEGETATION1/VEGETATION2 multitemporal registration accuracy: 210-m RMS, i.e., better than 450 m for 95% of the measurements. In conclusion, the geometric continuity between VEGETATION1 and VEGETATION2 is ensured.

20.5 Conclusion

The GCP database, generated through VEGETATION1 chips by a space triangulation method, substantially improved the VEGETATION1 image location. It has also provided a very fine VEGETATION2 geometric calibration. For both sensors, the absolute location accuracy and the multitemporal accuracy were very high compared to the pixel resolution, and fully compliant with demanding user requirements. Moreover, all the results presented in this chapter confirm the very good quality of the GCP database, estimated to 100-m RMS. Another benefit is that the GCP database can be used to register images from other sensors. For example, it was successfully used to evaluate the absolute location accuracy of images acquired by the POLDER 2 (POLarization and Directionality of the Earth's Reflectances) instrument on the ADvanced Earth Observing Satellite (ADEOS-2) and the third POLDER instrument on the Polarization & Anisotropy of Reflectances for Atmospheric Sciences coupled with Observations from a Lidar (PARASOL) satellite.

References

Barba, J., Clauss, A., Coste, G., and Durieux, A. (1994). VEGETATION cameras and imaging electronics. *IAF-94-B.3.085, 45th IAF International Astronautical Federation Congress*, Jerusalem, Israel.

Gleyzes, J. P., Meygret, A., Fratter, C., Panem, C., Baillarin, S., and Valorge, C. (2003). SPOT-5: System overview and image ground segment. In *Proceedings of the IEEE International Geoscience and Remote Sensing Symposium*, Toulouse, France, Vol. 1, pp. 300–302.

Henry, P. (1999). The VEGETATION system: A global monitoring system on-board SPOT 4. In *Proceedings of the 1998 Euro-Asia Space Week on Cooperation in Space – 'Where East & West Finally Meet'*, Singapore, ESA SP-430, pp. 233–239.

Pulitini, P., Barillot, M., Gentet, T., and Reulet, J. F. (1994). The VEGETATION payload. In *Proceedings of the International SPIE Symposium on Space Optics*, Garmisch, Germany, Vol. 2209, pp. 126–136.

Sylvander, S., Albert-Grousset, I., Henry, P., and Rollin, J. (2003). Geometrical performance of the VEGETATION products. In *Proceedings of the IEEE International Geoscience and Remote Sensing Symposium*, Toulouse, France, Vol. 1, pp. 573–575.

Sylvander, S., Henry, P., Bastien-Thiry, C., Meunier, F., and Fuster, D. (2000). VEGETATION geometric image quality. *Société Française de Photogrammétrie et de Télédétection*, **159**, 59–65.

21

Accurate MODIS global geolocation through automated ground control image matching

ROBERT E. WOLFE AND MASAHIRO NISHIHAMA

Abstract

A global network of ground control points (GCPs) is being used to maintain the geolocation accuracy of terrestrial remote sensing data from the two Moderate Resolution Imaging Spectroradiometers (MODIS) on NASA's Earth Observing System (EOS) Terra and Aqua spacecrafts. Biases and trends in the sensor orientation determined from automated control point matching are removed by updating models of the spacecraft and instrument orientation in the MODIS geolocation software. This technique has been used to keep the MODIS geolocation accuracy to approximately 50 m (1σ) at nadir. This chapter overviews an approach to automated matching of global GCPs and summarizes eight years of geolocation analysis. This approach allows an operational characterization of the MODIS geolocation errors and enables individual MODIS observations to be geolocated to the subpixel accuracies required for terrestrial global change applications.

21.1 Introduction

Two Moderate Resolution Imaging Spectroradiometer (MODIS) sensors (Salomonson *et al.*, 1989) have been launched as part of NASA's Earth Observing System (EOS). The first was launched in December 1999 on the Terra platform and the second in May 2002 on the Aqua platform. The observations from these sensors need to be geolocated to subpixel accuracies for Earth science research and applications (Townshend *et al.*, 1992; Roy, 2000). This chapter discusses the approach to obtaining and maintaining this accuracy through the use of finer-resolution Landsat Thematic Mapper (TM) and Enhanced TM+ (ETM+) GCPs.

Each MODIS instrument observes the daytime portion of the Earth in 20 reflective and 16 thermal spectral bands once every two days at the equator and at least once a day above 30°N and below 30°S latitude. In addition, MODIS similarly

437

observes the nighttime portion of the Earth in 16 thermal bands. Since first light, the acquisition time of the first on-orbit science data in February 2000, over eight years of data have been acquired by MODIS/Terra. Over five and a half years of data have been acquired since MODIS/Aqua first light in June 2002. MODIS is a cross-track whiskbroom scanner with two MODIS bands at 250-m nadir resolution (ground sample distance), five bands at 500 m, and 29 bands at 1 km. The 250-m bands were designed to help understand changes in vegetation (Townshend *et al.*, 1992). To enable this, the MODIS science team set a goal of obtaining nadir equivalent geolocation accuracy of 50 m (1σ), i.e., 20% of the 250-m pixel.

The approach to obtaining this accuracy was to use a geometric model of the MODIS instrument, spacecraft motion and orientation, and a digital elevation model of the Earth (Nishihama *et al.*, 1997). Before launch, detailed geometric characterization was performed for both MODIS instruments to measure their pointing. Also, a detailed preflight error analysis was done to help understand expected magnitude of pointing changes due to launch shifts and the on-orbit thermal environment. Both the Terra and Aqua satellites have very accurate onboard orientation (attitude). Terra and Aqua also have very accurate ephemeris (position and velocity), with the Terra ephemeris available in real time and a definitive Aqua ephemeris available with 24 hours of image acquisition. This allows the use of a set of globally distributed control points with a finer 30-m resolution to remove pointing biases, long-term linear trends, yearly cyclic trends, and within-orbit thermal variations. The control point location in the MODIS image is found by cross-correlating the MODIS 250-m red band with simulated 250-m resolution pixels obtained by aggregating Landsat 30-m red band GCPs to the MODIS resolution using a first-order MODIS *point spread function* (PSF).

21.2 The methodology

21.2.1 Geometric model

The MODIS geometric model has three main components: Sensor, spacecraft, and Earth. The geolocation algorithm implements a mathematical model of these components and solves for the terrain-corrected location of the center for each MODIS 1-km observation. The locations of the center of the 500-m and 250-m observations are then interpolated from the 1-km observations. A brief summary of this model is given below. For more detail see Nishihama *et al.* (1997) and Wolfe *et al.* (2002).

The MODIS geometric sensor model is represented by a paddle-wheel whiskbroom sensor model. The main component is an ideal array of ten 1-km detectors that is swept by a double-sided paddle-wheel mirror to scan the Earth, with

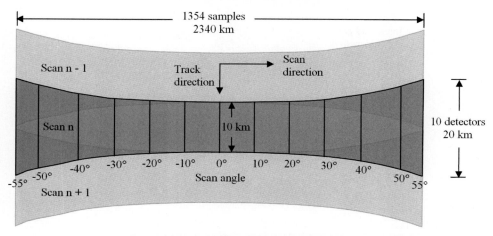

Figure 21.1. Ground projection of three consecutive MODIS scans illustrating the scan geometry and "bow-tie" effect (track direction dimensions are exaggerated). The darker areas near the left and right edges illustrate the area of overlap between scans.

1354 samples of each detector in a scan covering a scan angle of ±55°. The five 500-m bands have twice as many detectors and samples (20 detectors and 2708 samples), and the two 250-m bands have four times as many detectors and samples (40 detectors and 5416 samples). The scan pattern, shown in Fig. 21.1, has a bow-tie effect caused by the change in range of the observations, which increases from nadir to the edge of the swath. The pixel size at the edge of the swath is 4.8 km in the along-scan direction and 2.0 km in the along-track direction. This causes a 50% overlap between adjacent scans at the edges of the swath. The scan mirror's motion is determined by using encoder data that give the relative time the mirror passes each encoder position. The angle of the start of scan is measured prelaunch and the time of the start of each scan is measured on-orbit. Small misalignments of the mirror sides and axis of rotation are important geometric parameters that are measured prelaunch and then refined on-orbit based on control point measurements.

It is useful to define several terms that describe the spatial response characteristics of a scanning instrument such as MODIS. The PSF of an imaging system is defined as the system's response to a point source of light (Castleman, 1979). For a single sample of a specific detector, the detector's instantaneous PSF is the response of the static (nonscanning) detector to a point source of light. This response depends on the angular location of the point source and is primarily a function of the detector's shape and the instrument's optics. When the scanning motion of the instrument is included, the detector's integrated response is called

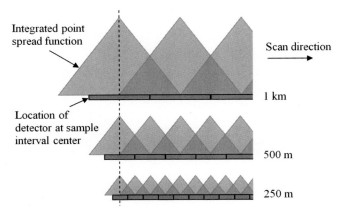

Figure 21.2. Relationship between the scan direction *dynamic line spread functions* (DLSFs) of the three MODIS spatial resolutions.

the dynamic PSF. Similarly, a detector's *line spread function* (LSF) is its response to light from a line light source, e.g., an idealized narrow slit.

The *instantaneous field of view* (IFOV) of the 1-km detectors is modeled with a 1-km ground projected area at nadir sampled to produce a nadir 1-km ground sample distance. This produces a rectangular instantaneous LSF in both the along-track (track) and cross-track (scan) direction. However, the integration of the detector signal in the scan direction during the scan causes the *dynamic LSF* (DLSF) of the detectors to be triangular. Ignoring the small readout time of the detector, the first-order 1-km observation's scan DLSF is a triangle with a base of 2 km. Figure 21.2 shows the first-order scan DLSF, as well as the relative location of the 1-km, 500-m, and 250-m DLSF.

When the MODIS sensor is mounted on the spacecraft, the alignment of the MODIS instrument reference axis to the spacecraft reference axis is measured and used in the geometric model. Typically there are small changes in this alignment caused by vibrations when the spacecraft is launched and by on-orbit thermal and other changes in the spacecraft and instrument. This alignment matrix models the initial on-orbit biases, as well as the long-term and within-orbit thermally induced changes in alignment.

The output of the sensor model is a line-of-sight vector $\mathbf{u}_{inst}$ and a sample time t for the center of each 1-km observation in the instrument reference frame. The components of $\mathbf{u}_{inst}$ are the direction cosines of the center of the 1-km observation DLSF. After the line-of-sight vector $\mathbf{u}_{inst}$ is first rotated to the spacecraft reference frame and then rotated to an Earth-fixed reference frame by using the spacecraft attitude, position $\mathbf{p}_{sc}$, and velocity at time t, it is called $\mathbf{u}_{ecr}$. The Earth intersection $\mathbf{x}_{ellip}$ of $\mathbf{u}_{ecr}$ is then found using an elliptical model of the Earth (Fig. 21.3).

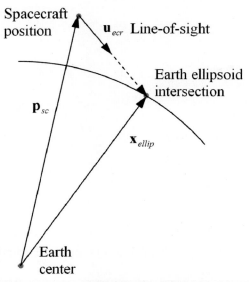

Figure 21.3. Intersection of the 1-km observation line-of-sight vector with the Earth ellipsoid.

Once the initial intersection with the Earth's ellipsoid is found, the location is refined using the terrain model of the Earth's surface to correct for off-nadir terrain parallax. This is done using an iterative search process to move down the observation vector from above the local terrain maximum to the first $\mathbf{x}_{terr}$ intersection with the terrain surface. The terrain model currently being used is a global 1-km model derived from the Shuttle Radar Topography Mission (SRTM) within $\pm 60°$ latitude (Hilland *et al.*, 1998) and other older models elsewhere (Logan, 1999).

21.2.2 Control points

Before the Terra launch, 417 GCPs were collected from precision-located and terrain-corrected Landsat-5 TM images (Bailey *et al.*, 1997). In 2005, this set of control points was refreshed by removing 40 control points that did not yield good performance and adding 837 new control points from Landsat-7 ETM+ images, for a current total of 1214 control points (Fig. 21.4). The majority of these control points ($\sim$75%) are in the Northern Hemisphere. The original control points were georeferenced to 15-m (1σ) accuracy. Each control point (Fig. 21.5) is an 800×800 subimage from the Landsat red band with a sample size of 30 m, and includes the fine-resolution terrain height at every pixel. At nadir, this represents a 96×96 sample area of 250-m pixels.

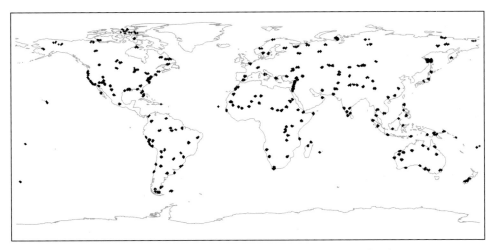

Figure 21.4. Global distribution of the ground control points.

Figure 21.5. Control point example: Landsat-5 TM band 3 (red) image of the control point centered at (44.2109 N, 88.4658 W). The image, sensed on November 8, 1999, is of Appleton, Wisconsin, USA.

21.2.3 Matching approach

This subsection first explains how simulated MODIS radiances are computed using Landsat data, and then how these simulated radiances are correlated to the actual MODIS radiances to compute the geolocation corrections.

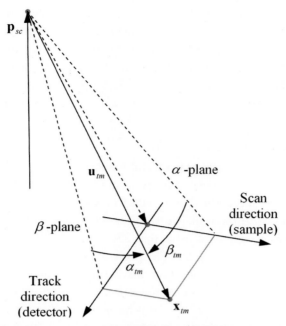

Figure 21.6. Geometry used in the control point matching algorithm to construct a simulated MODIS scene from a Landsat ground control point.

The control point matching process is used to determine the location of each control point in the MODIS image. Residuals between the expected location and the actual location are then used in the postlaunch error analysis. The matching algorithm computes the location of the best match between the MODIS scene and the Landsat control points by cross-correlating the MODIS 250-m red band (band 1) with simulated 250-m resolution pixels. A first-order MODIS dynamic PSF is applied to the Landsat 30-m red band (band 3) GCPs, aggregating them to MODIS resolution.

Initially, the expected location of the center of the control point is found by searching the MODIS image for the closest 1-km line/sample location. Only control points that lie within $\pm45°$ of nadir are considered because of growth of the ground pixel size near the edge of the swath. Since up to three scans may contain a control point image, each scan is treated separately to avoid crossing scan boundaries. For the center scan, a local coordinate system called the matching reference frame is set up, with the origin at the spacecraft position $\mathbf{p}_{sc}$ at the time of the expected location. The α-plane is defined (Fig. 21.6) as the plane containing the expected control point center location in the scan (s_{250m}, d_{250m}) and scan direction at the expected location, and the β-plane is defined as the plane perpendicular to the α-plane. The nominal angular separation for a 250-m sample α_{250m} (angle from the β-plane) and for a 250-m line (detector) β_{250m} (angle from the α-plane) is computed. Next, for the center of each pixel in the Landsat image, the Earth-centered coordinates $\mathbf{x}_{tm}$ are computed (including the terrain height) and then the vector in the matching reference frame $\mathbf{u}_{tm}$ is computed by subtracting the origin $\mathbf{p}_{sc}$ from the pixel center location $\mathbf{x}_{tm}$. Then, to obtain the fractional 250-m sample and detector numbers $(\Delta s_{tm}, \Delta d_{tm})$, the angular distances to the Landsat pixel from each of the planes $(\alpha_{tm}, \beta_{tm})$ are calculated and divided by the 250-m angular separation, i.e.,

$$\Delta s_{tm} = \frac{\alpha_{tm}}{\alpha_{250m}} \qquad \Delta d_{tm} = \frac{\beta_{tm}}{\beta_{250m}}.$$

For each of the (up to) two 250-m pixels (along-scan) that the TM sample intersects, a DLSF weight w_{250m} is calculated by evaluating the 250-m DLSF at the fractional location in the scan direction. The Landsat pixel radiance L_{tm} is then multiplied by w_{250m} and added to the radiance sum for the 250-m pixel. The weight is also added to a separate weight sum for that 250-m pixel. Of course, if the location is outside of the current scan, the TM pixel is ignored for that scan. Also, only intersections with the 250-m observations within -15 and $+16$ detectors (lines) or samples of the center pixel are stored (for a neighborhood containing a total of 32×32 pixels). Once all of the TM pixels have been processed for the current scan, the two adjacent scans (at most) are processed similarly, so that the intersections with all of the possible 250-m observations within the MODIS neighborhood are considered. Next, for each 250-m MODIS observation, the weighted radiance sums are divided by the sum of weights to produce the MODIS simulated radiance L_{250m}, i.e.,

$$L_{250m} = \sum_{\mathbf{R}_{250m}} w_{250m} L_{tm} \bigg/ \sum_{\mathbf{R}_{250m}} w_{250m},$$

where $\mathbf{R}_{250m}$ is the set of Landsat pixels that intersect the MODIS observation. If any simulated pixels have a zero weight, their value is set to zero. The cross-correlation example in Fig. 21.7 shows (a) the Landsat control point, (b) the MODIS scene, (c) the cross-correlation image, and (d) the simulated MODIS image.

Using normalized cross-correlation, these simulated radiance values are then correlated with the actual MODIS ones to obtain the cross-correlation coefficient at that location. This algorithm is then repeated over a set of angular displacements in the β and α planes, using a gradient search method, until the maximum cross-correlation is found (Fig. 21.8). Starting with the initial point, the cross-correlation coefficients for that location and the eight surrounding locations are computed. If none of the values is greater than the initial one, the location with the local maximum is found. Analysis has shown that this location almost always coincides with the global maximum as well. Otherwise, the new center location is set to the location of the maximum of the adjacent points, and the process is repeated until either the local maximum is found or the edge of the maximum search area is

(a)   (b)

(c)   (d)

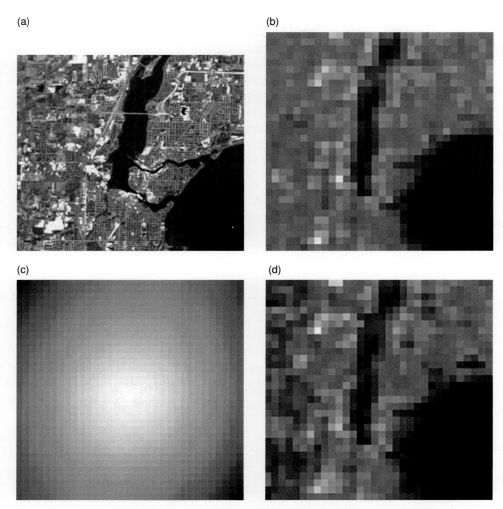

Figure 21.7. Example of control point matching: (a) Landsat-5 TM image of the control point (subset of Fig. 21.5), (b) MODIS 250-m band 1 (red) image centered on the control point, (c) cross-correlation image, and (d) simulated MODIS image at the location of maximum correlation. The MODIS/Terra image was sensed on July 18, 2006. This example uses a step size of 0.05 of a 250-m pixel (12.5 m at nadir).

encountered. Only matches with a maximum cross-correlation coefficient of 0.6 or more are considered good enough for use in the later error analysis.

21.2.4 Error analysis

After launch, the sensor orientation parameters that best fit the control point residuals were computed using a deterministic least-squares (minimum variance)

column

row	11	12	13	14	15	16	17	
15	0.8876	0.8881	0.8879	0.8870	0.8853	0.8828	0.8797	Search
16	0.8920	0.8926	0.8923	0.8913	0.8896	0.8871	0.8839	start
17	0.8962	0.8967	0.8964	0.8954	0.8936	0.8910	0.8877	
18	0.8999	0.9004	0.9001	0.8991	0.8972	0.8945	0.8911	
19	0.9027	0.9031	0.9028	0.9016	0.8996	0.8969	0.8934	
20	0.9045	0.9050	0.9046	0.9034	0.9014	0.8986	0.8951	
21	0.9049	0.9053	0.9048	0.9036	0.9015	0.8987	0.8951	
22	0.9043	0.9047	0.9043	0.9030	0.9009	0.8980	0.8944	

Maximum correlation

Figure 21.8. Normalized cross-correlation gradient search example (based on the correlation image in Fig. 21.7(c)). The search starts at row 16, column 16, and follows the maximum gradient (gray-colored boxes) to reach the maximum correlation at row 21, column 12. The locations containing the light-gray values are not computed during the search.

approach. In this approach, the errors were assumed to be randomly distributed with a mean of zero. In the error analysis, certain parameters were held constant while other parameters were determined to obtain a reliable solution. Estimates of long-term trends were made after obtaining a sufficiently long time-series. After a number of years, more dynamic changes in the parameters (e.g., those caused by thermal flexing) have been detected and modeled for MODIS/Terra. Lookup tables containing these parameters have been updated for both forward and retrospective processing.

21.3 Results

Using this automated control point matching and error analysis approach, the MODIS instruments went through the following phases of geolocation error analysis activity: (1) Initial on-orbit bias correction, (2) long-term trend, and, for MODIS/Terra, (3) a yearly cycle and Sun-angle corrections. The control point residuals have also helped in understanding the less-accurate geolocation near orbit and attitude maneuvers of the spacecraft.

21.3.1 Initial on-orbit bias correction

After the launch of MODIS Terra and Aqua, GCPs were used to remove the on-orbit biases from the sensor orientation parameters. These parameters include the instrument to spacecraft alignment (represented as the roll, pitch, and yaw angles), scan mirror coefficients (wedge angles in the scan α and track β direction, and the axis error γ in the track direction), and mirror axis tilt.

Table 21.1 *Last parameter update (seconds of arc) made during the initial on-orbit bias correction phase*

Satellite	Update	Start date	Instrument to spacecraft alignment			Scan mirror coefficients			Mirror axis tilt
			Roll	Pitch	Yaw	α	β	γ	
Terra	4	Apr. 2002	251.6	83.5	98.2	−4.1	38.0	−0.6	−180.7
Aqua	5	Jan. 2004	409.5	582.2	−76.3	−5.6	37.1	−6.4	−422.2

For MODIS/Terra, this first phase consisted of four updates (Wolfe *et al.*, 2002; Salomonson and Wolfe, 2003; Wolfe, 2006) and lasted 25 months from first light. Using an average of 82 control point matchups per day, the MODIS geolocation accuracy goal was reached within the first 13 months after first light with the third parameter update. The fourth update (Table 21.1), performed 26 months after first light, corrected an along-scan error involving a combination of the mirror axis tilt and pitch bias to further improve the accuracy.

The initial analysis of MODIS/Aqua showed on-orbit biases that were similar to MODIS/Terra (Wolfe, 2006). The second update at 4 months after launch was expected to reach the geolocation accuracy goal. However, it soon became apparent that there was an error in the Aqua flight navigation software (Glickman *et al.*, 2003) that caused large and varying within-orbit errors. The large number of globally distributed ground control points (average of 79 control point residuals per day) was a key element in uncovering this problem and verifying the correction. The geolocation accuracy goal was almost met with the fourth update to the parameters, which was made soon after the flight software fix, about 10 months after first light. The fifth update (Table 21.1) was performed 19 months after first light and further improved the accuracy.

21.3.2 Trend and cyclic correction

The second phase corrected for the long-term trend. Specifically, for the temporal parameterization, we used the following third-order polynomial combined with a sinusoidal equation for the yearly cyclic variation:

$$f(d) = u_0 + u_1 \frac{d}{P} + u_2 \left(\frac{d}{P}\right)^2 + u_3 \left(\frac{d}{P}\right)^3 + u_4 \sin\left(\left(\frac{u_5}{P} + \frac{d}{u_6}\right) 2\pi\right),$$

where the u_i's are the estimated coefficients, d is the number of days since 12:00 Universal Time Corrected (UTC) January 1, 2000 (J2000 epoch), and P is the number of days per year (365.242). A least-squares approach was used to fit the seven coefficients to the daily mean control point residuals. The temporal coefficients were estimated to determine the lowest-order polynomial that provided a good fit, but if the amplitude of the cyclic fit was small, the cyclic terms were not used.

For MODIS/Terra, a linear fit with a small yearly cyclic term in October 2002 (31 months after first light) was first used. This fit was based on the first 1.5 years of control point residual data and was used for the second MODIS/Terra reprocessing (called Collection Version 4, or C4). This fit was also used to extrapolate into the future for forward processing. Three additional updates were applied for C4 forward processing in March 2004, June 2005, and July 2006. In the June 2005 fit, all four terms in the third-order polynomial were first used. Then, in July 2006, even though a polynomial fit was performed, it was determined that for extrapolation into the future, a linear fit using the last two years of data was likely to produce the best results. These three updates also included adjustments to the yearly cyclic terms. Using this approach resulted in a C4 nadir equivalent geolocation accuracy of 38-m and 44-m *root mean square error* (RMSE) in the track and scan directions, when C4 ended in December 2006.

For MODIS/Aqua, a polynomial or cyclic fit (Wolfe, 2006) was not used at the beginning of the first reprocessing of MODIS/Aqua data, also called C4. The January 2004 bias continued to be used until the first linear extrapolation for forward processing was performed in June 2005 (three years after first light). A second linear extrapolation was then performed in July 2006. For the C4 time period, the yearly cyclic fit was too small to require modeling. Using this approach resulted in a C4 nadir equivalent geolocation accuracy of 43-m and 54-m RMSE in the track and scan directions. The scan direction residuals are larger than those of MODIS/Terra primarily because of a high-frequency jitter in the roll axis (scan direction) caused by another Aqua instrument, the Advanced Microwave Scanning Radiometer – EOS (AMSR-E).

21.3.3 Sun-angle correction

The third error correction phase was to model the within-orbit effect through a correction that depended on the solar angle. During testing for the MODIS Collection Version 5 (C5) reprocessing, approaches were considered to correct MODIS/Terra for within-orbit variations that were clear when control points in the Northern and Southern Hemispheres were analyzed separately. A clear relationship was apparent when the residuals were plotted against either specific instrument temperatures or solar angle (Wolfe, 2006). Because of this relationship, the effects were assumed to be primarily thermal distortions caused by the instrument heating up as it went through the sunlit portion of the orbit. During analysis, it was determined

Table 21.2 *Sun-angle correction coefficients used for MODIS/Terra Collection Version 5 (C5); distance units are meters at nadir*

	Term	Units	Track	Scan
s_0	bias	m	−58.5	−48.5
s_1	linear	m/deg	1.51	1.13
s_2	quadratic	m/deg^2	−0.0082	−0.0053
q_1	bias	m	36.2	40.4
q_2	linear	m/deg	−0.263	−0.247

that a correction that depended on the Sun angle would be more robust than a temperature-dependent correction.

The parameterization chosen for the Sun-angle correction is a combination of a second-order polynomial correction during the time the instrument is heating up in the sunlit portion of the orbit and a linear fit when the instrument temperature stabilizes and then decreases during the dark part of the orbit. This is expressed by the following equations:

$$h_1(\theta) = q_0 + q_1 (\theta + 360°), \quad \text{for } -180° \leq \theta < 0°,$$
$$g(\theta) = s_0 + s_1\theta + s_2\theta^2, \quad \text{for } 0° \leq \theta < \theta_{\max},$$
$$h_2(\theta) = q_0 + q_1\theta, \quad \text{for } \theta_{\max} \leq \theta < 180°,$$

where g is the second-order polynomial function, h_1 and h_2 are the linear functions, θ is the instrument solar elevation angle, and $\theta_{\max}$ is the transition angle between the heating and stabilization/cooling part of the orbit. For MODIS/Terra at the Spring or Fall equinox, θ is zero at the northernmost point of the orbit (near the North Pole) and close to 67.5° at the orbit descending node (Equator crossing at 10:30AM local time).

For the MODIS/Terra C5 reprocessing, a least-squares best fit was used to determine the coefficients for the Sun-angle correction. The transition angle $\theta_{\max}$ was then chosen to be where the polynomial fit approached its maximum and the curve began to decrease. The linear fit is the line between the polynomial values at the transition angle (120°) and at the start of the sunlit part of the orbit (0°). The fitted coefficients given in Table 21.2 produce a function that at nadir is 30-m peak-to-peak in the track direction and 35-m peak-to-peak in the scan direction (roughly 9 arcsec). Because some of the within-orbit variation was being captured in the yearly cyclic fit, a reduction in the yearly cyclic fit was made to compensate. These corrections were then applied in C5 and removed almost all of the hemispherical differences seen in C4.

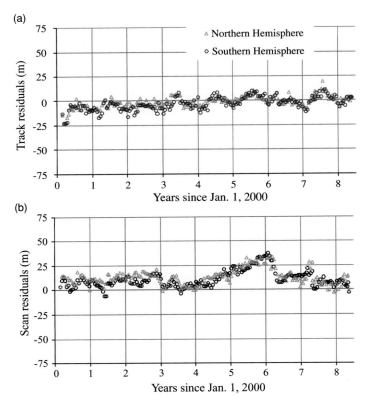

Figure 21.9. The 16-day means of the ground control point residuals for MODIS/Terra C5. Each hemisphere is shown for the scan (a) and track (b) directions in nadir equivalent units.

A similar analysis was done in preparation for MODIS/Aqua C5. However, the improvements were small and, at that time, were not considered large enough to warrant modeling.

21.3.4 Current results

As indicated, an updated set of GCPs was used in the third full C5 reprocessing that began January 2006. The updated set contained more than three times the original number of control points. The 16-day mean residuals for control points in the Northern and Southern Hemispheres are shown in Fig. 21.9 for MODIS/Terra and Fig. 21.10 for MODIS/Aqua. Overall, the residuals for the entire C5 period were good for both missions (Table 21.3). As expected, because of the larger library, the number of control point matchups for C5 was roughly three times that of C4. Note the smaller number of average daily matches for MODIS/Aqua (14% less). This is likely due to cloud-cover differences, as well as illumination and viewing geometry effects.

Table 21.3 *MODIS control point residuals in nadir equivalent units for C5, as well as a prediction of the residuals if a retrospective best fit (Table 21.4) were used*

Satellite	Matches per day	C5 results (m)		Best-fit prediction (m)	
		Track RMSE	Scan RMSE	Track RMSE	Scan RMSE
Terra	270	42	44	42	42
Aqua	232	47	53	46	51

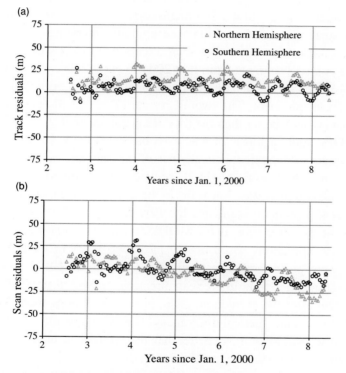

Figure 21.10. The 16-day means of the ground control point residuals for MODIS/Aqua C5. Each hemisphere is shown for the scan (a) and track (b) directions in nadir equivalent units.

Notice that compared to C4, the C5 track direction residual RMSE unexpectedly increased by 4 m for both instruments, though they were still below the overall geolocation goal. The scan residuals also did not decrease as much as expected. The reason for this can be explained by the accuracy of the ground control point source. When the at-launch library was developed, absolute GCPs were used to geolocate each Landsat scene before the control points were extracted. For the updated library, the Landsat scenes were corrected using a block triangulation approach that relied on absolute control points in some, but not all, scenes, and so these scenes might not be as accurately geolocated. Known issues have subsequently been encountered with some areas in the Global GeoCover Landsat data. In preparation for the mid-decadal Landsat dataset, these errors are expected to be corrected (personal communication with J. C. Storey; see also Chapter 18).

For MODIS/Terra, the initial polynomial fits and cyclic fit parameters were updated twice for forward processing, in April 2006 and May 2007, using linear extrapolations based on data of the prior two years. In these updates, small changes were also made to the cyclic terms. As mentioned above, the Sun-angle fit was used in C5. The C5 results in Fig. 21.9 show no significant hemispherical differences, demonstrating the success of the Sun-angle correction. In the track direction, there was a very small trend before 2005, and a small remaining yearly cyclic term that is most visible after 2004. The scan direction shows a small overall bias of 12 m. There is also a segment from mid 2004 to early 2006, where the trend was significantly undercorrected. The extrapolation starting in April 2006 for forward processing fixed this error. The overall results (Table 21.3) show a geolocation accuracy that is much better than the MODIS accuracy goal.

For MODIS/Aqua, a linear fit was used based on the C4 residuals without a cyclic fit. Because of apparent convergence of the onboard navigation data during the first three months after first light, these data were excluded from the long-term fit and, instead, two piecewise linear fits in the track direction and a separate linear fit for the first 45 days in the scan direction were performed. For forward processing, two updates at the same time as for MODIS/Terra were performed. Again, these were linear extrapolations based on the prior two years of data. No Sun-angle corrections were performed. The results for C5 (Fig. 21.10) show a small bias (12 m) in the track direction. In the scan direction, an overall trend of 6 m per year is still apparent after the correction. Also, in both directions there are remaining yearly cyclic and hemispherical differences. The overall results (Table 21.3) show a geolocation accuracy that is 3 m better than the MODIS accuracy goal in the track direction, but which misses the goal by 3 m in the scan direction.

In preparation for a possible future MODIS Collection Version 6 (C6), the C5 residuals are being analyzed to determine the overall trend and any additional corrections that may be needed. For MODIS/Terra, a small overcorrection that was found in the Sun-angle fit for C5 would be fixed for C6. Also, any biases and extrapolation errors would retrospectively be corrected with a new long-term trend correction (Table 21.4). By just updating the long-term trend an improvement in the scan direction is expected (see "Best-fit prediction" in Table 21.3), reducing the error from 44-m to 42-m RMSE.

Table 21.4 *Coefficients for the long-term and cyclic trends based on a retrospective analysis (cyclic terms are not used for MODIS/Aqua); distance units are meters at nadir*

	Term	Units	Terra		Aqua	
			Track	Scan	Track	Scan
u_0	bias	m	−47.9	32.4	−24.2	−31.9
u_1	linear	m/year	34.8	1.0	10.2	8.5
u_2	quadratic	m/year^2	−6.06	−1.89	0.74	0.86
u_3	cubic	m/year^3	0.362	0.171	−0.128	−0.137
u_4	amplitude	m	−9.9	−2.7		
u_5	phase	days	91.5	91.5	(not used)	
u_6	period	days	365.242	365.242		

For a possible MODIS/Aqua C6, a Sun-angle correction similar to the one performed for MODIS/Terra would be performed. This is expected to fix the hemispherical differences and a large portion of the yearly cyclic variation. As with MODIS/Terra, retrospective correction of the C5 biases, trends, and extrapolation errors would be performed. By just updating the long-term trend correction (Table 21.4), not including the yearly cyclic terms or the Sun-angle corrections, small improvements in the overall error (Table 21.3) are expected. With the other possible corrections for the yearly cyclic terms and Sun-angle corrections it is likely that the scan accuracy will also reach the MODIS geolocation goal.

21.3.5 Orbit and attitude maneuvers

Even though the overall geolocation accuracy is very good, there is degraded accuracy near orbit and attitude maneuvers. The Terra and Aqua spacecraft each perform an attitude maneuver for lunar calibration about once a month. In order to maintain the nominal equatorial crossing time, drag makeup maneuvers occur once or twice a year and inclination maneuvers occur once every few years. During and following all maneuvers, the attitude control system is in a low-accuracy pointing mode. In addition, the orbit must be recalculated following the maneuvers. The degraded periods are typically less than an orbit for attitude maneuvers and two orbits for orbit maneuvers. When processing orbit maneuvers, postprocessed ephemeris is also used for Terra geolocation (definitive ephemeris is already used for Aqua). Note that these degraded periods represent a very small portion (<0.3%) of the total data acquired. However, because large geolocation errors can have a significant impact on many higher-level products, particularly the higher-resolution Landsat products, data near maneuvers are currently excluded from the production of these products. The ground control point residuals have been used to help understand the extent of these periods of degraded geolocation performance.

21.4 Conclusion

We have explained the MODIS geolocation approach, described in detail the use of automated matching of global ground control points, and summarized eight years of geolocation analysis. Overall, a global network of ground control points is being used to maintain the geolocation accuracy of terrestrial remote sensing data from both MODIS instruments to approximately 50 m (1σ) at nadir. Biases and trends in the sensor orientation are determined from automated control point matching and removed by updating models of the spacecraft and instrument orientation in the MODIS geolocation software. This operational characterization of the geolocation errors has enabled MODIS observations to be geolocated to the subpixel accuracies required for terrestrial global change applications, and thus enabled the MODIS science team to concentrate their efforts on the accurate retrieval of biophysical parameters (Justice *et al.*, 1998).

For the future, we plan to maintain the control point library and improve our understanding and handling of data near maneuvers. We are also automating the long-term trend error analysis by the use of filtering techniques, such as a Kalman filter (Maybeck, 1979), to reduce the effort needed for the analysis, and to minimize extrapolation errors. This approach, which was developed for the MODIS instruments aboard NASA research satellites, is now being implemented for the Visible Infrared Imager Radiometer Suite (VIIRS) instruments aboard the National Polar-orbiting Operational Environmental Satellite System (NPOESS) Preparatory Project (NPP) and future operational NPOESS missions.

Acknowledgements

Accurate MODIS geolocation requires contributions from a large number of specialized groups, including the instrument and satellite builders, flight dynamics and attitude control groups, and other instrument teams. The authors acknowledge the contributions of these groups, the contributions of MODIS geolocation team members and the support provided by the MODIS science team. This work was performed under the direction of the MODIS Science Team in the Terrestrial Information Systems Branch (Code 614.5) at NASA Goddard Space Flight Center.

References

Bailey, G. B., Carneggie, D., Kieffer, H., Storey, J. C., Jovanovic, V. M., and Wolfe, R. E. (1997). *Ground Control Points for Calibration and Correction of EOS ASTER,*

MODIS, MISR and Landsat 7 ETM+ Data, SWAMP GCP Working Group Final Report. USGS EROS Data Center, Sioux Falls, SD.

Castleman, K. R. (1979). *Digital Image Processing*. Englewood Cliffs, NJ: Prentice-Hall, Inc.

Glickman, J., Hashmall, J., Natanson, G., Sedlak, J., and Tracewell, D. (2003). Earth Observing System (EOS) Aqua launch and early mission attitude support experiences. In *Proceedings of the NASA Flight Mechanics Symposium*, NASA Goddard Space Flight Center, Greenbelt, MD.

Hilland, J. E., Stuhr, F. V., Freedman, A., Imel, D., Shen, Y., Jordan, R., and Caro, E. (1998). Future NASA spaceborne SAR missions. *IEEE Aerospace and Electronic Systems Magazine*, **13**(11), 9–16.

Justice, C. O., Vermote, E., Townshend, J. R. G., Defries, R., Roy, D. P., Hall, D. K., Salomonson, V. V., Privette, J. L., Riggs, G., Strahler, A., Lucht, W., Myneni, R. B., Knyazikhin, Y., Running, S. W., Nemani, R. R., Wan, Z., Huete, A. R., van Leeuwen, W., Wolfe, R. E., Giglio, L., Muller, J.-P., Lewis, P., and Barnsley, M. J., (1998). The moderate resolution imaging spectroradiometer (MODIS): Land remote sensing for global change research. *IEEE Transactions on Geoscience and Remote Sensing*, **36**(4), 1228–1249.

Logan, T. L. (1999). *EOS/AM-1 Digital Elevation Model (DEM) Data Sets: DEM and DEM Auxiliary Datasets in Support of the EOS/Terra Platform*, JPL Tech. Doc. D-013508. NASA Jet Propulsion Laboratory, California Institute of Technology, Pasadena, CA.

Maybeck, P. S. (1979). *Stochastic Models, Estimation and Control*, Vol. 1, New York: Academic Press.

Nishihama, M., Wolfe, R. E., Solomon, D., Patt, F. S., Blanchette, J., Fleig, A. J., and Masuoka, E. (1997). *MODIS Level 1A Earth Location: Algorithm Theoretical Basis Document Version 3.0*, SDST-092. Laboratory for Terrestrial Physics, NASA Goddard Space Flight Center, Greenbelt, MD.

Roy, D. P. (2000). The impact of misregistration upon composited wide field of view satellite data and implications for change detection. *IEEE Transactions on Geoscience and Remote Sensing*, **38**(4), 2017–2032.

Salomonson, V. V. and Wolfe, R. E. (2003). MODIS geolocation approach, results and the future. In *Proceedings of the IEEE Workshop on Advances in Techniques for Analysis of Remotely Sensed Data*. NASA Goddard Space Flight Center, Geenbelt, MD, pp. 424–427.

Salomonson, V. V., Barnes, W. L., Maymon, P. W., Montgomery, H. E., and Ostrow, H. (1989). MODIS: Advanced facility instrument for studies of the Earth as a system. *IEEE Transactions on Geoscience and Remote Sensing*, **27**(2), 145–153.

Townshend, J. R. G., Justice, C. O., Gurney, C., and McManus, J. (1992). The impact of misregistration on change detection. *IEEE Transactions on Geoscience and Remote Sensing*, **30**(5), 1054–1060.

Wolfe, R. E. (2006). MODIS geolocation. In *Earth Science Satellite Remote Sensing: Science and Instruments*, Vol. 1, Chapter 4, J. J. Qu, W. Gao, M. Kafatos, R. E. Murphy, and V. V. Salomonson, eds. Beijing: Tsinghua University Press and Berlin: Springer-Verlag.

Wolfe, R. E., Nishihama, M., Fleig, A. J., Kuyper, J. A., Roy, D. P., Storey, J. C., and Patt, F. S. (2002). Achieving sub-pixel geolocation accuracy in support of MODIS land science. *Remote Sensing of Environment*, **83**, 31–49.

22

SeaWiFS operational geolocation assessment system

FREDERICK S. PATT

Abstract

An automated method has been developed for performing geolocation assessment on global satellite-based Earth remote sensing data. The method utilizes islands as targets that can be readily located in the sensor data and identified with reference locations. The essential elements are an algorithm for classifying the sensor data according to the source, a reference catalog of island locations, and a robust algorithm for matching viewed islands with the catalog locations. This method was originally developed and tested for the Sea-viewing Wide Field-of-view Sensor (SeaWiFS) before its launch in 1997, and was refined using the flight data after launch. The results have been used for both ongoing assessment of geolocation accuracy and development of improvements to the geolocation processing algorithms. The method has also been applied to other moderate-resolution satellite sensors.

22.1 Introduction

The determination of geolocation for global Earth imaging sensors is generally performed in-line with initial (e.g., Level 0–1) data processing, using parametric algorithms based on navigation (satellite orbit and attitude) and telemetry data. The accuracy of the satellite navigation data may be adequate to meet geolocation requirements at the resolution of global sensors (250 to 1000 m) without the need for manual intervention or postprocessing corrections. However, this approach does need verification and feedback using the sensed image data. Both navigation data and sensor geometric models are subject to systematic and time-varying errors, and a means of ongoing assessment of geolocation accuracy is needed over the full range (temporal and geographic) of data collection. The assessment results are used to demonstrate that geolocation accuracy requirements are met, or alternatively to

456

develop refinements to the parametric algorithms to remove biases and systematic errors. Ideally, this assessment would be performed automatically, in-line with data processing.

A method for automated geolocation assessment was developed for the Sea-viewing Wide Field-of-view Sensor (SeaWiFS), a global ocean color sensor. SeaWiFS was launched in August 1997 on the OrbView-2 (OV-2) satellite. The SeaWiFS data have been collected and processed by the Ocean Biology Processing Group at NASA Goddard Space Flight Center (GSFC). The method uses islands as control points that can readily be located in the sensed image data and matched with reference locations in a catalog. It has run since the start of routine imaging to produce global geolocation accuracy statistics. The results have been posted for monitoring, and have been used to refine the sensor geometric model, calibrate the spacecraft attitude sensors, and evaluate the results of algorithm improvements. The method was developed for SeaWiFS prior to launch (Patt *et al.*, 1997) and refined using flight data; it was also used to assess and improve the geolocation of data from the Coastal Zone Color Scanner (CZCS) on Nimbus 7 (Gregg *et al.*, 2002), and was also proposed for the MODerate-resolution Imaging Spectroradiometer (MODIS) (Nishihama *et al.*, 1997), although this was never implemented.

This chapter presents the method for locating islands in the sensed images, and describes the reference catalog that was generated for this purpose. This is followed by a description of the approach used to develop corrections to the geolocation algorithms, a summary of the assessment results for the duration of the SeaWiFS mission, and some specific examples of improvements to SeaWiFS geolocation.

22.2 Methods

In developing a geolocation assessment strategy for SeaWiFS, we sought to develop a system that minimizes the need for manual intervention. However, an automatic geolocation assessment technique must meet stringent requirements. Earth features to be used for computing geolocation errors must be distinguished in the sensed image data with a typical error of less than one pixel, and must be readily associated with a reference location whose coordinates are accurately known. The processing must be efficient in its use of computer resources.

Islands represent a useful set of control points for this purpose, as they can be readily located in sensed image data as clusters of contiguous land pixels surrounded by water (assuming that land and water can be easily distinguished). The geometric centroid of an island defines a point that can be computed from the data and also from a map, simplifying both the identification process and the error computation. There are thousands of islands of a suitable size for SeaWiFS (roughly 1 to 100 km in the largest dimension), widely distributed in latitude and longitude.

Table 22.1 *SeaWiFS instrument characteristics*

Scan half-width	58.3° (LAC/HRPT)
	45° (GAC)
Ground pixel size at nadir	1.12 km
Tilt	±20°
Band	Center wavelength
1	412 nm
2	443 nm
3	490 nm
4	510 nm
5	555 nm
6	670 nm
7	765 nm
8	865 nm

To automate geolocation assessment using island control points, we needed to solve the four distinct problems:

(1) Reliable classification of viewed locations in the sensed image data as land, water, or clouds/ice; the first two to locate the islands, and the last one to ensure the data are unobscured.
(2) Location of contiguous land pixel clusters corresponding to islands.
(3) Generation of a catalog of accurate reference island locations.
(4) Robust matching of islands located in the sensed image data with the reference locations, allowing for initial errors which may be significant.

We implemented this approach at NASA GSFC in a geolocation assessment system for the SeaWiFS instrument. SeaWiFS has collected global ocean color data since its launch in 1997. The sensor measures radiances in eight visible and near-infrared bands; see Table 22.1 for a summary of its characteristics. The sensor resolution is 1.12 km at nadir, and the geolocation accuracy requirement is 1 pixel (2σ). SeaWiFS data are collected on the daylit side of the orbit, for about 40 minutes centered within the daylight period. The sensor is tilted aft and forward to avoid ocean surface Sun glint, with the change from aft to forward at the subsolar point in the orbit. The primary data recorded onboard is *global area coverage* (GAC), which is subsampled every fourth line and fourth pixel, with sample spacing of 4.5 km at nadir. A limited amount of full-resolution *local area coverage* (LAC) data is recorded; in addition, full-resolution data are direct-broadcast to ground stations in high-resolution picture transmission (HRPT) format.

The results of automated SeaWiFS geolocation assessment have been used to determine the accuracy of geolocation on an ongoing basis and develop improvements to the geolocation segment of the scientific data processing. In the early mission phase, the results were used to detect static offsets. Over longer timescales, the results were analyzed to characterize periodic and secular errors, which were incorporated as refinements to the geolocation system. The following subsections describe the methods developed for each of the problems listed above, and also summarize the approach to the interpretation of the results.

22.2.1 Sensed image data classification

The goal of the classification algorithm is to locate islands that are not obscured from the sensor view by clouds or ice. Therefore, we need a reliable technique for determining the source of the incoming radiance by analysis of the data from one or more of the sensor bands. The assumption is implicit that each source has a unique spectral signature that can be detected with the band complement of SeaWiFS, which would enable reliable classification. An additional requirement is that the algorithm be relatively insensitive to the effects of illumination level, which varies with solar zenith angle, and atmospheric path length, which varies with sensor scan angle.

Our method of data classification was originally developed using simulated SeaWiFS data (Patt *et al.*, 1997), and subsequently refined using the flight data. The overall approach is to investigate correlations between the band radiances for the viewed areas and utilize the difference in response of the blue and red bands for land and water. This difference results from the variation in surface reflectance with wavelength and surface type. However, first the effects of the atmosphere and illumination on the total (top of the atmosphere) radiances must be corrected for. The largest contribution to the atmosphere radiance is Rayleigh (molecular) scattering.

The classification approach for SeaWiFS uses bands 2 and 8 (443 and 865 nm). The first step is to subtract the Rayleigh contribution and normalize to a constant solar zenith angle, i.e.,

$$R_{2N} = (R_{2T} - R_{2R})/\cos\theta_0, \tag{22.1}$$

$$R_{8N} = (R_{8T} - R_{8R})/\cos\theta_0, \tag{22.2}$$

where R_{2T} and R_{8T} are the total radiances for bands 2 and 8, respectively, R_{2R} and R_{8R} correspond to the Rayleigh radiances of these bands, and R_{2N} and R_{8N} to their Rayleigh-corrected, normalized radiances, and θ_0 is the solar zenith angle at the viewed location.

A scatter plot of the corrected, normalized radiances for these bands from a typical scene illustrates the next step in the classification approach (Fig. 22.1). The

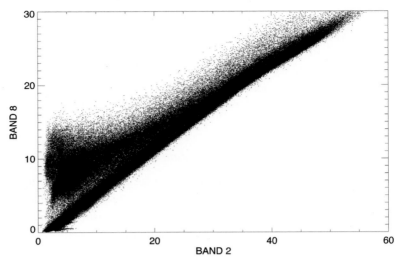

Figure 22.1. Rayleigh-corrected, normalized top-of-the-atmosphere (TOA) radiances: SeaWiFS band 8 vs. band 2.

overall distribution appears as a long diagonal streak, which is divided at the lower end. The divided region represents the land and water, with the points below the diagonal being the water. The points to the right of the divided region represent the clouds. These features show indications of a strong correlation between the two bands for these surface types.

The separation of the land and water data along the diagonal in the plot suggests the final step in the analysis. If the average slope is removed from the distribution, then the classification can be performed using simple thresholds. The slope of the distribution is estimated visually to be 0.6. We can remove this slope by computing the weighted difference of the normalized radiances,

$$R_D = R_{8N} - 0.6R_{2N}, \tag{22.3}$$

and by plotting this difference vs. R_{2N} (Fig. 22.2). This plot shows the desired result, with the division between land and water radiances distributed roughly horizontally, and with increasing cloudiness from left to right. The data can now be classified using simple thresholds, as shown in Fig. 22.2. As seen in this figure, there is no clean division between clouds and other data types; the cloud threshold for band 2 is set at the approximate point where the division between land and water disappears. Currently the following rules are applied: All data with $R_{2N} > 12$ are classified as "clouds," all remaining data with $R_D < 1.2$ are classified as "water," and the rest of the data are classified as "land."

The results of this classification are shown in Fig. 22.3, for a portion of a GAC scene that includes the islands of Borneo and Java. From left to right, the images

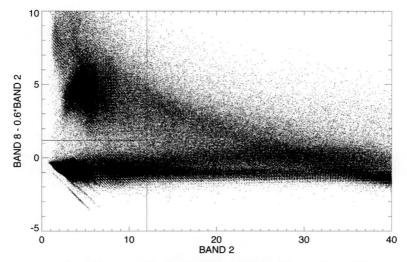

Figure 22.2. Weighted difference of SeaWiFS bands 8 and 2 vs. band 2.

(a)   (b)   (c)

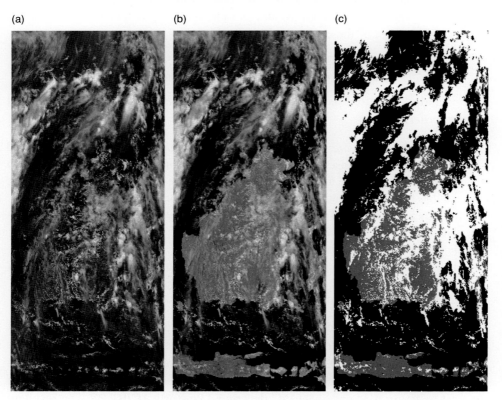

Figure 22.3. GAC of SeaWiFS scene and its classification to land, water, and clouds: (a) Band 2 TOA radiances, (b) band 8 TOA radiances, and (c) classification results.

show the sensed image data (top-of-the-atmosphere (TOA) radiance) for bands 2 and 8 and the classification results, with black being water, gray being land, and white being clouds. The land areas are clearly visible, including several islands not obscured by clouds.

22.2.2 Island location

Once the sensed image data have been classified, the next step locates contiguous clusters of isolated land pixels in the data that can be identified as islands. We also require that clouds do not contaminate the island view. For SeaWiFS, we have implemented an algorithm that identifies land pixel segments in series with the data classification, so that the number of pixels that need to be considered as potential islands is relatively small (Patt *et al.*, 1997). Each island is then identified as a group of contiguous land segments, completely surrounded by water pixels.

Once an island has been located, longitudes and latitudes for all pixels in the cluster are computed using the geolocation data for the pixels. The centroid coordinates (mean values of the longitude and latitude) are computed for the pixel cluster, along with the size of the cluster in each dimension.

22.2.3 Generation of the reference island catalog

The islands to be used as references for control point identification must meet several criteria:

(1) Fall within a useful size range. For SeaWiFS, the minimum island size was selected as about 0.5 km^2 in area. This value was selected to ensure that, for an island within a single 1 km pixel, the majority of the signal would be produced by the land surface, and thus be reliably identified by the classification algorithm.
(2) Be sufficiently separated from other islands and land masses to be readily distinguishable in the sensed image data.
(3) Be reasonably compact in shape.
(4) Have well-determined locations, compared to the required accuracy of the sensor geolocation.
(5) Be reasonably distributed worldwide for global assessment.

We generated the island catalog for SeaWiFS using the Global Self-consistent Hierarchical High-resolution Shorelines (GSHHS) (Wessel and Smith, 1996). This data set contains over 35 000 islands worldwide that meet the minimum size criterion (1). The islands were checked against the separation criterion (2) above, which was set to 10 km for SeaWiFS GAC data and 4 km for LAC/HRPT. The resultant catalogs contain 5882 and 15 376 islands, respectively. See Fig. 22.4 for the global distribution of the GAC island catalog.

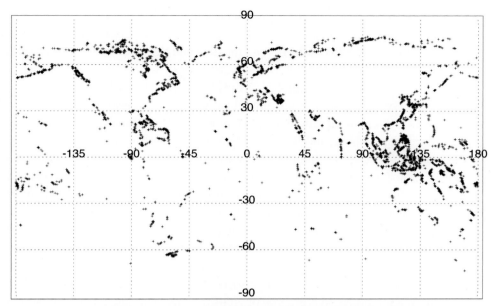

Figure 22.4. Island control point catalog generated from GSHHS data for use with SeaWiFS GAC data.

22.2.4 Island matching

We need a robust matching algorithm based on the expectation that the initial errors in the sensor geolocation might be large compared to the SeaWiFS pixel resolution. These errors could result from timing, orbit position, or instrument pointing errors, or some combination of these.

The simplest matching algorithm is direct, or "nearest neighbor" matching, in which each island located in the data is associated with the closest cataloged island, within a specified matching tolerance. This method is reliable if the expected geolocation errors are less than the separation of catalog islands, but it degrades if the errors approach the minimum island separation specified for the catalog.

If the errors are large, the probability of correctly matching the islands can be improved substantially by comparing the relative separations of the islands located in the data with those of the cataloged islands. Islands in the data can be matched with reference islands if their angular separations (measured on the Earth's surface) are consistent with those of the reference positions, even if the absolute location differences are not small. Our implementation of this approach is known as *triplet matching*, in which separations for groups of three islands located in the data are compared with the separations of cataloged islands. Triplet matching provides more robust matching of islands in the presence of large geolocation errors, although the computational requirements are correspondingly greater.

In addition to the locations or angular separations of islands, the sizes can also be used to aid the matching algorithms. This is mostly useful for the larger islands, since the vast majority of islands, in both the data and the catalog, are a few pixels or less in size.

This pattern-matching approach has already been applied in other situations. The GSFC Flight Dynamics Analysis Branch has developed algorithms and software to match star tracker observations to a reference catalog. These routines perform direct and triplet matching, and also use stellar magnitudes to improve the reliability of the matching. We have obtained the source code for the matching algorithms and incorporated it into our geolocation assessment software. We substituted island sizes directly for stellar magnitudes, and estimated the island position uncertainties based upon the position in the scan line and the number of pixels.

The details of the island matching are as follows. The first step for both algorithms is the selection of candidate islands from the catalog. For each island located in the data, candidates are selected from the catalog based upon a specified search radius. The size of each candidate is also required to match that of the located island within a tolerance, to eliminate obvious mismatches. The islands are sorted by number of candidates and position uncertainty. At this point the algorithm branches according to the matching method selected. For direct matching, each island with candidates is matched with the nearest candidate from the catalog.

If triplet matching is selected, the triplets are formed starting from the top of the sorted island list (i.e., the islands with the fewest candidates). The triplets are checked to ensure that all three islands have at least one candidate and have not yet been matched. Triplets are also required to pass a colinearity test, that is, the smallest angle of the spherical triangle formed by the three island locations must be greater than a specified minimum value. This check ensures that the angular separations for the triplet are geometrically independent measurements of the islands' relative positions. The angular separations for the three islands are then compared with the corresponding separations of all possible combinations of candidates. If the separations correspond within a specified tolerance, the islands are considered to be matched. The process continues until either all of the islands are matched or all possible triplets are checked.

As stated, the choice of matching algorithm depends upon the expected geolocation errors. If the errors are significantly larger than the pixel spacing, the triplet algorithm provides robust matching. However, it is more computationally intensive and may result in mismatches and unmatched islands. For small geolocation errors, the direct matching algorithm is both fast and accurate. The reliability of these methods depends upon choosing values for the parameters (candidate search radius, matching tolerance, and triplet minimum colinearity angle) according to the characteristics of the data. Initial parameter values can be selected based upon, for example, the assumed uncertainties in the located island positions, but these values are best refined empirically by running tests on actual sensed image data.

22.2.5 *Interpretation of geolocation errors*

Once the matches have been found, the coordinate differences (in terms of latitude and longitude) between the island and reference location are computed and stored, along with the location and scan position. By themselves, these results provide an indication of the overall geolocation accuracy for the particular instrument. We can also use these results to characterize the errors and develop improvements to the geolocation processing. For this purpose, we need to be able to interpret a set of geolocation errors in terms of possible sources, for example, spacecraft navigation and instrument geometry, and (if possible) to distinguish among the various error sources. This requires the calculation of partial derivatives of the island locations with respect to each error source under consideration.

For SeaWiFS, the potential error sources considered are the spacecraft attitude angles (i.e., roll, pitch, and yaw), the sensor scan geometry, specifically the scan angle per pixel and the out-of-plane angle, and the scan line time. Although the partial derivatives can be computed analytically, a numerical approach has been used for simplicity. This is performed in two stages. First, lookup tables are generated, which contain partial derivatives of SeaWiFS observed latitudes and longitudes with respect to each error source as a function of observed pixel number and latitude. Then, the latitudes and pixel numbers for a set of island matches are used to interpolate from the table values to generate the partial derivatives for every island in the set.

The lookup tables are generated as follows. First, a nominal set of observed locations is generated using a simulated orbit and nominal attitude. Then, the parameters are each perturbed by small amounts, and the observed locations are regenerated. The differences in observed locations, which represent the numerical partial derivatives, are computed and saved for each parameter, along with the nominal latitudes and pixel numbers. Since the SeaWiFS tilt angle can be either forward or aft, two sets of tables are generated, one for each tilt.

After the tables are generated, the partial derivatives for each island are computed. For each island, the associated pixel number is used to select a table column. The row is selected by finding the tabulated latitudes within that column which bound the island latitude. The partial derivatives are then interpolated to the island location. The output is a set of partial derivatives of the island latitude and longitude with respect to each parameter.

Once the partial derivatives have been computed for a set of island matches, they can be combined with the island location errors for analysis. For example,

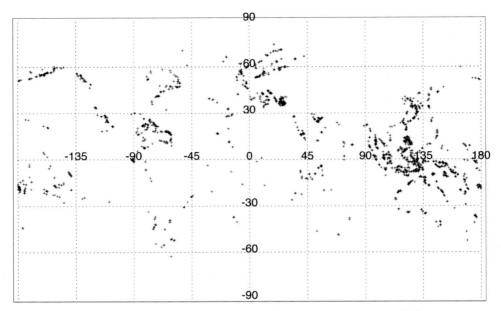

Figure 22.5. Island matching results for GAC data from May 2006.

the errors can be transformed from geographic to image coordinates (along- and cross-track) for visualization, or be used to estimate the errors from the various sources (either the full set or any selected subset) using the least-squares method.

22.3 Results

In this section we present the overall results of SeaWiFS geolocation assessment over the entire mission, followed by specific examples of how these results have been used to improve the geolocation accuracy or detect problems.

22.3.1 Routine geolocation assessment results

The SeaWiFS data are collected and processed automatically by the Ocean Discipline Processing System (ODPS) at NASA GSFC. The sensed image data are located by postprocessing of the spacecraft navigation telemetry data (Patt, 2002). The geolocation assessment processing has been routinely performed on every SeaWiFS scene produced by the ODPS. For GAC data, typically 200 island matches are produced per day. An example of geographic coverage for one month (May 2006) of GAC island matches is shown (Fig. 22.5); it consists of roughly 6600 total matches of about 1700 different island control points.

As shown in the figure, the matches span the full range of global longitude, and the latitude range (about −61° to 75°) represents the full range of SeaWiFS

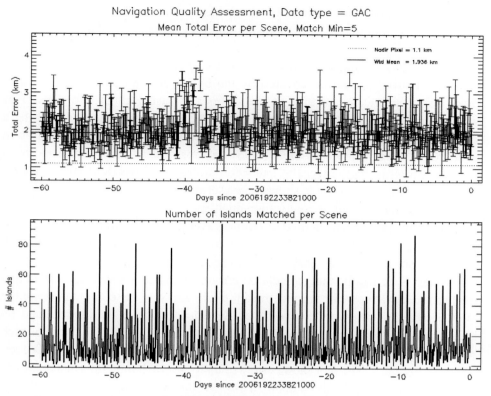

Figure 22.6. Example of SeaWiFS website display of geolocation error statistics; GAC data for the most recent 60 days as of July 11 (day 192), 2006.

data processing for this period. Clearly, however, there are areas which are not well represented in the matches, particularly in the southern oceans and over large continents; this is also seen in the island catalog distribution (Fig. 22.4). It would not be possible in these regions to compute geolocation errors or corrections for individual scenes of limited geographic extent. Nonetheless, for the stated purposes, i.e., global geolocation assessment and identification of system errors, the geographic distribution of the matches is sufficient.

The geolocation error results are analyzed and displayed on the SeaWiFS Mission Operations website (http://seawifs.gsfc.nasa.gov/plankton2/d3/userweb/ navqc). For each scene, the results are summarized as the mean and standard deviation of the geolocation errors (km) and the number of islands matched. These statistics are displayed on the website over short (last 15 days), medium (last 60 days), and long (mission) timescales, and are separated by GAC, LAC, and HRPT scenes. A typical plot for 60 days of GAC data is shown in Fig. 22.6. The upper plot shows the error statistics for each scene. The midpoint of each bar shows the mean of the absolute geolocation errors for the scene, and the bar height is the standard deviation. The bottom plot shows a histogram of the number of islands matched in each scene.

These plots are a fairly sensitive indicator of problems that affect geolocation accuracy. During the period covered by the figure, starting at about −40 days, there was a noticeable increase in the errors that lasted about two days. This was observed and investigated at that time. The cause was determined to be a shift in the spacecraft time code, which in turn was caused by an interruption in onboard Global Positioning System (GPS) tracking. The shift was estimated from the geolocation assessment results, and incorporated as a correction to the SeaWiFS data processing. This example shows how useful the automated assessment is as a tool for daily monitoring of geolocation quality.

The results have also been used to estimate average spacecraft navigation errors. The primary source of errors on SeaWiFS has been the spacecraft attitude (roll, pitch, and yaw angles). The analysis of these errors has been performed on monthly timescales. Specifically, the geolocation errors for a month are accumulated, and the partial derivatives with respect to roll, pitch, and yaw are computed using the lookup tables as described above. The island locations are then partitioned by latitude (typically $10°$ zones), and the average attitude errors are computed for each zone using the least-squares method. The results for one month (May 2006) are shown in Fig. 22.7. The plot shows errors of up to several hundredths of a degree in roll and pitch, and up to about $0.14°$ in yaw. At the SeaWiFS altitude ($\sim700\,$km), $0.1°$ of pitch or roll error is equivalent to 1.2 km of location error at nadir, or slightly more than one pixel. The effect of yaw error is small at nadir and increases with scan angle. The results show that, on average, the 1-pixel geolocation accuracy requirement is being met, with no large biases or systematic errors.

22.3.2 Improvements to geolocation

In addition to routine assessment of geolocation accuracy, the island matching results are also well suited to estimation of systematic errors in the navigation (orbit and attitude) data and sensor alignments. Early in the mission, the average roll, pitch, and yaw errors were incorporated into the spacecraft-to-instrument alignment matrix, which is a fairly standard improvement to satellite sensor data processing systems.

The geolocation error results are useful for evaluation of systematic navigation errors as well. The coverage provided by the island control points (geometric, geographic, and temporal) allows the errors to be interpreted in terms of various types of source errors. Depending on the error source, the effects on geolocation errors may vary with scan angle, sensor tilt, latitude, or season, or some combination

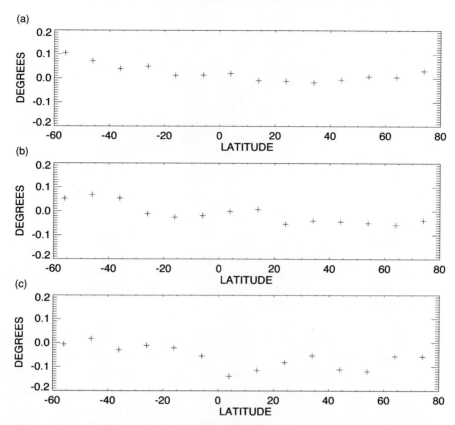

Figure 22.7. Average spacecraft attitude angle errors vs. latitude, determined from island matching results for one month (May 2006) of SeaWiFS GAC data: (a) Roll, (b) pitch, and (c) yaw.

of these. The basic approach, as described above, is to compute the partial derivatives of the geolocation errors with respect to the suspected error sources, and then solve for one or more of these sources using least-squares regression. The solutions can then be incorporated as improvements to the satellite navigation or sensor geometric model.

An example of such an improvement is a combined pitch and time correction that was incorporated into the SeaWiFS navigation processing during the first year of the mission. After the average errors were removed, the geolocation errors in longitude varied systematically with scan angle. The sense of this scan angle dependence changed depending on the sensor tilt angle (recall that SeaWiFS is tilted 20° forward and aft to avoid Sun glint). This suggested a pitch attitude error. Indeed, a plot of the longitude errors versus the partial derivative with respect to pitch for the month of February 1998 shows a high correlation (see Fig. 22.8). However, a pitch error

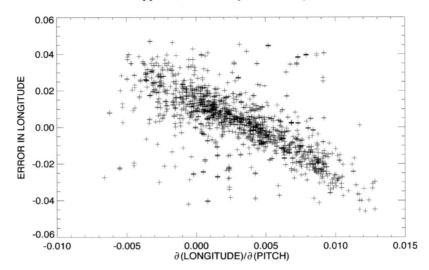

Figure 22.8 Geolocation errors in longitude vs. the partial derivative with respect to the spacecraft pitch angle for SeaWiFS island matches for February 1998, prior to combined pitch/time adjustment.

would also be expected to produce an average error in latitude. Since the average pitch error had previously been estimated and incorporated into the sensor alignment, the average latitude errors were close to zero. It was determined, though, that a combination of pitch and along-track orbit position errors could offset in latitude, but still produce the longitude errors that were observed. Since orbit along-track position is correlated with observation time, a combination of time and pitch adjustments was incorporated into the spacecraft navigation processing; specifically, a 1-sec adjustment to the sensor scan line time tag was offset with a pitch adjustment of 0.6°. This had the desired effect of removing the scan angle dependence of the longitude errors, while maintaining the average latitude errors near zero.

Numerous additional corrections have been developed and incorporated into the SeaWiFS geolocation processing by similar analyses of the island matching errors, although a full discussion of these is beyond the scope of this chapter.

22.4 Conclusion

A system has been developed for automated assessment of satellite-based Earth sensor geolocation using island targets, and applied to SeaWiFS data. This approach involves filtering the sensor data to locate islands, and matching them with reference island locations in a catalog. Algorithms were developed before launch for classifying the sensor data and locating the islands, and refined using the postlaunch radiometric data. A global catalog of island locations and sizes was generated based on a precise vector shoreline database. An existing pattern-matching algorithm was adapted to perform robust matching of located and reference islands. Additional methods were developed to interpret the geolocation errors in terms of the spacecraft navigation (orbit and attitude) and sensor geometric parameters.

This system has been in continuous use since the SeaWiFS launch in 1997, and is run as part of the routine scientific data processing. During that time it has produced a continuous record of the sensor data geolocation accuracy, which has been publicly displayed on the project website. The results represent a significant number of daily target matches that are well distributed, geographically and temporally, and thus provide a valid indication of global geolocation quality.

The method has also been very successful in identifying and testing potential improvements to the SeaWiFS geolocation processing algorithms. Systematic or correlated geolocation error trends have been associated with specific spacecraft navigation or sensor geometric parameters, and have been used to refine the models or algorithms used to compute geolocation. These have been incorporated into the routine data processing and product generation, and have been applied to the entire mission dataset during reprocessings. The result has been to produce SeaWiFS geolocation that is consistent and verifiable across the full geographic and temporal range of the mission dataset. The methods described here can readily be applied to other global sensors of comparable resolution.

References

Gregg, W. W., Conkright, M. E., O'Reilly, J. E., Patt, F. S., Wang, M. H., Yoder, J. A., and Casey, N. W. (2002). NOAA-NASA coastal zone color scanner reanalysis effort. *Applied Optics*, **41**(9), 1615–1628.

Nishihama, M., Wolfe, R. E., Solomon, D., Patt, F. S., Blanchette, J., Fleig, A., and Masuoka, E. (1997). *MODIS Level 1A Earth Location Algorithm: Theoretical Basis Document, Version 3.0.* NASA Technical Memorandum SDST-092, Laboratory for Terrestrial Physics, NASA Goddard Space Flight Center, Greenbelt, MD.

Patt, F. S. (2002). Navigation algorithms for the SeaWiFS mission. In *SeaWiFS Postlaunch Technical Report Series*, S. B. Hooker and E. R. Firestone, eds., NASA Technical Memorandum 2002-0206892, Vol. 16, NASA Goddard Space Flight Center, Greenbelt, MD.

Patt, F. S., Gregg, W. W., and Woodward, R. H. (1997). An automated method for navigation assessment for Earth survey sensors using island targets. *International Journal of Remote Sensing*, **18**(16), 3311–3336.

Wessel, P. and Smith, W. H. F. (1996). A global self-consistent, hierarchical, high-resolution shoreline database. *Journal of Geophysical Research*, **101**(B4), 8741–8743.

Part V
Conclusion

23

Concluding remarks

JACQUELINE LE MOIGNE, NATHAN S. NETANYAHU,
AND ROGER D. EASTMAN

The field of satellite sensing/imaging is still in full expansion. The new millennium will see developments related not only to scientific missions, but also an explosion of commercial satellite systems providing data that will have economic and sociopolitical implications, with telecommunications taking a large place in the space market. In space, returning to the Moon and Mars, as well as exploring distant planets will see a growing number of distant satellite systems providing unprecedented amounts of data to analyze. The Lunar Reconnaissance Orbiter (LRO) is merely one recent example of such systems. Automatic, accurate, fast, and reliable image registration will increase the success of these future endeavors by providing data products that will foster interdisciplinary research and fast turnaround of information for applications with societal benefits.

This book has brought together invited contributions by 36 distinguished researchers in the field to present a coherent and detailed overview of current research and practice in the application of image registration to satellite imagery. The contributions cover the definition of the problem, theoretical issues in accuracy and efficiency, fundamental algorithms used in its solution, and real world case studies of image registration software applied to imagery from operational satellite systems.

As the field keeps evolving, we anticipate that new research will deal with combining multiple band-to-band registrations, extending 3D medical registration methodologies (Goshtasby, 2005; Hainal et al., 2001) to the registration of hyperspectral data, and automatically extracting windows of interest (Plaza et al., 2007) to guide more refined registration techniques. We also foresee that all the methods described in this book will be extended to many other data sources, especially those generated by future planetary systems (e.g., Troglio et al., 2009), but also to such problems as the verification of optical systems. Back on Earth, new constellation systems will require real-time integration of products onboard the spacecraft, and methods will be needed to combine imagery obtained by three or more sources, and to implement onboard registration on specialized hardware such as field programmable gate arrays (FPGAs).

We anticipate additional future directions in image registration research. Ground satellite teams and those focusing on specific satellite sensors depend on research and development to extend the success of automatic registration algorithms to more challenging imagery. For example, they require better cross-band and cross-sensor registration to successfully calibrate single instruments and multi-instrument platforms, as well as more robust registration algorithms for imagery acquired from highly off-nadir sweeps, or very high resolution imagery that reveals more distinctly distortions due to significant elevation variation.

On the other hand, Earth science end users need flexible, robust, and adaptable software solutions that include and integrate a number of registration algorithms suitable for different sources of data. They may need to register satellite imagery, vector map data, airborne sensor data, ground station data, or even historical aerial photographs and hydrology studies, so their software must also support evaluation and choice of the underlying algorithmic registration engine. This community also could use expert system methods to choose optimal algorithms for their datasets, metrics to assess the reliability and accuracy of a software registration technique, and scripting tools to assist them in quickly automating procedures for use in a new project.

A number of trends in the research literature support the creation of solutions for the above requirements. One such trend, that addresses some of the needs of end users, as well as the needs of ground satellite teams for successful performance on more challenging imagery, involves the development of well-engineered, multistage registration algorithms that combine basic principles and different approaches by exploiting the various strengths of these elements. Such algorithms may use initially complex interest operators or statistical similarity measures to successfully search for a large transformation in a simple domain, and then refine the results by computing, in a robust manner, an optimal correlation with respect to a more complex transformation model for achieving precise subpixel registration.

There are many other trends in the literature, involving theoretical derivation of error bounds, analysis and impact of error propagation, growing innovation of statistically-sound and robust similarity measures, large degree of integration of domain and semantic knowledge into the algorithms, and more robust search strategies less prone to local minima. It is very likely that the solution to image registration for remote sensing will not emerge from a single innovative algorithmic principle, but rather from an interdisciplinary systems engineering approach that will integrate a number of research fields into a cohesive set of knowledge.

References

Goshtasby, A. (2005). *2-D and 3-D Image Registration for Medical, Remote Sensing and Industrial Applications.* Hoboken, NJ: John Wiley & Sons.

Hainal, J., Hawkes, D. J., and Hill, D. (2001). *Medical Image Registration.* Boca Raton, FL: CRC Press.

Plaza, A., Le Moigne, J., and Netanyahu, N. S. (2007). Parallel morphological feature extraction for automatic registration of remotely sensed images. In *Proceedings of the IEEE International Geoscience and Remote Sensing Symposium*, Barcelona, Spain, pp. 421–424.

Troglio, G., Le Moigne, J., Moser, G., Serpico, S. B., and Benediktsson, J. A. (2009). Automatic extraction of planetary image features. In *Proceedings of the Third IEEE International Conference on Space Mission Challenges for Information Technology*, Pasadena, CA, pp. 211–215.

Index

Printed in the United States
By Bookmasters